REFLECTIONS & CONNECTIONS

Personal Journeys Through the Life Sciences
Volume II
Healthcare Economic, Environmental &
Medical Scientists

Editors

Otto J. Crocomo

Julius P. Kreier

William R. Sharp

ISBN: 1500465755
ISBN 13: 9781500465759

ACKNOWLEDGEMENTS

The editors are grateful to the following colleagues, family members and friends for their contributions and support during the two year book development journey.

Consuelho Baehr, Author and One of Four National
   Content Providers, Named by "Fast Company".
Dr. Roy S. Chaleff, Biotechnology Scientist and Author
Lauren Chomiuk, Associate Editor
Rebecca Butcher, Production Editor
Carla Maisa Crocomo, Educator
Ruth O'Heron, Publishing Assistant
Sally Sharp Holland, Educator and Author
Dr. Rachel Kreier, Healthcare Economist & Author
Walid Lofty, Engineer and Pilot
Rosa Shine Raskin, Microbiologist, Library Information
   Scientist and Author
Jeffrey W. Sharp, Entrepreneur and Film Producer
Edward Ramos Sousa, Attorney and Author
Dr. Douglas S. Steinbrech, Surgeon and Author

COVER DESIGN AND PHOTOGRAPHY

The editors provide special thanks to Yuan Hong Lin, Photographer, for providing the magnificent cover photograph entitled "Water Lily Reflections" and to Mauricio Diaz, Book Cover Designer, for his brilliant cover design.

Yuan Hong Lin resides in Taipei, Taiwan and Mauricio Diaz resides in New York City.

DEDICATION

The authors dedicate Reflections & Connections to the
legendary author Maya Angelou, our mentors, and our
successors.

"When You Learn, Teach, When You Get, Give"

Maya Angelou

PROLOGUE

R. S. Chaleff

We are taught from an early age about science; that it is an analytical process by which observation and experimentation increase human knowledge. It is a large and intimidating concept, but we get a sense of it early on because we are a science and technology-driven society. We are bombarded from birth about the importance and achievements of, and our dependency on, science. But who are the practitioners of that array of disciplines we call science? Yes, we learn the names of the most famous, those who made extraordinary discoveries or advances. But we don't learn who they were, or are, as individuals or how they became the monoliths we study and venerate. We are taught about the development of their accomplishments, but not about their personal development. We study science, but not scientists. Yet scientists are not like Athena who sprung fully formed and armed from the forehead of Zeus. What motivates people to become scientists? What is it like to labor in that profession? What are the challenges, frustrations, and the rewards? And how does one become a scientist? So while it is only natural for many young people in our science-dominated society to entertain the notion of becoming scientists, we provide little or no information on which to base such a life-forming decision.

This collection of autobiographical essays by scientists strives to address that void. Admittedly, it displays a bias toward the life, rather than the physical sciences, and within that domain the plant sciences were represented more than the medical sciences, but the the lessons are universal. In this book, we are offered insights into the lives of scientists and what influences motivated and shaped them and why they chose their particular fields. Moreover, we find how they were trained and about their failures and successes. The practice of science can be punishing: experimental designs may be flawed, long arduous hours are spent alone in the laboratory, and results can be inconclusive, misleading, or contradictory. Moreover, unconventional results and new theories that challenge established dogma can be met with brutal rejection. Yet the sense of personal fulfillment and achievement, and the awareness of making a meaningful contribution to humanity can be glorious. Open this book and you enter the lives of scientists, turn the pages and their lives and careers will unfold before you. Herein readers considering careers in science may find inspiration and acquire first-hand knowledge of what it is like not so much in the headlines, as in the trenches of science. Those already in the profession may find guidance for managing their careers, for knowing when to pursue and when to abandon an experimental strategy or interpretation, and perhaps most importantly, find courage and stamina to weather the inevitable storms and uncertainties of their ever challenging journey.

In American (and, as far as we know, elsewhere as well) graduate and postdoctoral programs train students and novitiates in experimental design and execution and in the analysis, interpretation, and presentation of results. However, it is only later, when they venture from beneath the protective wing of a mentor that these fledglings discover that additional personal skills are as important as skills in science for a successful career. Means and strategies for procuring funding, navigating departmental politics, promoting one's program, advancing one's career, and winning support for innovative ideas, are omitted from graduate school curricula. Many a promising career has been shipwrecked on these treacherous shoals. This collection of autobiographical essays hopes to rectify this lamentable deficiency in our system of graduate education. There is an art to the successful practice of science: an art that is essential, yet is not taught, that may come naturally to some, but that is acquired all too painfully by others.

It has been said that experience is something acquired only after it is needed. But it is also possible to learn from the experiences of others who have traveled the same path upon which one is embarking and to adopt methods employed by those predecessors to facilitate success, avoid pitfalls, and overcome obstacles. This book contains a trove of such experiences. It is not alone in that regard. Many valuable and informative books – some by individual authors and others comprising collections of personal narratives - provide lessons and advice for managing one's career in academia (e.g., *Rhythms of Academic Life: Personal accounts of careers in academia*, by Peter J. frost and M. Susan Taylor, 1996; *An Academic Life: a handbook for new academics*, by Robert H. Caldwell and Jill J. Scevak, 2010; *Academic Entrepreneurship: university spinoffs and wealth creation*, by Scott Andrew Shane, 2004). *Reflections and Connections* is distinct in that it chronicles career journeys in the natural sciences and is not a handbook, but a collection of intimate histories of the challenges, frustrations, and triumphs encountered on those journeys. It relates personal ordeals that inevitably obtrude on one's career, including bullying, competition, political infighting, racism and sexism as well as the human misfortunes of death, divorce, a family conflict, the Holocaust, medical disabilities, sickness and war. It also recounts the encouragement, inspiration, support, and counseling of colleagues, family, friends, and mentors that proved crucial in prevailing over these challenges. But above all, *Reflections and Connections* strives to capture and communicate the passion that drives scientists to succeed in the bittersweet, lifelong struggle of scientific research.

INTRODUCTION

Otto J. Crocomo, Julius P. Kreier & William R. Sharp

The book is organized in two volumes: Volume One - Agricultural Economic & Plant Scientists, Volume Two - Health Care Economic, Environmental & Medical Scientists and There is a back section which includes tributes to our influencers and mentors, biographical sketches about the authors and editors and an appendix with information about the long-term collaboration between The Ohio State University and the University of Sao Paulo.

The book chapters were written by various people in various stages of their lives. They were asked by the editors to reflect on how they got into their careers. What they were asked to write was loosely described. It was requested that they describe their family background and how they felt their families affected the careers they chose. They were asked to describe not only the effects their families had on their choices but also how encounters and connections with other people influenced what they chose to do. They further were asked to describe the results of their choices and how they handled these results. And to reflect on any other aspects of their personal and professional lives that they considered significant to their development.

The resulting chapters are quite heterogeneous as they are strongly autobiographical. There are also quite similar in many respects. This is because not only are job searches inherently similar but much of human development also is quite similar.

The editors of this book are all biologists but we are to some degree from different fields. Julius Kreier is medically oriented. His field was parasitology and he has a degree in veterinary medicine. He worked in animal disease control for some years before he returned to school to obtain a PhD in microbiology. After he obtained the PhD degree, he was hired as assistant professor of microbiology at the Ohio State University from which he retired in 1989 as a full professor. His colleague and friend Rod Sharp was trained as a plant cell biologist. They met in 1969 when Rod joined the department of microbiology at OSU because of his interest in cell biology and plant cell and tissue culture, a field not originally much thought of in botany. The third editor, Otto Crocomo, an agronomist and biochemist was also like Dr. Sharp a plant cell biologist but with a strong interest in agriculture and plant biology. Otto Crocomo and Rod Sharp have collaborated in research and teaching for almost 42 years since meeting in 1971. Dr. Crocomo recently retired as a full professor from the department of chemistry and from the position, Director of CEBTEC, the Biotechnology Center of the University of Sao Paulo. He was the founder of CEBTEC. Dr. Sharp has served as a full professor at Ohio State University in the microbiology department and at Rutgers University in the department of plant science along with stints in the food and biotechnology industries.

The result of the differences in the fields of the editors is that Sharp and Crocomo primarily picked the authors with an agricultural and plant biology slant while Julius Kreier picked primarily the authors who wrote chapters with a medical slant. A few of the chapters were authored by individuals with degrees in plant biology who later pursued medical research. The people we chose were people who had been our colleagues, collaborators, graduate students, and others whom we met at scientific meetings. The author Rachel Kreier is the daughter of Julius Kreier and the chapters describing growing up with a life scientist are authored by the editor's children.

The leaders of various fields of science usually consist of a fairly small group of people who get to know each other as a result of reading journal papers and attending national and international scientific meetings in their respective fields. The subject matter of the chapters included in the book is largely determined by the choice of the contributors. What is really of interest to us is whether we created a book that will be useful to the reader. It is the hope of the editors that the chapters will be read and will aid some young people in their struggle to develop careers as well as secondary school and university career counselors in the provision of advice to students. It is always useful to learn about what others have experienced. It may help young people find that their unpleasant as well as pleasant experiences were not unique but similar to other's experiences. Some of us, already established or even retired, on reading these chapters may be comforted by finding that they were not unique in having gone through struggles and having encountered setbacks while trying to develop careers. Also the hope is that the reader will gain a better understanding of research and teaching institutions and the lives of career scientists.

There is something positive we would like to say about our experience while reading and editing the chapters in this book. Some of the authors were our students and they have made us happy when they said that our teaching and mentoring helped them in their careers and even in their personal lives.

CONTENTS

CHAPTER 1.

A Checkered Career

Rachel Kreier

Introduction

Having looked through a number of the chapters in this book, I have to warn prospective readers that my own rather checkered career is more scattered than those of their authors. Nevertheless, perhaps some may be interested in the winding career path of a woman who has been a health rights activist, union staff member, editor, health reporter, "non-traditional" graduate student, health economist, and wife and mother.

Early Days

My earliest professional ambition was to be a novelist. I used to read novels hidden on my lap in elementary school. More than once, when the teacher called on me, I was so engrossed in the story that I did not hear her. I have a distinct memory of the disorientation I would feel as I was pulled up to the surface by hoots of laughter from my classmates. But though I spent hours every day reading, I never spent more than a few hours a month writing fiction. By the time I graduated high school, I had decided that my true vocation was as a reader, rather than a writer. It doesn't pay the bills, but I think I have gotten something out of reading all those books. It requires an effort of the imagination to gain even an imperfect sense of the realities that others live. Good novels help with that effort.

While I was an avid reader, I was not an enthusiastic student as an adolescent. I did well in the classes that came to me with little effort (English and social studies), and not well at all in the math and science classes that would have required a modicum of sustained attention. As I neared high school graduation in 1974, I told my friends and parents that I did not want to go to college. Instead, I said, I planned to hitchhike around the country. My parents were naturally worried about this. Joseph Hamburger, an Israeli scientist working with my father, suggested an alternative adventure – attending Hebrew University in Jerusalem. Long past the regular deadlines, I submitted an application, and was accepted.

Even if the option of going to Israel had not been brought to my attention, I think that I would have been afraid actually to carry out my hitchhiking plans. As a young woman, I felt ashamed to admit that fear. I've changed my mind. In 1976, a man who picked her up hitchhiking murdered my close friend Debbie Linton. Over the years, two friends have been raped, and several others attacked or threatened. Such occurrences are not rare. I now believe that it is a mark of self-confidence for a young woman to treat her misgivings about danger with the respect that they deserve.

College and graduate school: Round I

Somehow, between high school and college my attitude towards school changed. I enjoyed my classes at Hebrew University, and I worked hard at them. In addition to intensive Hebrew, my classes included a course on the Cold War, a course about the Jewish experience in America, and a course on the Holocaust. Perhaps unfortunately, I did well enough on the math placement exam to avoid being required to take any math classes.

The Holocaust course, in particular, made a lasting impression on me. While I was working my way through the textbook, silent nightmares full of trains and paperwork and shuffling crowds disturbed my sleep. The nightmares receded only after I read Elie Wiesel's memoir, *Night,* which was also on the syllabus. The personal narrative calmed the anxiety that the textbook's dispassionate documentation of the scope of the horrors aroused. An observation I read somewhere while I took that course (I don't remember the title or author) has stayed in my mind over the decades. There is a limit, the author said, to how much pain any individual can experience. No single person experiences the pain of millions.

I came away from that course with the conviction that the Holocaust says less about the nature of Germans than about the nature of human beings. The bottom line for me is that if they could do it, so could we. I also came away with a mistrust of ideology and an overly neat worldview. Remaining open to doubt, I believe, makes it harder to be ruthless.

Learning opportunities that year were by no means limited to my course work. The people I knew in Israel were quite different from the people I had grown up with in Ohio. I had had only limited contact with other Jews before. For example, I had never known anyone who kept kosher, but in Israel I had a roommate who followed the dietary rules strictly. She was a very nice woman, three or four years older than me, who finished school and moved away in the middle of the year. After her, I had a South African roommate whose unselfconscious racism was a new experience for me. I also got to know another South African, an unusually mature young woman whose commitment to fighting apartheid was a revelation of a different sort.

Jerusalem was beautiful and exciting. I still feel a surge of nostalgia when the scent of pine trees brings back the hillside on Har Hatzofim where I used to sit looking out over the walls of the

Old City. However, living in Jerusalem was also stressful. It made me feel like what I was -- a sheltered child of the American midwest. I knew no Hebrew at the beginning. As I started to learn, I found that Israelis had little patience for my efforts to speak their language, preferring to use their English, which I found frustrating and insulting. (My interactions with Israelis were mostly with the working class people I encountered in stores and restaurants. I think now that they probably welcomed a chance to practice their English, and also that their attitude may have contained an understandable element of envy and resentment.) I also found it stressful dealing with the predatory attitude of many of the men, Israeli and Arab alike. It seemed to me that whenever I walked down the street, I came under siege.

A pair of odd experiences that occurred early that year has stuck with me as a lesson about taking risks. The first experience was very nice. While walking through the Old City with my American boyfriend, we were approached by an Arab man with a note asking us to donate blood. We went with this man to a clinic. In the process of giving blood, we fell into conversation with the young Palestinian doctor staffing the clinic. We talked about belief in God. He told us about sitting out at night under a sky full of stars, watching his family's sheep. How could one witness that and not believe in God, he asked us.

With that pleasant experience in mind, we didn't hesitate the next time we were approached about giving blood. We went to the same clinic, but a different doctor was on duty. The doctor directed an older man – an orderly perhaps – to insert the needle in my arm and begin collecting the blood. I was lying down, with my arm extended through a small opening in a partition topped by a glass window. The doctor took my boyfriend to another room. I was left alone, while the older man and my arm -- with the needle in it -- were on the other side of the partition. The man began massaging and kneading my arm in a distinctly unwholesome way. I was afraid to jerk my arm back through the hole in the partition, because I was worried about the needle breaking. After a few minutes, the doctor returned, saw what was going on and yelled angrily in Arabic at the older man, who dropped my arm and scuttled off. It was all very surreal. The next day, my entire arm was black and blue. It was painful to move for a month.

I had gone to Israel intending to complete college there. Instead, I came back to the United States at the end of my freshman year and enrolled at Cornell University. I remember standing at the top of the hill near the library that fall, looking out across the green landscape to Lake Cayuga -- so different from Jerusalem's bone-dry, sun-drenched sand and stone -- and feeling every muscle in my body relax. It was as though I had been holding my breath for a year, and finally could exhale.

I spent a large portion of the next 10 years in Ithaca. I hiked in the gorges. I swam in the reservoir. I developed my musical tastes and political convictions. I met the man who would become my husband, Steve Peggs, as well as the half dozen people who remain my closest friends. I live now on Long Island and drive upstate two or three times a year. The upstate New York landscape still feels like home to me.

In 1978, I graduated with a double major in near-eastern studies (mainly Hebrew and Arabic language studies) and government, and with no clarity about what I wanted to do with my life.

A chance encounter set my direction for the next few years. As I was studying for final exams in one of the public rooms Cornell set aside for that purpose, I fell into conversation with a young man working in Cornell's agricultural school. When I told him that I had no post-graduation plans, he suggested I speak with someone he knew in the Cooperative Extension program who was looking for a research assistant. I followed up on that suggestion, and landed a job with an evaluation project for a state agricultural districts program. Within the districts, property taxes were levied only on the agricultural value of land so long as the land was used for purposes related to agriculture. The idea was to relieve farmers of financial pressure to sell their land for development.

I really enjoyed that job. I spent the summer driving around Onondaga County conducting interviews with farmers, agribusiness owners, real estate developers and local government officials and planning staff. It introduced me to a world that, in its own way, was as foreign to me as Jerusalem had been. In the fall and spring, I participated in the design and distribution of a questionnaire and in the analysis of the responses. I got an introduction to the rudiments of zoning and land use planning, and gained some insight into how decisions made by local governments on matters like sewer and water lines shape communities. I became aware of the many complications arising from our system of financing schools through property taxes. I also got my first inklings about how the interests of knowledgeable insiders influence the workings of representative government, and of the problems that arise when there is a tension between the concentrated interests of a few and the diffuse interests of the community as a whole.

Although I had held a number of part time jobs while I was an undergraduate, this was the first time in my life that I was financially independent of my parents. In addition to supporting myself that year, I was able to save enough money to finance a summer backpacking trip around Europe with two friends. That trip included quite a bit of hitchhiking (with no frightening incidents to report). We met a nice young man from Belfast, on his way to find seasonal work in the Dutch tulip fields and greenhouses. This was in the midst of the Troubles, but what I remember is that his Belfast accent gave him the most pleasing speaking voice I had ever heard. We lost our deposit money when our rental bikes were stolen in Amsterdam. We slept at a youth hostel within the city walls of Bruges, which were like a jewel box enclosing the exquisite little medieval city.

When I returned from Europe, I started a graduate program in the Department of Food and Resource Economics at the University of Florida in Gainesville. The department enrolled me in a number of undergraduate courses to fill in my yawning deficits in mathematics and economics. Rather to my surprise, I loved the calculus course. I had always been intrigued by the paradox that only "now" appears to exist, and yet "now" is a slice of time of zero duration. As I tell my math-econ students when I am trying to get them past viewing calculus as a series of arbitrary magic formulas that they need to memorize, calculus is a methodology for dealing with this paradox.

I also found that economic analysis came to me easily, especially microeconomic theory, which to a large extent consists of teasing out the logical implications of a simple and explicit model of human psychology. The master's thesis that I wrote for that program was an evaluation of the merits of using cost-benefit analysis to guide policy. In it, I started to explore the questions about how markets operate, and under what conditions they serve human well-being that continue to be at the center of my interests.

I left Gainesville before I had finished my thesis to join Steve in Geneva, Switzerland where he was working at the European Center for Nuclear Research (CERN). I finished the thesis during the seven months that I spent in Switzerland. When Steve returned to Ithaca for a job at Cornell's Wilson Synchrotron Laboratory, I followed him there. I initially enrolled in Cornell's PhD program in Agricultural Economics, but left after one semester, having decided that the PhD was not the right direction for me. The next fall, I enrolled in a registered nursing degree program at Tompkins-Cortland Community College. I thought a nursing degree would be more portable than the PhD. I also thought it would give me a skill of undoubted value for others in the world – something I worried would not be true of the PhD.

In retrospect, that year in nursing school was a valuable part of my education, although for many years I viewed it as a wrong turn down a blind alley. The rotations I spent in a community hospital and a skilled nursing facility were eye opening. I remember caring for an elderly man in the hospital and learning the meaning of the letters "DNR" (do not resuscitate) on his chart. The use of such orders was routine long before all the nonsense about "death panels" splashed across the media. I also remember confidently responding, "Yes, I have," to a patient who asked somewhat anxiously if I had ever given an injection before. I had indeed done so – once, when I gave an injection of saline solution to a fellow nursing student the previous day.

I remember conversing with a patient with dementia in the nursing home. There was nothing wrong with the way her brain worked in the present tense. Her words and intonation were perfectly appropriate, and perfectly identical, each of the three times I tried to introduce myself – but each time, her brain retained no shred of the conversation the moment after it was over. In a normally functioning brain, consciousness makes it possible to operate in a now of zero duration by linking past and present. Her dementia had severed that link.

I also remember the woman who was sent to the nursing home for a short stay following surgery to repair a broken hip. She was terrified by her situation. The nursing home was not a badly run place, as nursing homes go. It wouldn't have been selected as a placement for nursing students if it had been. The nursing aides who worked there were by and large conscientious people, though underpaid and overworked, as are all nursing aides to this day. But they weren't used to dealing with a resident whose mind was intact, and they didn't have the time or patience to acknowledge her as an autonomous human being. She clung to me like a drowning person clinging to a log just because I recognized that she was mentally competent. The memory still disturbs me because I should have done more to help that woman.

Towards the end of that year, Steve took a job at Lawrence Berkeley Laboratory in Berkeley, California. He returned in May 1985 for our wedding in Ithaca, after which I joined him in California.

My years on the Left Coast

By the time I got to California, I had started to doubt whether I was suited to being a nurse. Instead of completing my nursing studies, I looked for work in the social justice sector, which was very active in the San Francisco Bay Area, the heart of the region sometimes referred to as America's Left Coast. After an unhappy year as a fundraiser for another organization, I found a job as media coordinator for the Committee for Health Rights in Central America (CHRICA) in San Francisco. I worked for CHRICA for three satisfying and productive years.

The organization was composed of a half dozen or so semi-autonomous subcommittees, each focused on a specific area, such as maternal and child health, behavioral health, and physical rehabilitation services. Some subcommittees supported projects in specific countries, including Guatemala, El Salvador and Nicaragua. Several worked with Central American refugees in the Bay Area. Health professionals made up most of the active membership. As media coordinator, I was responsible for editing the monthly newsletter (which usually included writing most of it) and for working with the subcommittees on publicity and press coverage for their projects.

One project focused on raising donations of funds and professional services to meet the mental health care needs of Central American survivors of torture. One of the two psychologists coordinating the project was a gentle, soft-spoken man. I'm sorry to say that I've forgotten his name, although I recall his face quite clearly. He had worked with victims of torture, and also with perpetrators. He held an optimistic view of the rehabilitative capacity of the human psyche, but even he had his limits. I remember a conversation after a horrific local crime involving the murder of several children. I said something about how difficult it was for me to conceive of reconstructing anything like a worthwhile life after having done such a thing. He replied that a meaningful rehabilitation requires coming to terms with the reality of what one has done, and he thought there were circumstances in which that would imply committing suicide.

I got my introduction to some of the less attractive aspects of left-wing factional politics while working for CHRICA. Our staff included an adherent of a Maoist group, the League of Revolutionary Struggle (LRS), and an adherent of a group aligned with the Soviet Union, Line of March (LOM). Both of my colleagues were hardworking, dedicated, and intelligent, but they devoted significant efforts to undermining each other's influence among the mostly non-communist membership. The rivalry between them was almost comical, but the distrust it engendered was detrimental to the organization and the work we did.

The years I worked for CHRICA, 1986-1989, were the end of the line for that sort of American communism. Gorbachev took office in 1987 and the Soviet Union collapsed four years later. Communist China, of course, is still very much in existence, but there are no longer significant numbers of Americans able to maintain the pretense that it is an unambiguously progressive force in the world. I remember speaking with my Maoist colleague after the Tiananmen Square massacre in the spring of 1989. It was as though someone had died. I suppose it was a crisis of faith for her not so different from when people lose their belief in God.

My years with CHRICA overlapped the years in which the United States under the Reagan administration funded the Contra insurgency against the Sandinista government of Nicaragua. CHRICA -- which sent medical volunteers to Nicaragua, raised funds for rural health clinics and provided training and support for Nicaraguan health *brigadistas* and midwives -- played a role in documenting a systematic pattern of Contra attacks on health workers. From the standpoint of the insurgency, the attacks made sense. Whether insurgents are left wing or right wing, impeding a government's ability to meet the needs of its people furthers the goals of the rebellion. The Sandinista government had promised to improve the delivery of desperately needed social services to the Nicaraguan people. If the Contras could prevent the government from delivering on those promises, they would undermine support for the Sandinistas. Indeed, Contra disruptions contributed to the electoral defeat of the Sandinistas in 1990.

In 1987, the Contras murdered a young American engineer named Benjamin Linder and his Nicaraguan colleagues Sergio Hernandez and Pablo Rosales, while they were working on a small hydroelectric dam in the village of San Jose de Bocay in northern Nicaragua. I didn't know Linder personally, but other people in CHRICA did because he volunteered his considerable talents as an amateur clown and juggler to entertain children during vaccination campaigns with which we were involved.

The autopsy established that Linder had been wounded by a grenade and then shot at point blank range in the head. The two Nicaraguans were also shot at close range. Unfortunately, people who were sympathetic to Linder and the Sandinistas circulated false reports that the Contras had tortured Linder before they killed him. I don't know who originated those false reports, but I saw at first hand the damage they did. Journalists lost interest in further reporting about Linder's death once they realized that Linder's friends had given them incorrect information. (CHRICA had not been part of circulating the reports, because the Linder family did not confirm them.)

Towards the end of my time in the Bay Area, I also became involved as a volunteer with the innovative organizing campaign of Local 2 of the Hotel Employees and Restaurant Employees (HERE) Union. Influenced, perhaps, by my background in economic theory, I increasingly saw domestic and international issues as parts of an integrated whole. I disagreed with the tendency of some activists to view the interests of American workers as being in fundamental conflict with those of poor people in other countries. On the contrary, it was clear to me that the logic of market forces in an increasingly globalized economy dictated that the standards of living of American workers and workers in other parts of the world would be determined jointly.

Baby boomers like me grew up in a truly remarkable period of rapidly increasing mass prosperity. As a young woman, I took it for granted that this pattern would continue into the foreseeable future – that the irresistible tide of economic growth and rising equality would gradually chip away at the remaining pockets of hardship and poverty. Instead, as we now know, the year I graduated high school (1974) marked the turning of the tide for that particular version of the American dream. Since the end of World War II, American wages had closely tracked gains in productivity, but starting in 1974, the share of wages in GDP started to fall. As I write this in 2013, the concentration of income and wealth in the United States has regressed to levels last seen in the 1920s.

When I teach introductory economics courses, I explain globalization in terms of the "law of one price." As national markets become integrated into a global whole, wages – that is, the price of labor – will inevitably converge across nations. It is in the interest of workers in the poor countries that this process happens quickly, and in the interest of workers in the rich countries that this process happens slowly. But it is in the shared interest of workers everywhere that wages converge at as high a level as possible.

Balancing work and family

My older son Michael was born on May 31, 1988. [1] Steve and I moved to the Chicago area the following year, when Steve took a job at Fermilab in Batavia, Illinois. We were sitting in front of the TV in our living room in Illinois, preparing to watch the Oakland A's take on the San Francisco Giants in the 1989 World Series, when the Loma Prieta earthquake struck the Bay Area. It felt very odd to be so far away.

I found work in downtown Chicago as media coordinator for AFSCME (American Federation of State, County and Municipal Employees) Council 31, the largest public employee union in Illinois. I worked for AFSCME for three years, from 1989 to 1991.

At AFSCME, the majority of my time was devoted to producing Council 31's monthly tabloid newspaper. As a rule, the lead article each month was a profile of one of Council 31's local unions. I travelled throughout the state to interview and photograph local union members. I interviewed state nuclear safety inspectors, civilian employees of the Chicago police department, sewer district workers in Aurora, nursing home workers in Galena, and social workers with the Department of Children and Family Services, among many others. I became familiar with struggles over privatization of public services, and over the deinstitutionalization of psychiatric patients and clients with developmental disabilities.

I also witnessed the relentless pressure that rising health care costs put on wages. This was the era when managed care was rapidly gaining market share. Union and management negotiators alike hoped it would be possible to avoid difficult tradeoffs by moving employees from traditional indemnity insurance into HMOs and PPOs. There was a great deal of confusion about what the changes meant for costs and quality of care, and a great deal of over-promising. The changes affected many of our members not

---

1 Michael was killed in a car accident on February 11, 2012. I don't feel ready to write about this in more detail here, but do think I need to mention it, since raising a family has clearly been an important factor in my career choices.

only as consumers of health care, but also as providers of care, often at safety net public institutions, such as Cook County Hospital, that served the poorest and most vulnerable populations.

The job at AFSCME was extremely demanding. Producing the paper was a fulltime job by itself, but I also was pressed into service for public relations and electoral campaign activities. The media work, in particular, was unpredictable. More than once or twice, I found myself with an extra 20 hours of work dumped in my lap with no advance notice the week that the paper was due to go to press. As primary care giver for a small child, I found it hard to cope. When I became pregnant in 1990, I was forced to admit to myself that I simply did not want to continue working such long hours while taking care of a newborn and a toddler.

I spoke with my boss, prepared to tell her that I needed to quit. Instead, she offered me the opportunity to become an independent contractor, paid at 80% of my previous salary, with my responsibilities limited to producing the paper. The council hired a fulltime employee to handle media relations, which perhaps gives some indication of just how demanding my job had been. (After I left, the council hired another full time employee to edit the paper.) Under the new arrangement, I was permitted to do much of my work from home, only going into Chicago once or twice a week, which made it much easier to care for the new baby.

One often hears about extraordinary women who raise wonderful families while achieving remarkable things in their professional lives. In fact, I personally know and admire a number of women who belong in this category. I also know women, including myself, who find that balancing the demands of fulltime work and family life exacts a significant toll. I never wanted to be a stay-at-home mother (although I have nothing but respect for the women who choose that path), but I counted myself lucky to have a husband who earned enough that I could work part-time while the children were young. For me, combining a demanding fulltime job with raising small children was not a recipe for happiness. Many women, of course, do not have the options I had.

My younger son Simon was born on March 19, 1991. The following year, Steve took a job at Brookhaven National Laboratory on Long Island, NY. We moved into a house in the small town of Port Jefferson in December 1992. By coincidence, my predecessor as media coordinator at CHRICA had left the Bay Area for Long Island, where she had worked as a reporter and editor for a chain of community newspapers. She gave me the name of one of the editors she had worked with at the Times-Beacon-Record Newspapers, whose offices were only a couple of miles from Port Jefferson. Once we were more-or-less settled into our new house, I contacted this editor about doing freelance reporting for the papers.

The first story I wrote was about local reaction to the March, 1993 murder of Dr. David Gun, a Florida abortion provider. I soon took on a number of regular assignments for the papers, covering local Port Jefferson politics and producing a weekly health column. I also started reporting about Long Island's changing health care industry, focusing on the formation of hospital-based health care systems in reaction to the increasing penetration of managed care, and on local angles to the national debate over health care reform. Within a year, I was freelancing on a regular basis for the Long Island section of the *New York Times*, and for *American Medical News*, a weekly news publication of the American Medical Association. I also became editor for the lifestyle section that was common to the six weekly newspapers in the Times-Beacon-Record chain.

I worked as an editor and freelance reporter for the next six years, averaging 30 to 35 hours a week. The lifestyle section that I edited went to press on Tuesdays, so that was always a long day for me. In compensation, I took Wednesdays "off," meaning that I was able to exercise, clean the house, and do the big weekly grocery shopping while the kids were at school or in childcare. That was also the day that I could use for doctors' appointments, meetings with teachers, and similar chores.

On an hourly basis, I earned much less than I had in Chicago, but I enjoyed the work and appreciated the flexibility it gave me to care for my family. If I got a call from school saying one of the kids had just thrown up, nobody looked at me cross-eyed when I left the newspaper's office to go collect my sick child. For a while, I had my name on a freelance list for the *New York Times*, which I think might have been a path to a better-paid career as a journalist. Unfortunately, the calls always came at the least desirable times – on weekends and in the middle of the night when the regular reporters didn't want to cover the story – and I finally told them to take my name off the list. The point of the freelance career was to achieve a reasonable work-family balance – that meant being able to hang out with my kids during the evenings and weekends.

Back to academia

As time passed, I started to toy with the idea of going back to school. My editing work, which had been challenging and interesting at first, came to feel routine. The freelance reporting on the health care industry remained fascinating, but I found myself frustrated with repeatedly turning a story around in a week or a month, and then moving on to a new topic. I felt the need to take a more systematic approach to understanding the issues posed by the changes roiling the health care industry, and the national debate about how to reform our systems for delivering and financing health care.

I considered pursuing a public health degree, but that would have required commuting into New York City, which I decided was unworkable given my family responsibilities. The State University of New York at Stony Brook was less than 15 minutes' drive from my house. The economics department had recently hired a young professor with an interest in health economics named Debra Dwyer. We met and hit it off. I started in the economics PhD program in September 1999, when Michael and Simon were 10 and 7 years old, with Deb as my advisor.

In my arrogance or ignorance, I had thought that the PhD course work would be easy – I thought of myself as a person to whom school came easily, and I had not found the course work in the Cornell agricultural economics PhD program difficult. In fact, I was quite unprepared for the level of mathematical sophistication. I spent the first semester, in particular, barely holding my nose above water. I had also thought that if I found some of the math challenging, I could always ask my physicist husband for help. I realized how wrong I was about that when I sought his help on the first problem set. I am sure if he had been enrolled in the program, he would have mastered the math more easily than I was able to do, but he certainly was not able to understand the material off the top of his head. I was going to sink or swim on my own – or, to be more precise, as it turned out, to swim with the help of several close study partners among my fellow graduate students.

Eventually, I found my feet. Microeconomic theory, again, was my strong suit. The department had only one course in health economics, but I arranged to do two independent studies with Deb – reading courses on U.S. and international institutional structures for the delivery and financing of health care. During my last two years in the program, I also designed and taught an undergraduate course entitled, "Health Economics in International Perspective" – and learned an enormous amount in the process.

Indeed, one of the most important discoveries I made in graduate school was that I loved teaching. I held teaching assistantships for four years, beginning by leading recitation sessions for introductory economics, and progressing to teaching introductory economics, intermediate microeconomics, and – as mentioned above – a health economics course that I designed.

My dissertation evaluated how the release of information about the quality of cardiac surgeons – specifically, risk-adjusted mortality rates for surgeons performing coronary artery bypass grafts (CABG) --affected the allocation of surgeons among different socioeconomic groups. Using 12 years of New York State hospital discharge data, I was able to show that disparities in surgeon quality across socioeconomic groups widened after the Department of Health made the mortality rate information available. This was true across patients from low- and high-income zip codes, across racial and ethnic groups, and also across groups with different types of insurance coverage. A large body of earlier work had established that the mortality rate reporting program had contributed to a substantial improvement in New York's CABG mortality rates. But while mortality rates for the state as a whole fell quite dramatically over the 12 years I analyzed, they barely budged for the surgeons treating patients from the lowest income zip codes.

I spoke above about my interest in understanding the circumstances under which markets promote human wellbeing. A major portion of economic research and theory is devoted to devising means to correct so-called market failures. Economists come up with policies to increase competition, internalize external costs or benefits through appropriate taxes or subsidies, or correct for informational asymmetries. The CABG reporting program, itself, falls into this category. By providing information about the quality of surgeons, it empowers consumers (patients) to make market choices (pick surgeons) that serve their interests. Competition for customers can then be relied on to give surgeons and hospitals incentives to improve their performance. As I mentioned, the available evidence supports the view that the reporting program did, as predicted, contribute to reductions in CABG mortality rates.

What my research showed was that providing consumers with information about the quality of a commodity (the surgeon, and, by implication, the surgery he or she performs) also may facilitate the allocation of the highest quality commodities to those with the highest ability to pay for them. This is not, in fact, a market failure in the standard sense in which economists use the phrase. Textbook "perfect" markets allocate commodities to the uses with the highest economic value. The economic value of a commodity to any given consumer is measured by the maximum amount of money that consumer is willing to pay for it. *Ceteris paribus* for individual variation in preferences, higher income will increase a consumer's willingness to pay for increments in quality. Stated baldly, it is to be expected that even textbook perfect markets will tend to allocate the highest quality health care to people at the top of the income scale, just as markets tend to allocate the nicest houses, best cars, and best wines to the wealthy.

Many people would argue that that is fine. People with the highest incomes will be the most productive people (at least at the margin, if not on average), and it is only fair that they be rewarded for their productivity. Even those who don't think such inequality is fair, may accept that providing sufficient incentives to motivate people to work and to develop the skills that society needs is impossible without a certain level of inequality. I count myself among that latter group, although I strongly feel that the level of inequality in contemporary American society is way too high for a variety of reasons.

However, from a normative point of view, I consider health care to be a special case, in which resources should be allocated so as to accomplish the largest possible improvements in health – for example, the best surgeons should treat the most complex cases, where their skills will make the biggest difference to health outcomes. My reasoning for making health care a special case is straightforward. I consider it a minor "bad thing" that rich people drive nicer cars than poor ones. If that is the price that must be paid for the "good thing" of providing incentives to work and build skills, I can live with that. But I consider it a major "bad thing" if rich people live longer or healthier lives than poor ones. The life of a rich child should not matter more or less than the life of a poor child. However, speaking now from a positive point of view, allocating health care resources to achieve the largest possible improvements in health is not as a general rule consistent with allocating them to the uses with the highest economic value.

That basic perspective continues to inform my views on health policy. One of the concerns I feel about the growth of managed care is that it has facilitated economic segregation in health care. In managed care, when an individual selects a health insurance plan, he or she receives care from the doctors and hospitals that are in that plan's network. Health plans charging different premiums have provider networks catering to patients from different socioeconomic backgrounds. The plans that charge enrollees the lowest premiums will naturally tend to pay providers the lowest rates. Medicaid, the public program that provides coverage for low income Americans, pays the lowest rates of all.[2] Of course, there are excellent doctors and other health care providers who choose to work at public clinics or safety net hospitals from altruistic motives. In general, however, it is likely that provider quality will correlate pretty closely with the rates the health plans pay. Increasingly, specialist physicians with the best reputations refuse to take insurance at all. They set their fees, and are able to find enough patients willing to pay them that they avoid haggling over payments with the managed care companies altogether.

Life after Graduate School

Given the geographic constraints imposed by Steve's job and my family responsibilities, I was worried about being able to find

---

2 One very good provision of the Affordable Care Act raises Medicaid payment rates for primary care doctors to parity with Medicare rates. Disparities in specialist rates remain unaddressed.

work after I finished my graduate studies. I was very lucky to find a job with the economics department at Hofstra University, about an hour and a half's drive from my home. I was a fulltime member of the Hofstra economics faculty for seven years. Although I am no longer with the economics department, I continue to teach at the university on a part-time basis.

My research since finishing the doctoral program has maintained a focus on the role of socioeconomic status in health care markets. I have written about the economic and political implications of a body of economic theory known as managed competition, and the influence of managed competition ideas on Clinton's failed health reform initiative, and on changes that were made to the Medicare program during the Clinton and Bush administrations. With my Hofstra colleague Bhaswati Sengupta, I have modeled the effects of income and health status on consumer choices in markets offering two types of managed care contracts – HMOs that only cover enrollees for care from network providers, and PPOs that cover out-of-network care, with higher cost sharing. With Swiss health economist Peter Zweifel, I have evaluated the Swiss health insurance system that was established in 1996, and have compared and contrasted it with the provisions of the Affordable Care Act.

My approach to health policy has always been interdisciplinary. I teach regularly in Hofstra's Masters of Health Administration program. I also have developed a close working relationship with two non-economist colleagues: Janet Dolgin at the Law School and Corinne Kyriacou at the School of Health and Human Services. Together, we organized several speaker series. In March, 2010, we served as co-directors for a major health policy conference entitled, "New Directions in American Healthcare: Innovations from Home and Abroad." Johns Hopkins sociologist Vicente Navarro and Swiss health economist Peter Zweifel were the keynote speakers. I am working with them now to establish an interdisciplinary clinical program for public health and health law students.

Teaching and health policy have been the intertwined concerns of my professional life. While I enjoy teaching a variety of courses, I have poured my intellectual energies into developing my undergraduate and graduate health economics courses. These courses, and my views about how to achieve high levels of population health, equity, and reasonable cost control, have evolved together over the years.

The core of the material I teach attempts to answer the question, why are American healthcare costs so much higher than costs in the other wealthy democracies. Many students start the class with the mistaken idea that our costs are high because we use more healthcare that people in other countries. In fact, that understanding is incorrect. Americans visit the doctor and are hospitalized less frequently than the citizens of most other wealthy countries. When we are hospitalized, we are released from the hospital more quickly. Our system of malpractice insurance also accounts for only a small portion of our excess costs. And our aging population certainly doesn't explain why our costs are higher, because our population is significantly younger than the populations of most of the other wealthy countries.

In fact, there are three major factors that account for the vast majority of the cost differential: 1. Controlling for type of healthcare service, prices are higher in the U.S. than in other countries, 2. Our service mix tilts heavily towards expensive, high-tech curative services but skimps on primary care and preventive care for chronic conditions, and 3. Our administrative costs are very high.

Main stream health economists have, unfortunately in my view, focused on the idea that people use a lot of health care because insurance pays for it, and that the way to control costs is therefore to reduce insurance coverage through large deductibles, copayments and coinsurance. This is an example of a phenomenon known as moral hazard, which occurs when insurance causes changes in the behavior of the insured that raise the insurer's costs.[3] The move towards so-called "consumer-directed care," basically, a high-deductible health plan coupled with a health savings account, is the logical outgrowth of this point of view.

America's high and still increasing level of patient cost sharing has had the unfortunate effect of warping the service mix towards high-tech, high-cost procedures whose costs are way above conceivable deductible levels. Simultaneously, because they bear a large share of the costs, patients tend to skimp on primary care and routine preventive care for chronic conditions. This is particularly unfortunate as chronic conditions come to account for an increasingly large share of the nation's morbidity burden.

Despite all the attention that has been focused on the demand-side effects of moral hazard, what might be called "supply side moral hazard" has received much less attention from economic theorists. What I mean by supply side moral hazard is that health insurance removes the constraints that the demand curve places on the prices that doctors and hospitals charge – to the extent that insurance pays for health care, providers can raise their prices without losing customers. Without enunciating this theory, the practical policies of every wealthy country but the U.S. acknowledge this aspect of health insurance by creating an all-payer system of regulated and/or negotiated prices for doctor and hospital fees, pharmaceuticals and medical devices.

Whether for political or ideological reasons, the U.S. has so far been unwilling to embrace government price regulation. Instead, we have tried a sort of private sector approach to controlling prices, empowering private managed care organizations to use their market power to negotiate discounted prices from health

---

3 Moral hazard played a role in the 2008 financial crisis. Key players at large financial institutions understood that no government could permit them to fail because the consequences of their failure would be disastrous for the national economy. In other words, they took on excessive risk in their pursuit of high profits because they knew that they enjoyed an implicit insurance against that risk from taxpayers.

care providers. For a while, this approach seemed to be quite effective. In the early 1990s, the rate of growth of U.S. health care costs slowed dramatically. In my opinion, this slowdown in costs undermined political support from large corporate employers for Clinton-era health care reform. They thought perhaps the move to managed care made government reforms unnecessary to control costs. However, the dominance of managed care created strong incentives for health care providers to band together in hospital-based health care systems, so that they could counteract the market clout of the health plans with their own market power.

History in the Making

Clearly, the issues I teach about excite controversy, often quite passionate controversy. To cite just one example, as I write this in the autumn of 2013, the Republican majority in the House of Representatives has recently forced a shutdown of much of the federal government in an unsuccessful effort to defund the Affordable Care Act -- the health care reforms that constitute the signature legislative accomplishment of the Obama administration.

I am conscious in my teaching that I deal with issues about which reasonable people may disagree. In fact, I increasingly see that teaching and journalism have a great deal in common. I recognize that neither a teacher nor a journalist can be entirely objective. Indeed, it is too often the case that teachers or journalists adopt the trappings of objectivity to provide cover for their own biases. I have opinions, and I do not keep them a secret from my students. But I make a point of exposing students to the work of health economists and other experts who disagree with me, and with each other. And I make clear to students when I hold an opinion on an issue about which there are other intellectually respectable views, and when I am describing a settled scholarly consensus.

Like many other health services researchers, and, indeed, Obama himself, my first choice for U.S. health care reform would have been a single-payer, "Medicare-for-all" approach. Nevertheless, I view the passage of the Affordable Care Act as an historic achievement. The malfunctioning of the federal health benefits exchange website is infuriating, but I do not share the view common among supporters of a single payer approach that it is the inevitable consequence of fundamental failings in the law itself, and I fully expect that the website will function reasonably well within a year or so.

I do, however, have two important concerns about the fundamentals of the law. The first relates to cost control. This concern is not limited to the share of health care costs that are covered by public funds. Instead, it concerns the 18 percent of the economy's output that we currently direct towards health care, whether those dollars come from private pockets or tax dollars. Due partly to the recession, the rate of growth in costs has moderated since 2008 – but I'm old enough to remember that we went through a similar experience a couple of decades back, and then costs picked up again.

The ACA includes a number of payment reform initiatives – programs to encourage bundled payments and the formation of Accountable Care Organizations with a financial stake in reducing the medical procedure and hospital admission rates for their patient populations. Maybe these reforms will be the silver bullet that bends the cost curve – but I am skeptical. We keep trying to reinvent the wheel. Diagnosis related groups (DRGs), resource based relative value scales (RBRVS), capitation, bonuses and withholds: all have had their modest successes, but none of them have been silver bullets. And the ACA locks in place the fragmentation and very high administrative costs of our patchwork system for paying for care.

My second concern about the fundamentals of the ACA relates to equity. I spoke above about my worry that managed care has facilitated economic segregation in health care. The overwhelming majority of the health plans being offered through the health benefits exchanges place stringent restrictions on enrollees' choice of doctor and hospital. The stage is set for an acceleration of the development of provider networks catering to different socioeconomic niches – "Walmart" doctors and hospitals for the working class, and "Macy's" doctors and hospitals for the middle class, and "Neiman-Marcus" doctors and hospitals for the wealthy. And, of course, an expanded Medicaid program for the poor. (I find myself unexpectedly sympathetic to proposals from several Republican governors to use funds allocated for Medicaid expansion to subsidize purchase of private coverage through the exchanges.) Subsidies for people with low and moderate incomes to purchase coverage will mitigate this trend to some extent, but the subsidies are consciously structured to leave individuals bearing costs at the margin if they choose more expensive coverage.

All of these misgivings remain highly speculative. The truth is, only time will tell whether the ACA will be successful in reining in health care costs, and whether it will mitigate or widen class disparities in health outcomes and access to care. I expect the next couple of decades in my professional life to be extremely interesting.

Conclusion

I am a conscientious teacher, and I am prouder of my accomplishments in the classroom than of anything else I have done in my professional life. In academia, good teaching is a notoriously under-appreciated skill. And yet, for many of us, it has enormous intrinsic rewards. For me, at least, the satisfaction of watching as the proverbial penny drops, and a student who was lost suddenly "gets it" is very potent. I think that for many academics, perhaps for female academics especially, teaching is almost a guilty pleasure. We know that our careers would benefit if we invested less intellectual effort in teaching – but teaching is so much fun!

I am often shocked by how little understanding health professionals have of the economic forces and policy choices that shape the financing and delivery of care. In my teaching, I strive to do my bit to remedy this deficit, because I believe society benefits when the individuals who work on the frontlines of our health care system are able to bring a broader and more sophisticated understanding to their professional lives.

CHAPTER 2.

A Lifetime of Opportunities and Decisions

Robert M. Pfister

The following short paper is a synopsis of my career. This is not meant to be inclusive of all the 80 years that I have enjoyed life. The following pages merely indicate some of the decisions and opportunities that I have had during that time. The paper represents what I think were major crossroads and opportunities for me, which resulted in the career that I had and actually still have. I will include a short section on some of the scientific areas that I have worked in. This will not explain any of the science that had been done in my laboratory, but merely make reference to a number of the different science opportunities that I've had.

This chapter is not meant to be a guideline for any other individual, but merely an examination of some of my philosophy about how a person might help themselves in gaining a career.

I think the following comments should appear at the beginning of this chapter. I believe that modesty requires me to explain that I do not feel as though I have done anything extraordinary or unusual, that anybody with reasonable intelligence could not also accomplish. The reader will learn by going through all the chapters in this book that the eventual success or failure of any individual depends on both nature and nurture and to some degree chance.

Living in this country we have been given the opportunity to develop our lives in a direction that we choose suitable. That does not mean that we automatically reach these goals. I would've loved to have been an astronaut flying through space when I was young, or perhaps a cosmologist uncovering the great mysteries of our universe. Perhaps to a certain extent the position we find ourselves having been born into does in fact partially dictate the eventual success or failure of our efforts. Trying to write this document has brought out to me the complexities of the normal activities of life and how they affect us. I have commented in this chapter, on the importance of the decisions we make in our lives and how they shape our career development. I think that is easy to get caught up in thinking about the roles of nurture versus nature in our development. The roles of nature and nurture are nearly insoluble when trying to have a discussion about, or to give advice regarding how to achieve a career. That debate does not help. Biologists will quickly point out that genetics plays a major role in anyone's existence. I doubt that there's much argument about that fact. Modern biologists will also comment on the value of upbringing (nurture) and its effect on the eventual life or career of any individual. As a consequence of these facts it is nearly impossible to give advice or to predict in any way how a young person will end up in their career. The only thing that is certain is that it will be a result of a confusing jumble of genetic segregation, nurturing, and luck. One could argue that the environmental nurturing is also partially a result of genetic segregation. What I mean by that is that of course our parents were results of the genetic segregation of their parents and so genetics in effect carries through all prior generations. That of course excludes the concept of bad luck, accidents or tragedies that befall any given family and how that then affects the direction of one's life

In my own case my parents were both educated and led exemplary lives up to the point of their early demise. No one could predict that my father would die at age 47 of heart disease because looking at his parents they lived well into their 80s. My mother's early demise was in much the same fashion, her parents and great-grandparents were reasonably long-lived. One has to question whether nature or nurture habits had any affect. Of course with genetic segregation it's difficult to say because of the complexity of gene movement. It is also very difficult to say that our modern environment had in some way altered or affected their lives. This holds true even more so today because living has become more complex. We are getting more technological achievements, and probably now suffer a great deal more from exposure to complex chemicals and other environmental effects then did earlier generations. I don't really mean to say that life is just a crapshoot and that we simply turn out as a result of all of these complex involvements and that you just have to live with whatever happens.

The number of decision-making events, (junctures) in one's life is enormous. No one can keep track of all of these decisions nor should I think really try to. You can always point to situations which are identifiable as having a positive effect or potentially a serious effect on one's life. For example whether you go to the store in the morning at 10 o'clock or later in the evening is a very small decision. I'm not saying that it can affect your life; in fact it could end it if one had an accident. But the decision to leave home and set out on one's own path or the decision not to go to college or the choice to join the military are all very important junctures. There are myriads of minor decisions too many to count, and probably hundreds, and perhaps more that could be very important in producing the final result.

I'm reminded of the well-known quote of Yogi Berra, the famous Yankee catcher, "when you come to a fork in the road take it." What this statement really suggests is that there are three decisions to be made at any time you reach a fork in the road. You could take the left or right fork or decide not to make any choice at all. That really results in three chances for you to do or not to do something. The decision to not do anything is in fact a decision. I think it's quite easy to envision the many opportunities and situations one would find oneself in, by using or not using these three points. Unfortunately, many of these choices are never fully recognized or realized. If you decided to go to the store and had a flat tire on the way you would've had to either fix it yourself or call a service to have it fixed. I suppose you could actually decide not to do anything and walk away from the car, or just drive home and ruin the tire and the rim. If you had made an early decision to have a (AAA) service that could come and help you it may have been helpful in many ways or you could've also made the decision to fix it yourself. Then you have to be careful that the car doesn't fall off the jack or you become injured by a passing car. Perhaps you're even lucky and a passerby stopped and helped you fix it. It is becoming evident that when one thinks about these things, the number of choices one makes in one's life are way too numerous to count. Perhaps you would've stayed home and not picked up the nail and not had the flat tire in the first place.

One can never know how these situations turn out. I realize, and I'm certain the reader recognizes, that a person cannot always follow the directions or dictates of another, especially where success or failure are concerned. Furthermore, the particular time in history in a person's life is certainly a factor in the development of anyone, and eventually what direction he or she may take.

In my case my parents died while I was very young, I had very little money, the Korean War was underway and the United States of America had a military draft for young men. When I first started college at Hofstra, I realized that when I reached 18 and registered for the draft that I was then in a position to eventually spend some time in military service. That was the conclusion of any healthy normal male at that time in history. The draft had been in effect during World War II and was not rescinded after the war, but continued on into the Korean conflict. The draft of young men turned out to be an extremely valuable part of my development.

It is nearly impossible, in a few lines, to explain the history of any individual. This becomes particularly difficult when trying to examine one's own life. My attempt in this document is to simply try to explain to a generally diverse audience, how I arrived at my career. I would very much like to help young people make their way in the direction that would point toward a valuable and productive life.

I have to admit at the outset that I was very lucky to have been born into a family with devoted parents. My mother was born in Kankakee, Illinois in the last year of the 19th century. My father was born in the first year of the 20th century in Switzerland. How did these people meet? That story unto itself might be an interesting romance novel. I will only say that after my mother had received her college education in America. She spent some time in Switzerland with her mother, who was recovering from a divorce. During that time my mother to be took a job as a dental technician in Zürich. One of her customers was a young woman who she became friendly with and who suggested that she should meet her brother who was about the same age. They met and became romantically involved.

I'm going to interject at this point that, in anyone's life, as discussed in an earlier paragraph, there are huge number of" forks in the road" that have to be navigated and which at the time seem inconsequential. It is unfortunate that one has to pass so many years in order to establish at what point significant decisions were made. For example, the trip my mother made to spend time with her mother in Switzerland, certainly affected my being here at all. The fact that my father fell in love with a foreign woman and left Switzerland, much to his family's displeasure, to chase her back United States and propose marriage was certainly unusual in itself. My father was a graduate of the University of Zürich. My mother was a college graduate from the University of Illinois. My mother had a liberal education. My father was trained as an accountant. In those days in the early 1920s, jobs were hard to come by. I know my family did not have much money and they, like many families, worked hard to establish themselves.

During the early years of their marriage the economy in this country was not good and my father worked at whatever job he could get. The depression was having its effects and I know he had a job in a factory in PA making some type of stove. I was told later on his salary was approximately a dollar a day. My father arrived in this country. In the latter part of 1924, reunited with his girlfriend and proposed marriage. She accepted and they were married in 1925 and I was born in 1933. At some time after that he got a job with the Chrysler Corporation in New York City and held an office position with them during the latter part of the 1930s. Chrysler wanted to establish a dealership in Zürich, Switzerland and he was a logical choice to go to Switzerland to help set up that dealership. Unfortunately, that was about 1939, which was not a particularly good time to go to Europe.

My family arrived in Zürich with my father at his new position, only to face the threat of world war. During the six months that we lived in Switzerland. I started school in Zurich. We spent less than a year there, living with my grandparents. The Swiss government suggested that we return to the United States because we were all US citizens at the time and they could not guarantee our safety. I also was registered with the Swiss government and hold dual citizenship with that country. As history notes that was a very difficult and dangerous time in Europe. Even though I was quite young I have vivid memories of traveling from Zürich, eventually to France and Holland, to catch a trip back to the United States. During that trip while we were on board several trains, I remember my family insisting that I do not speak German or Swiss German because it was clear that the German people were not appreciated in much of Europe at that time. While we were American- Swiss, the language resembles that of German and could easily miss identify us as possible German citizens. The trains were darkened and had shades over the windows so that no light could be seen from the outside.

I am not sure of the exact timing during the period we were in Switzerland but we visited my father's, sister, (Aunt Millie) who lived in Holland in the small town of Dalfsen. She had married a Netherlander who ran a chicory factory in Dalfsen. Somewhere during that time I remember seeing trees along the highway with sticks of dynamite wrapped around them so they could be exploded to drop the trees over the highway and stop the German advance. Many older people and historians will clearly remember the Maginot line, which was set up in France to interfere with the German progress. History has shown the German army simply went around the flank of the Maginot line and made it totally ineffective and an expensive defense which was worthless. Apparently, the blitzkrieg military technique that the Germans devised had not been recognized prior to its use and so most of Europe was quickly overwhelmed. These are more appropriately stories for the history books. Suffice it to say we made it safely back from Europe after a rough crossing on the Atlantic, with the concern that German U-boats might accidentally sink an American vessel. The United States at this time was not in the war and I clearly remember a giant American flag painted on the side of the ship with big lights at night illuminating the flag so the Germans could presumably not torpedo or take a shot at that liner. We returned to the United States during the latter part of 1939, and were caught in a hurricane at sea. Our ship was damaged and had to be repaired by a diver who worked on some aspect of the propulsion system at the aft part of the ship. I think he may have been working on a damaged propeller. I do not remember what the reason was, but I do clearly remember seeing a diver working at the stern of the ship.

My father was a citizen and completely Americanized by the time the job with Chrysler fell apart because of the war. He eventually he took a job with the Rural Electrification Association (REA), as an accountant and we traveled to Glencoe, Minnesota where we lived for several years. It is interesting to note that only 70 years ago most of rural America was not electrified. Farming was still very primitive and done without the help of electric power. After working for the REA we returned to New York City where he got a job working for the Airco Air Reduction Sales Company. He worked for them in New York City until his passing in 1949. He was 47 years old. There was no job security in those days, and there was no insurance program so my mother found herself with a teenage boy and no income. The company my father worked for was a good company, and offered her a job so that she could support her family. Thus, after being a stay-at-home mother for quite a few years she returned to work traveling to New York on a daily basis. We were living in Hollis, Long Island at the time, and she worked for the company for about a year when she became ill and developed a cancer which resulted in her passing in 1950.

I had started going to Hofstra College, just before she became ill. I found that I did enjoy learning and in particular was comfortable in science. History reared its ugly head again, and difficulties in Asia resulted in the Korean War. After my mother's passing, I lived with a friend whose mother offered me an attic room in her house. I worked for the Wall Street Journal for short while and copied all customer records to be placed in the vault in case of atomic war. It was a good job for a young person and I enjoyed working on Wall Street for the short time that I was with them. I decided to volunteer for the Army and was inducted in 1951.

As it turns out the decision to enter the United States Army was inadvertently one of the most important decisions I ever made. At that time in history the United States was still making use of the system of drafting young men at the age of 18 into the military service. Those of us who wanted to complete some sort of advanced education were somewhat in fear of being snapped up right during the time of our educational process. This did not appear to be a good idea. So we had many conversations, my friends and I, about the nature of the future for ourselves and for our country. This period in history was during a time when there was great fear that we would be bombed with a nuclear device .We all lived under a cloud of concern as to whether we were going to have any future at all. In my own case I honestly thought I would never reach age 21. Not primarily for health reasons but cause I probably was going to be killed along with many others due to an atomic war. During that particular time many people were building bomb shelters along the East Coast and in other places in our country.

To my knowledge at the present time, no one has been able to predict the future with any degree of accuracy. Sometimes even the past is difficult to recognize or analyze until a great deal of time intervenes. Thus, a decision which is made today may not even be recognized as a significant decision, but in future years will become a major factor in determining the direction an individual or in fact the country may take. I believe that this is not a discouraging factor, but actually an encouraging one. I think one can examine decisions that have been made and use them as a guideline for future choices. Of course, in some cases, actions are irrevocable. There are however, many decisions or options which, by using your native intelligence can be a huge factor in your future life. For example, not to continue your education, because you might have to borrow some money, is a very serious one. In my personal opinion the borrowing of money for future education is a no-brainer. In my own case the decision to go to college and eventually complete my doctorate was perhaps the single most important decision I ever made. Obtaining my doctorate did not guarantee that I would achieve success, win the Nobel Prize, or become a nationally known figure. It did however open many doors for me, which would've not been opened, had I not had that degree.

Jeanne D. Nordheim (1933-2008). Most of us who make our way to an advanced stage of life will recognize that you really have very few true friends. I realize that this goes against the current concept of" Face book" in which people seem to have dozens if not hundreds of friends, or when I speak with younger people who assure me that they have many friends and that their lives will be complete by knowing them. I'm willing to make a statement which may not be agreed by all to be true. If you have one or perhaps two true friends during your lifetime, I think you've done very well. We all have many acquaintances, colleagues, and people that we are very friendly with. But the kind of friend for whom you would give up your life, does not occur very often.

I am a strong believer in family and while some people say you can choose your friends, but you're stuck with your family, I must comment that there perhaps is certain amount of truth to that, but by and large, in general you can depend upon your family. I have been asked many times what I have missed in life. Sometimes, to the consternation of whom I'm speaking with I say I wish I had a brother or sister. I am quite serious about that. It's possible that perhaps because I was an orphan at age 16 that I have been sensitized to the need for family. Perhaps there's some truth in that statement, but in my own heart of hearts, my family is my most important strength. The name at the beginning of this paragraph became "Jeanne Nordheim Pfister" quite clearly the most important person in my life and my best friend for life.

We met at the Hollis Presbyterian Church when we were about 15 years old. I am happy to say that I have been in love with her ever since. She stood by me during the difficult year of my mother's illness and alongside me when I buried her. I am very lucky that not only for my relationship with Jeanne but my relationship with her family which was extremely strong. During her years Jeanne became valedictorian in our high school and was always a very smart young lady. She received a scholarship to go to Cornell but because of family financial difficulties could not take it. She eventually went to Barnard College for several years and studied Spanish as her major. Jeanne eventually completed her education at The Ohio State University. After our children were grown and out of the nest, she applied for law school at Ohio State. She was accepted and completed the program, passing the bar exam on her first try and becoming a full-fledged practicing attorney for about 10 years.

I volunteered for the draft, at age 18, took my training as an infantry man, at Indiantown Gap, Pennsylvania, and completed a course in leadership school. I was sent to cooking school at Fort George G. Meade near Washington DC. And became a member of the 101st Airborne Division stationed at Camp Breckinridge Kentucky. After some training with the Airborne Division, they assigned me to cook for the soldiers in training. I spent the rest of my time during that enlistment cooking for and training troops.

I was very lucky during that period and was able to drive to New York frequently and spend time with my girlfriend Jeanne. When my Army experience was finished Jeanne and I were married and set off in a "New Moon "house trailer to attend college at Syracuse University. We overwintered in the house trailer, but experience and wisdom got the better of us and we applied for married students housing at Syracuse University. Jeanne got a job with the engineering department at the University and I was accepted as an undergraduate in the Department of Microbiology for a Bachelor of Arts program. After graduating from Syracuse with a B.A. degree in microbiology, I applied for graduate school and was accepted. By the time I graduated with my doctorate. Jeanne and I had four children, a dog, a parakeet, and a junker car. My children Ellen, Alison, Valerie, and Robert were all born in the Memorial hospital in Syracuse. Our whole family certainly became used to snow and cold weather.

Jeanne passed away in 2008, having contracted lung cancer.

My son Robert lives in Puerto Rico owning his own business. My daughter Valerie has her doctorate from the University of South Florida and resides near Gainesville, Florida. Alison has worked for nationwide insurance for 25 years, having completed her degree in communication at The Ohio State University. Ellen is a successful l artist, specializing in wildlife and animal Art (artbyellen.com). She has survived lung cancer and currently is doing very well.

The fact that I was working on my doctorate allowed me to get a job with the Syracuse University Research Corporation while still going to school. In this job I learned to do basic microbiology for the paper cup industry. One aspect of the job at the research corporation was to inspect the factories of all members of the paper cup industry who participated in the program, at least once a year for microbial sanitation purposes. In other words, to determine that the cups and paper products were being produced properly, in a sanitary fashion and in general to determine how paper products were handled. When I visited these plants and I visited all of them in the United States during my graduate career, the doors of the corporation's presidents were open to me and I was treated in a professional fashion. I never anticipated meeting the presidents of corporations such as the Lily Tulip Cup Corporation, the Maryland Cup Corporation, the Dixie Cup

Corporation and so forth. These experiences were of enormous help to me, because I was a bit shy and retiring and found it difficult to relax in that type of situation. I learned a great deal in those years, and that personal development served me well for the rest of my life. While the decision for me to work at the Research Corporation, which was an important one for my family (we had four children by then) to provide income, It turned out it was extremely important for me in a professional and personal developmental sense. Actually, while not to be discussed in this chapter it is true that I could write a lengthy treatise on my particular time with the Research Corporation and the various projects that we had while I was there. So once again the decision to take that particular job, while I did not realize its eventual effect on my expertise and personality, was extremely important in my personal development.

When I decided to go to Hofstra College my mother was still alive, working for Airco Air Reduction Sales in New York City and was most pleased that I had made the commitment to go to college. The money for the first year college came from a savings account that I had made from my work as a paperboy for the Long Island Daily Press. I worked for quite a few years for the Long Island Press and delivered papers after school and on Sundays. The paper route that I worked on was made up of about 120 -130 customers and for its time was very lucrative for a young man. I remember toward the end of my newspaper career. I received $100 in tips at Christmas time, which for me was a lot of money. My father was still alive when I first got the job and was upset that I had to get out early on Sunday morning and go to the office to fold and deliver the papers. But, I think he must've recognized it would be a good thing for me to put an effort into something which could be helpful in my own life. So in essence that time and money that I earned was put toward my eventual education, as I paid for my first year at Hofstra College with the money I earned as a paperboy. The summer after my first year in college I got a job as a construction worker and in those days I earned two dollars an hour which was very good pay and which I saved for my second year at Hofstra.

I would not be writing this chapter if it had not been for the United States military. I had saved some money during the first two years of my college education. I found myself in limbo at the end of that time and by my volunteering for the draft I eventually received enough support for about four years of college. The G.I. bill in my opinion was one of the most successful government programs ever attempted. The number of soldiers who received benefits after World War II was huge. Many men returned from the military and got their education, thanks to government help. I remember as a freshman at Hofstra I had to wear a beanie during my freshman year, which was embarrassing but necessary because that was what was done at the time. However, after the veterans from World War II had their influence the wearing of a ridiculous hat ceased. No man in military service and perhaps in armed conflict was going to put on this beanie so he or she could walk around the campus. Needless to say that practice disappeared. I was very fortunate in the military because of my size on loud voice. They sent me to leadership school and so I started training troops to go to Korea. During that time they needed cooks in the Army and so it was decided to send me to cooking school at Fort George G. Meade.

Some Accomplishments

The following paragraphs have been added to explain a number of the projects that I have been involved in during my career. There will be no detail as to results or papers produced.

When I entered graduate school at Syracuse University in the Department of Microbiology, I was given the opportunity to work at the Syracuse University Research Corporation to help support my family. I was also a teaching assistant in the microbiology program. . My job was to supervise the testing of paper cups and paper products, according to a contract that the corporation had with the paper cup industry. We tested random samples of consumer products for microbial contamination on a continuing basis. We also were required to inspect the factories that produced the paper products at least once a year. My job was to make sure that the microbiology of the paper products was accomplished and to travel to all of the paper cup producer's factories and do an inspection for sanitary conditions. I was also involved with a number of other projects, including the evaluation of chemical toilets onboard recreational boats.

When I joined the research program of Dr. Donald G Lundgren,(my academic advisor) he assigned me the task of assembling a used" RCA" electron microscope that he had purchased. I used that equipment to do an ultra-structural evaluation of the formation of spores in the genus <u>Bacillus</u>. At the conclusion of my doctoral program, I received a postdoctoral fellowship at the Lamont Geological Observatory in Rockland County, New York. Lamont Observatory was part of the research facilities of Columbia University.

I worked with Dr. Paul Burkholder, doing electron microscopy for him, and established a sponge collecting program in Puerto Rico. One of our missions was to isolate microorganisms which might have antimicrobial activity. I also wrote a computer program to do numerical analysis for the identification of microbes in Antarctica.

I received a call from my friend and colleague, Dr. Patrick R Dugan, who was a professor at the Ohio State University. He told me there was a position of Assistant Professor available in the newly formed College of Biological Science.

I presented a seminar there and was offered a position as Assistant Professor with the faculty of microbiology. Most of my research involved ultrastructure of microorganisms, and I taught a course in electron microscopy as well as beginning microbiology for many years. My research always had overtones of environmental microbiology. Three colleagues, Dr. Patrick Dugan, Dr. James Frea, Dr. Chet Randles and I carried out a research program in Lake Erie for four years, evaluating the serious concerns of microbial contamination in the lake. I also worked on the migration of heavy metals in the environment and the involvement of microbial activity with those metals. These heavy metals were primarily Mercury and Cadmium and we looked at how certain microorganisms could help the translocation (movement) of these elements in the environment.

In conclusion, I will say that when starting this chapter, I was given a set of criteria to guide me as to its contents. The criteria were to encompass the entire book and hopefully encourage younger people to continue their education and to realize the absolute value gained by doing so. I remember my parents saying to me "go to school and learn, no one can take that away from you". The idea of this particular chapter and book is to give the reader a chance to examine how various individuals made themselves successful either by pure luck, stubbornness or unusual skill. In general it is my opinion that luck does play a role but the individual has to be able to recognize opportunity when it presents itself. That recognition expertise is part of the inherent nature of any individual. In essence what I'm saying is that in our entire lives we are given a huge number of opportunities (decision points) or some say" forks in the road" and our choice at these times, determines what we become.

There is no substitute for hard work and determination. Have a great career, live long and prosper. Best wishes. Bob Pfister

Robert M. Pfister Photograph

CHAPTER 3.

Jeff Alder's Story

Jeff Alder

Prelude

Today as I write this, I'm on a 777 heading to Beijing. We're somewhere over the North Pole and the exact time is debatable. I know its +12 hours to China (all China has one time zone, and there is no daylight saving time in China). This is my 7th trip to Beijing in the first half of this year. I'm going to help develop an anti-infective drug for Chinese patients.

Before the flight, at the Newark airport, I quickly progressed through the special security line that benefits the frequent travelers. I got to bypass the long line of people waiting to be searched, x-rayed, scanned, and possibly patted down. That was me a few years ago – standing in the long lines, looking at the frequent travelers whisked through security. If you ever felt the urge to become intimate with a total stranger, then a trip through airport security can be just the ticket. My status is currently called "Global Platinum Elite" – how many different words can be strung together to convey the message of special? The only qualification to become a Global Platinum Elite is to spend (or have your company spend) a lot of money on airfare.

There was a time when I gazed with great envy at the special airline travelers. Those people who sat in first class, with choice of movies, real food and all comforts.

"Shiraz or chardonnay, sir?"

Now this relative comfort seems commonplace, routine, expected. If something does not go smoothly with air travel, I'm experienced enough to accept it. Anything that arrives the same day – and the definition of "day" gets muddled during a 12 hour 12 time zone flight. Still, no first class airline flight compares with the simple pleasure of coming home to cook a nice dinner with my spouse. And all this comfort occurs due to drug money.

In human endeavors, the majority of people must work hard to maintain the status quo of society – to keep civilization going and fend off the darkness. These are the people that really make society work – fill in the blank with any job that is associated with the middle class that the politicians all seem to be forever praising and promoting. This is the glue that holds the societal thing together. From the water that comes out of the tap to the food in the store, and the general good order of traffic on the roads to get to the store – it takes many to maintain a working society.

Relatively few have the great fortune, and perhaps some level of determination and skill, to devote their careers to the creation of something entirely new. To the creation of something that did not exist before. In the past, this was known as an invention or a scientific discovery. Invention is associated with repeated failure. To be an inventor, you must first be prepared to be a complete and utter failure.

I have had the good fortune to work in the business of anti-infective drug discovery and development. This means creating new substances, some of which have never before existed, and taking them from test tube through human clinical trials to produce safe and effective medicines. In the field of drug discovery, where failure is the norm, I've also had the good fortune to contribute to the discovery and development of multiple anti-infective drugs. One drug (daptomycin) was for drug-resistant MRSA bacterial infections of the heart, blood, and deep skin wounds. Another (clarithromycin) is for upper respiratory infections. I had a small role in helping with the first HIV protease inhibitor, Norvir.

Currently, I am the Chief Clinical Microbiologist for Bayer HealthCare (the aspirin people), where I work on ciprofloxacin dry powder for inhalation (ciprofloxacin DPI) for bronchiectasis, and COPD, amikacin aerosol for serious ventilator associated pneumonia in ICU patients, and on tedizolid for MRSA infections in pneumonia and deep skin wounds.

Anti-infective drug therapy is a very old and established field of medicine. Humans have been trying to find cures for the sudden mysterious ailments that afflicted individuals for a long time. Different natural medicines have been in use well before Alexander Fleming initial discovery and later Howard Florey's rediscovery and commercialization of penicillin.

Something my graduate school advisor, Julius Kreier, said once has stuck with me. To paraphrase, he said that the really outstanding individuals in any field don't have to act like prima donnas or constantly remind everyone that their work is considered remarkable. It's the second tier (or lower) of individuals who seem to be forever self-promoting; always reaching to get to where the great ones reside. Being bombastic is never attractive. Self-promotion, perhaps the new norm in our society, used to be considered oafish. The really outstanding individuals are typically humble, accessible; readily acknowledge their own limitations, and their good luck in their modest achievements.

Early Years

There was not much in my early years that would have lead anyone to predict my future role as a scientist and drug developer. I was raised in a lower middle class family in Toledo Ohio, the son of an accountant and a secretary (administrative assistant in the current lingo). My dad was the first Alder known to graduate from college. My mother was raised on a farm in Oakland, Nebraska, and graduated high school. We were raised with an attitude I would later come to call "Toledo" – the belief that all the "common working people" were going to get the shaft (cheated) sooner or later, and that anyone who has made it big must have done so by unfair or illegal means. In Honoré de Balzac's tragicomic novel "Le Père Goriot", he states (to paraphrase), that "the secret to a great success, for which you are unable to account, is a great crime properly executed."

My parents used a catch-phrase when asked about how some situation was progressing: if asked how their (example) sore back was, the response was "just as bad, if not worse". This phrase sticks with my brother to this day; I've banished it from my vocabulary. But my brother is still in Toledo, I left long ago.

We had a healthy contempt for anyone in authority – bosses, politicians, public figures. They were all considered arrogant, pompous, and condescending. We poked fun at them. Perhaps this is a bit of the American attitude; we bow to no royalty. However, we were also resentful of great success, and while offering short term satisfaction of one's own failings, it was not a healthy attitude.

There was also a type of puritan attitude that having great fun might just be mildly wrong in some way. I can't really remember my parents ever coming home from work happy. Work was not fun for them, especially for my father. Work was something to be endured, not enjoyed or appreciated. It was a source of money, and the more money the better. But to be fair, I heard a saying later in life (from someone who had clearly made it) that rings true: "Money is not that important – unless you don't have enough of it!"

My parents presented a view that the bosses were stupid and inconsiderate, business companies were bad, and coworkers were occasionally slackers. This is similar to the view shown in most popular books and movies. The workers are good; the bosses are not. I think this helped shape an early view of work, and probably helped me to seek a different situation for myself. In case this makes it seem as if my parents were liberals, it should be stated that their view of government was even worse.

But this probably paints way too negative a picture of my childhood. It was a different time than today. We kids actually went outside and had unsupervised play. We did not eat from the refrigerator at will, and were not allowed to consume candy, soda pop, or watch TV whenever the mood struck. Control of such behavior at the time, was not considered abnormal.

Both my parents eventually became alcoholics; I've wondered if their attitudes led them down the alcoholic path, or if their alcoholic path led them to bad attitudes. But they genuinely tried to provide a good home for my brother and me. The popular portrayal of alcoholics as violent, emotional, unhinged people is not necessarily the norm. Our parents provided for my brother and me, and there are no horror stories of abuse to be told. All of us were very unhappy at parts of our lives, and each of us learned to cope. Both of my parents sacrificed themselves to try to raise my brother and me; at some point the pain became too much and they turned to instant relief.

School, Growing Up, and the Neighborhood

I was certainly nothing special in grade school. In fact, I was well on the road to a "Toledo" attitude. I began to mirror my parent's attitudes: the teachers were all bad and unfair, and I looked for the worst in people. I did excel somewhat at childhood games. Our neighborhood in the 1960s was Polish and Irish, and very homogenous. I was a middling student in grade school and was getting into some minor trouble; basically tussling and fighting with other kids. I liked baseball, kickball, and became a Detroit Tigers (and to my lasting detriment) Detroit Lions fan.

Then integration happened to our school and neighborhood. I was still a little young to understand the concepts, but I did recognize that my parents became worried and afraid. The idea of integration in Toledo was to build low cost housing units with minimal financing needs around established older middle to lower middle class neighborhoods, encourage minority buying, and then let the magic of integration happen. I remember when our older neighborhood of modest, brick houses on small tree lined boulevards became surrounded on three sides by the first housing projects. They were single story, slab housing in a former farmland and woods where I had played as a younger child. A housing unit of 200 homes went up in just over a year. To my parent's dismay, a large billboard proclaimed that "$200 dollars moves you in!" This was in the late 1960s and early 1970s.

I don't recall overt racism from my parents. They were not the kind that truly believed in racial inferiority. It was more that they were worried about the value of their home falling (it did), worried about crime (it increased), and the safety of my brother and me; just worried about everything. This concern was based on life's experiences, and a firm belief that things would *not* work out for the best.

Both of my parents have passed away. My dad died young at age 60, when I was 25. He had a 10 year gradual decline of emphysema, cirrhosis of the liver and general bad health. My best years with my dad were his last few, when he had quit drinking and some of his sense of humor had returned. I think we related better to each other as adults, and I grew to see the good qualities in him. My mom liver to be 80 and she died when I was 51. After my dad died, I helped her move back to her hometown of Oakland, Nebraska (population < 1,000). She lived 16 happy years in retirement in rural Nebraska, with county fairs, local coffee hour, and local news. Rural life is not for everyone, but the desire for a rural life seems to also be in my genes – I also look forward to a retirement in a rural setting.

At school, I went from being the one of the best athletes (in 6th grade terms) to not even being close to the top. It was quickly apparent that the new kids were older, tougher, and willing to use more force. This has had a deep influence on me throughout my life. Flash forward from grade school to high school, and I worked a summer in a Black-Hispanic majority factory making fiberglass shower and bath stalls. (Toledo is the fiberglass capital of the country!) There was a definite self-imposed racial separation of the workers, but also casual respect between groups. None of this silly pandering that you see from some people, but rather an acceptance and respect that everybody would defend themselves, nobody would get a break based on race, but if someone got hurt on the job, everybody would help. With patronizing removed, we were all on equal footing on the factory floor. Nobody claimed to hold racially superior views. Here I learned to respect, appreciate, but also evaluate others based on what they could do, independent of their category. In some odd way, it has helped me in my work with broader cultural categories across the globe.

A big break came my way during high school. It was unexpected, at least by me.

I had been working summers as a caddy at the local country club. The Inverness Country Club brought me into contact with successful people who in Toledo terms had "made it". I wish I could say there was someone – some local bigwig - who had a profound impact on my life, some mentor who changed my worldview. But it didn't happen that way. If anything, my experiences with the rich local golfers tended to reinforce views of bosses and the rich as typically thoughtless, condescending, and somewhat pompous. There were notable exceptions, and in retrospect, some of the golfers were actually quite decent. The caddies collectively called the "Jobs", as in "I got a bad Job today; only a one buck tip." "Job" was not an insult, it was a functional description that also reinforced the division between employer and employee. The summers spent as a caddy have given me an appreciation of workers who earn part of their income by tips.

What caddying did get me in addition to a little money was an opportunity at a scholarship. Over 50 years ago, the golfer Chick Evans established a scholarship fund for deserving caddies who merited help based on financial need and outstanding scholarship. This fund has grown through the donations of the generous golfers (This generosity was another hidden trait that I did not appreciate in the "golfers") and also from Evans scholar alums. Most any caddy at Inverness would qualify based on financial need. But what about the academic qualifications?

I started to blossom academically in 7th grade, and this continued through high school. I can't explain exactly what happened to change me. It was almost as if a switch was thrown and I suddenly found the hard sciences interesting and a challenge in an exciting kind of way. In grade school, math and science seemed babyish; in junior high and high school they seemed to have much more application to the real world.

I still remember one of my first physics experiments. It dealt with light behaving as both energy and particle. The experiment was to have two slide projectors cross beams at some angle and observe if the images on the two screens were whole or disrupted. I was pretty sure I knew the answer, but still the moment of doing the actual experiment to prove or disprove a concept involving electrons too small to see stayed with me. (Of course, light behaves like energy under these conditions and the two images appear complete).

I got 7 A grades in my first report card in 7th grade. I was on my way! Throughout high school, I was in the top classes and had a high grade point average of 3.8 in the days when an A grade was not easy to come by. But I had to work for my grades and was not the top student in my class of 550 graduates (I edged into the top 10). But this peer pressure helped me strive to improve.

I was awarded an Evans Scholarship to the Ohio State University in my senior year of high school. This was a life-altering experience that put me on the way to my career path. I had not traveled much, and had not spent much time in the company of strangers. I would now do both.

College

At Ohio State, there was a significant culture change. I think that young people benefit from early broadening experience, and an opportunity to lead. At Ohio State, I certainly experienced the broadening.

Academically at Ohio State, I was solid, mostly B grades and some A grades (3.2 Final GPA). But what I did learn is the value of rational thought, hard work, and a good emotional quotient. It was the norm in the 1970s and early 1980s for the really smart people to be portrayed as brilliant, but eccentric. The thinking was that really smart people could be arrogant, difficult, and pompous. It still surprising how many scientists and physicians seem to think that being pompous must make one exceptional (all it accomplishes is making one exceptionally pompous).

Drug and excessive alcohol use were common among students newly away from home, and many really smart people destroyed themselves by trying to ape the popular images. Being away from home is a good growing experience. There appeared to be differences between young college studies who left home and those who stayed in their hometown. Growing up and gaining some level of maturity early is a good thing. Keep your ties and friendships from home, but be ready to move and grow.

I had to study hard to make grades. My major was microbiology and the coursework was mostly challenging. This constant challenge to make grades would help me in my career. Those who had it easy seem to be easily frustrated when a problem is not quickly solved. Kind of like a strongman who is frustrated by the first thing he finds too heavy to move. I learned that the best problems do not yield easily, and there is a better satisfaction from solving deep problems. This has also taught me a suspicion of people who claim to instantly see the solution. Few people are as smart as they think or as they project themselves. There are many new things that we encounter for the first time each day. The people who ultimately succeed in science are those who are able to stay with difficult problems and come up with the difficult answer. This skill set seems to be getting much more-rare.

As I was nearing the end of my college career, I had the opportunity to work as an undergraduate in the research lab of Julius Kreier, who was a well-respected malaria researcher. This chance encounter would shape significant portions of my career.

The experience was similar to that of 7th grade. Here was real science; not the baby stuff, but real discoveries. I was impressed with the people and the science. My initial projects were really just helping the graduate students with their projects. The projects involved mouse and rat models of disease, and experimental vaccines.

This early exposure to animal models had a lasting positive impact on my career choices. Experiences with animal models would be reflected in the rest of my career. While my contemporaries were focused on molecular biology applications, pure in vitro research, and pieces of DNA, I was learning about whole organ systems, physiology, and the interactions of elements at the body level. This training would serve me well.

I decided to go for graduate school instead of medical school. The idea of doing research and making discoveries was more attractive than treating sick people one by one. This would have consequences for my career path. In the USA, physicians have been held in high esteem – some merited, and some not. In most of the rest of the world, physicians do not hold such a vaulted place by dint of degree. As the USA moves to a more socialized form of medicine, a big change will be that physicians transform from being employers to employees. Physicians will become company employees who will be expected to follow procedure and will be accountable for costs and results. This will have both positive and negative consequences – in net centralized control of complex processes is negative. Medicine will become more standardized in the USA, to the benefit of cost containment and to the detriment of quality.

Graduate School

Graduate school was a tremendous learning experience. I also learned the value of people and thought. Today, we have become completely enamored of technology, the quick analysis, the instant Google-supported answer and the quick comeback.

An amusing antidote is that some people blame Frank Sinatra for the fall of our education system and the decline of the USA lead. Frank made it acceptable to be cool rather than smart, to be cagey rather than intelligent, streetwise instead of thoughtful. After Frank Sinatra, the heroes were no longer the Einstein, Salk, or Edison. People of quick action, the quick comeback, and a crafty, shrewd approach to life replaced these heroes. These are the people who have "the angles" rather than the deep thoughts.

This probably puts gives too much blame on a single person. But our culture does seem to glorify the action hero. Quick – who's a famous scientist in the world right now? Most people might come up with something along the lines of "the guy in the wheelchair" (Stephen Hawking), but could easily name a dozen or more athletes.

In graduate school, it was important to persevere through negative results, think about the data and then plan the next step. A good and qualified scientist is able to plan an experiment, execute, analyze and interpret the data, and then plan the next experiment based on the results of the first. I learned in graduate school to improvise, plan new approaches, and adapt to changing conditions.

My Advisor Julius Kreier was instrumental in guiding these approaches and I am forever grateful. Julius viewed graduate students as people to be developed to their fullest capability, rather than instruments to advance his own research. This allowed them to develop more fully and had the effect of multiplying the influence of Julius through the impact of his many students in the world.

My thesis research certainly changed during the course of the years of work, and the finished product was nothing like the original plan. The exact topic (immune complex mediated immunomodulation) and perhaps the scientific advances (to be honest – minimal!) of my thesis are perhaps not really that important compared to the goal of training a scientist to plan, execute, analyze, interpret, and then start planning again.

My graduate work involved classification of malarial antigen classification and the effect of antigens upon the immune response in an animal model system. I fully doubt that my thesis has been read by anyone after I graduated (including me). But I learned a lot from the process and the thesis was training for a career in drug discovery and development. The goal of training a budding scientist was accomplished.

Similarly, the exact ranking of the college in which you train is less important than the level of training and achievement of the student. Ivy League schools such as Harvard, Yale, Columbia, and some of the other elite biomedical schools (Stanford comes to mind, but there are about a dozen that would rank highly) do have a certain cachet. But a poorly adapted scientist from a top school will not fare as well as a well-rounded person from a middling school. Certainly my Schools, Ohio State for graduate school and then University of Wisconsin for post-graduate work, would be considered middling to upper middle. But these rankings are artificial; each person is a product of their environment, work habits, and approach to life.

Post graduate school took me to the University of Wisconsin. I underwent some stress in my personal life that made some days very bad, but the work environment was excellent. My advisor, Dr. Ron Schell had striking similarities to Julius Kreier. Ron clearly cared about his students and post docs as people, and he gave each person latitude to work and develop. Ron was also dedicated to developing students in an environment where many other advisors used students to advance their own careers.

I was again working with animal models, this time with syphilis and Lyme disease. The approaches to my thesis work with malaria were somewhat similar, and I again worked on immune responses to complex host-pathogen interactions. The work involved incremental advances, and I learned that there was not to be a great final eureka moment when everyone runs down the hall with a new scientific discovery. But once again the latitude given in planning through interpretation of scientific results reinforced a self-motivated work ethic. Ron Schell, like Julius Kreier, is another developer of scientists, and we need more people like them.

My professional career has now involved four biopharmaceutical companies and 25 years. I have worked on antibacterial drug discovery and development, with a few years of oncology and anti-viral research only. I've had the pleasure of working at two large pharmaceutical companies and two small biotech companies. The environments are very different; but both can be very rewarding.

I have been fortunate to contribute to the research and development of five drugs that have been great successes, and to one that was a complete failure.

Work in the pharmaceutical industry teaches the great interconnected nature of research and development, and the extreme importance of quality in work. Academic research tends to isolate scientists who are expected to become great experts in a fairly narrow field. The pharmaceutical industry teaches that to be successful, a scientist should be well rounded and conversant in multiple fields. I started out in a very basic research pharmacology position, similar to my graduate work. I now am in a pure clinical development (human medicine) role, including designing, analyzing, and interpreting the clinical development of an antibacterial drug portfolio.

How does one progress from mice and rats to humans? Several quips come to mind, but in the interest of this article, we'll stay on task. The teaching and training from graduate and post graduate school proven invaluable. Ability to plan, analyze, and move to the next step is a key attribute. This training seems to becoming more-rare in the current crop of scientists. The ability to be responsible and accountable for work, in a positive sense, is a skill that will make a scientist invaluable in all situations. Planning a clinical trial in 1,000 people is obviously a different kettle of fish than an experiment with 20 mice, but the principles of following sound scientific, medical, and ethical principle remain.

Early stage research is about overcoming failure. Nearly everything fails initially. The successes in drug discovery are few and very far in between. Some scientists have worked their entire careers without the fortune to be associated with a successful drug. There are many times when the laboratory work seems futile. However, simply staying on task represents a kind of victory. Some of my best memories of lab work are coming home after a very long day (and sometimes night) in the lab and knowing I had done everything I could possibly do to advance the project.

Later drug development is about minimizing risk, especially risk to the patient. The clinical design changes from that in experimental animal studies. In clinical research, generally every test must be pre-specified, every data analysis plan must have statistical backing. There is no "try and see what happens" in clinical research. Data mining a clinical trial to look for interesting data is certainly done, but the results (especially if positive for the drug) must be confirmed in a preplanned trial. This is a key difference between academic study and pharmaceutical study.

I've had the pleasure to work on one of the first successful HIV drugs, several drugs to treat serious MRSA infections, serious pneumonia infections, and in general drug-resistant bacterial infections. I've been able to interact with the FDA and the major FDA equals (the EMA for Europe, the PMDA for Japan, and the cFDA for China) for gaining approval for these life-saving medications. I've also worked closely with key opinion Leaders (termed KOLs in the business) in these fields. All of this is extremely gratifying. My career highs have almost all come as a result of a successful trial and final approval of a new drug.

It was mentioned above that I've had worked on a drug that was a failure. The study of failure is usually an interesting topic and a brief analysis may add some insight.

First, the pharmaceutical industry does not try to develop toxic, ineffective drugs. A drug that successfully survives all the clinical trials and is finally approved by the FDA (and the European, Japanese, and Chinese equivalents) has consumed 500 million to close to 1 billion dollars in research and development money. A typical anti-infective drug must be sold on the market for 5 or more years *just to break even*. True profit from a program standpoint occurs in the later years.

The biggest financial disaster for a drug company is to have a drug progress all the way through approval (500 million to 1 billion spend) and then fail due to an unexpected lack of efficacy or safety. This results in a huge loss.

But there is another aspect that is not mentioned. The scientists and clinicians in a pharmaceutical company are trained in the same places as academics – Universities. The ethical standards of the pharmaceutical drug discovery and development people are underappreciated. In 25 years in the business I have not once encountered a researcher, developer, or commercial person who suggested something that would deliberately put a patient at risk. There are in fact debates about the optimal dose, best drug, acceptable adverse events, and all other aspects the balance the risk and benefit of patients who are often desperately ill.

Of the five successful drugs in my career, all of them had very dark moments when people inside and outside the company were convinced that it was not going to work, and the drug was either ineffective or toxic. I've never worked on a drug that had no issues or any factor that did not require careful thought and application of some sound judgment of difficult situations arising in its development.

It is amusing to compare the portrayal of a pharmaceutical company in the popular media compared to the reality. In the movies, the pharmaceutical company is (for some reason) pressing ahead with the development of a drug which has some horrible side effect. This side effect has escaped the notice of the hundreds of scientists and clinicians who have developed the drug. The only persons aware of this unpleasant fact is the company CEO, identified by wearing an expensive suit, and perhaps his thug bodyguard. A brave reporter, perhaps added by a University scientist break into the company at night, enter a data room, and by selecting a file at random, or perhaps logging onto a computer, discover the hidden side effect data with about 30 seconds of effort. A climatic chase results, with the company SWAT team trailing the heroic reporter and professor, who escape and foil the evil company.

By the way, I've never once seen a company bodyguard, and while a SWAT team would come in handy on some days, I think the budget justification would be a bit difficult.

The reality is that in drug develop, every little or big adverse event is analyzed and debated. Every drug is toxic if the dose is escalated to a certain level. This level can even vary from patient to patient. But the idea of a drug in development with a known, horrible toxicity (my favorite in the movies on is the one that turns people into some type of monster!), with the toxicity covered up and concealed, with 100s of people, most of whom have never met, all conspiring to keep this a secret is laughable. And to what purpose; the secret would not remain secret for long once the drug is released for sale!

In the first few months on my career, I worked on an antibacterial drug that was in the final stages of approval. My work was unrelated; I was just starting out and worked on animal models looking for possible new indications. The drug was approved by the FDA based on data from about 5,000 patients. In the first months after approval, a serious renal adverse event was noted in approximately 49 patients out of about 50,000 treated. Statistically, it is possible to have an event rate of 0 per 5,000, when the true rate is approximately 1 per 1,000. Thus a spectacular failure was recorded, people left the company in disgrace, and lawsuits (from many more than the 49 people) followed.

In your career, your scientific confidence and perhaps your integrity will be challenged. There will be dark moments when your work will look wrong. Perhaps it is wrong – don't personalize your work inappropriately. Try to look at all work objectively, most especially your own. Avoid unproductive burning of your time and energy. Keep a positive attitude, learn how to truly consider other people's views, but also learn when to stay true to your own ideas and ideals.

Also important is the ability to work with people. Many skilled scientists see their progress retarded by a lack of people skills. The ability to influence others at all levels of an organization will serve you well. Start with the receptionist. This person sees all who enter and leave. Learn this person's name, greet them each day, and perhaps remember a small gift on appropriate holidays or birthdays. Also show great consideration to your Administrative Assistant if you have one. I have been bailed out of difficult situations many times by a good Admin. This person becomes your wingman through all situations. And learn to work well with your colleagues. My darkest professional moments have come about through a failure to work with certain people. Learn to work well at all levels – those viewed as subordinates, peers, and superiors. Most everyone was a subordinate at some time. Try to be a person that everybody wants to have on a team. Look and observe – are people happy to see you in the hall? If not, consider changing something about yourself.

Finally, become well rounded. Have hobbies and friends. I took up marathon running and triathlon competition, which are appropriate metaphors for research and development. Much time is spent in preparation to gain the few, wonderful moments when the finish line of a race is finally crossed. I've made great friends in running whom I am not really sure what they do professionally. A spouse can be a great addition to your life, and you will live longer and happier if you are in a trusting relationship. A spouse will complete an important part of your life. Spiritual fulfillment has become important to me, and can be a great help during a difficult time, and importantly keep you grounded when things are going well. May God bless you and your dreams, live happy, laugh often and be joyfully. As long as you're here, you might as well enjoy the view.

Early Years - I'm second from the left, on my
Grandpa's lap, age 4 (Nebraska, 1963).

A happy moment with my Dad– Birthday,
age 12, Toledo, Ohio 1971.

I remember taking this picture of my Mom and Dad – I
was about 10 years old - before they had a big Saturday
night out. I've learned to appreciate all they did and
tried to do for me. Rest in peace Mom and Dad.

With my Mom, High School graduation, 1977, Toledo, OH

The Author is wearing the hat of his beloved Detroit Tigers with the view from the deck of his Mountain House in the Catskills. This is where I plan to live a rural lifestyle in happy retirement.

CHAPTER 4.

Guiding Angels

Carolyn Brooks

It seems my life has been charmed. I have known so much happiness; I've loved and been loved, and admired and been admired by such amazing people that I marvel at how my life evolved – almost like I was always guided by those who saw something much more special about me than what I perceived of myself. I have had so many guiding angels in my life, leading me toward very rewarding opportunities. One of my weaknesses I believe is that I am neither ambitious nor competitive, yet I have risen in the world of academics, to some pretty lofty places, and when I look back in perspective, I realize I owe so many so much, that it amazes me.

Let me start by describing what I remember about my childhood. I grew up in the segregated society of Richmond, Virginia. I remember living on the north side in the 1950s where there were African Americans who were doctors, lawyers (even a judge – Judge Oliver Hill whose home I was invited to because of my friend Nannette), college professors (at Virginia Union University), and business owners. My parents, my sister and I lived in a nice house that belonged to my great-grandparents who allowed us to pay them rent until my parents could save up enough to buy their own home. I do not think it was a pleasant place to live for my parents, but my great-grandparents doted on me, so I was really happy there. My friend Claudette and I played quietly on her porch every day and it was kind of obvious that we were a little 'different' than the other kids who were more physically active. I could read before I started kindergarten, but I do not remember anyone ever teaching me. Education was not revered by my family – working hard was. My father was a truck driver and my mother worked in an antique store doing all sorts of things. I guess she was supposed to be a maid because she certainly kept everything clean, but she was very smart, so oftentimes she did clerical-type things for the owners. I guess that is where I got my love for beautiful things. My mother used to bring me some unique gifts from the store because the family of the owners thought I might like them. My mother had a high school degree but my father, who was from 'the country' did not have very much schooling.....but both of them were very smart and were always working sun up to sun down. My father was extremely personable and friendly. People always liked him. My mother was feisty.

The segregated black schools on the north side of Richmond were thought to be best because the children of the black professionals would have it no other way. So when my parents finally were able to buy their own home, it was in the blue collar west side of town. We moved to a new little house in a lovely neighborhood. One of my aunts lived nearby and told my mother to leave me in the north side schools and I walked to her house every morning and she drove me to the north side school where she taught. I was an excellent student. I guess you can say I was often the teachers' pet. No one begrudged me that and I was always helping my friends with their school work. I cannot remember getting anything but A's (until I got to high school and got a B or two) but I never remember studying. I just loved learning and could remember everything I was taught. Once I had a teacher who was boring and that was my wake-up call. He said nothing worth remembering so I had to read more.

My parents thought I was 'smart' but did not dwell on it. They never mentioned college and I remember my dad saying, "You are a smart gal. I bet you will get one of those big jobs at Phillip Morris or Reynolds Aluminum." My mom would let me buy records at the grocery store, and I bought classical music. I loved listening to the *Triumphal March from Aida* and then she started bringing me other records, especially Beethoven. She would also encourage me to check out books from the library and I remember borrowing thick novels like *Gone with the Wind and Jane Eyre and mythology, etc.* No one paid any attention to my listening to classical music or my love for classical novels though because I was also like other young girls. I was busy being in social clubs, busy in Girl Scouts, being president of the Student Government Association and president of academic clubs, being majorette captain and high school queen, and being a part of the most popular group in school. Life seemed pretty good to me. I did not know about the mean and unfriendly world of bigotry nor did I pay much attention to the fact that blacks had facilities that were separate and most definitely not equal. I started paying more attention though when our counselors started talking to us to see if we were interested in attending the white schools nearby. The integrating of schools was beginning in Richmond. I knew a few students who were brave enough to transfer. Life was tough for them there though and I was not brave enough to give it a try. My teenage years were terrific! I am somewhat ashamed to say, I was too selfish to be a martyr. Also, I had never forgotten seeing the gruesome newspaper pictures of Emmett Till, the young boy who had been lynched for what was reported as whistling at a white woman. I could not understand such hatred and it made me fearful of the white race.

My experiences in the 1950s and 1960s were as a happy black child and teen who did not know that I was lacking for anything. I was old enough now to pay a little more attention to the world around me though. My parents took the morning and evening newspapers and we did have a little television. My mothered struggled to find money to pay for me to take dance lessons like my more affluent friends. I took these dance lessons at the Chapman's School of Dance and the dance lessons on Saturdays were at a dinky recreation center in an impoverished neighborhood. The white girls took their lessons in an impressive facility on Monument Avenue. But we all had our recitals at the mosque (a very grand place) with beautiful costumes that my mother struggled to buy. So, I noticed that dance school inequality stuff. I also did not know I was supposed to ride at the back of the bus and from the time I was an elementary student I used to board the public buses, pay my money and sit right up front and no one ever told me to move. When I heard people talking about what had happened to Rosa Parks, I wondered if that could happen to me. So my friends and I started walking miles and miles and stayed away from the buses for a while. Then in high school we left a football game one night and went into White Castle (fast food restaurant) to buy hamburgers – they told us to get out. When one of my affluent friends got a car, we went riding in new parts of the city and saw a put-put golf course. We excitedly went to pay to play and were surprised when they told us "We do not cater to blacks yet." We were shocked.....where had we been? In Never-Never-Land? I had been selected by my church to attend a Presbyterian summer camp. Everybody was so nice to the three of us African Americans. The young boys asked me to square dance with them. The girls were curious about my hair because it was kind of straight. They all seemed surprised that I was just as smart as they were. I thought I had made some new friends and convinced myself that if the two races were not segregated, we would find out we had a lot in common and could even be friends. But....I soon changed my mind about that. I saw some of them in Thalheimers and Miller and Rhodes Department stores on a few occasions and they would turn away so I would not speak to them. I then decided, segregation was just fine with me.

I was in a trigonometry class when we learned that President Kennedy had been assassinated. For some reason images of Emmett Till resurfaced in my mind. I thought of George Wallace and the other governors who said under no circumstances would blacks be allowed to enroll at their universities. I knew these people were called conservatives and to this day anyone who says he is a conservative comes across as an enemy in my mind. I cannot shake this because it was the 'conservatives' who were adamant that blacks had to stay second class citizens. I wanted to stay in my segregated world and never be exposed to this direct hatred and evilness. Although I was greatly saddened by the death of President Kennedy, it had always been his brother Bobby that I regarded with much more interest and admiration.

While all of this was taking place, I was about to graduate from high school. There were more than 350 students in my senior class, and my grades resulted in my being salutatorian. Education in Richmond for blacks had been unique it seems. I took French for 11 years. In elementary school I was taught these French classes by a real Frenchman who seemed out of place in our segregated system because he was white. I also took German in high school from a real German and there was also a Russian at our school teaching the Russian language. I had college preparatory classes and my teachers saw to it that I attended any special programs at universities for those in high school who excelled in science. At one of these programs a medical microbiologist came to talk in the career exposure sessions and I was hooked. After hearing him talk, I knew I wanted to become a microbiologist. Yet college was never even discussed at my house. But I had guiding angels who were my teachers and guidance counselors. When I stood up at graduation to have my scholarships read by the principal, I was shocked myself. I had ten full or partial scholarships to Historically Black Colleges and Universities (HBCU) and know that the paperwork had to have been done by my high school guiding angels. As college forms kept arriving at my house I guess I responded most to Tuskegee Institute (University), because in the end that paperwork was finalized first. My affluent friends were mostly attending Howard, Hampton, Morgan, VA Union, VA State, or Norfolk State and I wanted to be different and be the one to go furthest from home.

Life in the small town in Alabama was foreign to me. College was a shock to me. The Civil Rights movement was strong at Tuskegee and black pride was everywhere. I felt ashamed of myself that I had been living such a self-absorbed life and now, at 18 years old I was paying attention and I wanted to do my part. I met the Tuskegee student, Tally, who eventually became my husband and he was close by when a Tuskegee student was shot and killed by a gas station attendant for something minor that he said to him. The whole campus marched on the town. Then there was the Montgomery March. I went with Tally (his nickname which is short for his hometown of Tallahassee) and oh how frightening the experience was. The police on horses threatened to harm to us. We sat on the streets for hours and had no possibilities for restroom usage and had been told by the march organizers – no matter what, do not move....not from the police, not from the horses, not from the possibility of water hoses or whatever they came with......and I really started to mature during this time. It was a very rude awakening to the cold world that it was (and is) for African Americans. It seemed strange that I knew so little about the real world. Fighting for our civil rights was becoming more and more ingrained in me.

Tally was my first college mentor I guess. He came from a family that did value education. Both of his parents had college degrees. He noticed that I took my academic skills for granted and it was he who made sure I would not skip classes and that I would stay focused on my studies. I was a freshman and he was a fifth year student finishing his student teaching and taking graduate courses. We married when I became a sophomore. By the time I was a junior I had my first son, Charles. Soon thereafter I was pregnant again with my son Marcellus. I was ready to quit school. Tally was employed as a County Agent (for black farmers) by Auburn University. I saw no reason why I couldn't just be a wife and mother. Tally was livid and wouldn't hear of it and my mentor at Tuskegee, Dr. Howard Carter who was a microbiologist himself and who had given me lab jobs and treated me like a daughter since he had become my advisor, made me ashamed I had even considered quitting. Tally was both mother and father to our sons so I could study (taking 22 hours one semester to include physics, biochemistry and an advanced biology course). During my senior year, in order to get a B.S. degree in four years, my mother-in-law and my father-in-law took my little sons for nine months so I could graduate. As a result, I did graduate in four years, with honors.

Since we lived in rural Alabama, with Tally working as a county agent, there was no job possibility for me but to teach high school. I had not been an education major so I taught with a special certificate. I taught biology, environmental science, physics and chemistry. It was depressing to teach these wonderful young teenagers who wanted to learn but were in a school that had no microscopes, hardly any chemicals and glassware...almost nothing but me.....struggling to teach in a school that had very few resources and where students were taught by teachers who lived fifty miles away and were teaching in areas outside of their expertise, I was depressed by it all, remembering how I had been blessed with a pretty good secondary education. Only the black students attended the public schools. Rather than endure integration, all of the white kids attended the county's private schools. My guiding angel at Tuskegee, Dr. Carter who was also my older son's godfather, asked me to come back to Tuskegee to get a master's degree and with the fellowship I got, I rushed back. Tally was delighted. Again, he did so much to help with my sons, and guess what, I was soon pregnant again. I had labor induced so I would not miss the beginning of classes and Alexis was born. No problem, in two years I had my MS degree in biology.

Tally decided that if one of his friends from college could get a PhD, he could too. So he applied to the Ohio State University to major in Ag. Extension and was accepted, received a Farm Foundation Fellowship and eventually an OSU assistantship as well. I was so happy….the five of us could move to the great city of Columbus. There would be so much more to offer my children there. Tally told me that I could not come unless I applied to graduate school too. I was so disappointed, but did apply and shortly thereafter I heard back – I was accepted into the PhD program in microbiology with a full fellowship. Darn!!!! I was tired of school. But we loaded up the U-Haul and off we went. I loved Columbus!

I received instructions that I was to go to the Microbiology building (it was about 5 floors) and ask to meet with the professors so I could find my major professor. Each professor I met kindly discussed the research going on in his lab, but all of them ended with – "Aren't you going to be Dr. Kreier's student?" I would say no, I have not heard of Dr. Kreier. No one I talked to, ended the conversation with an invitation to join his research lab. I noticed that there were very few females in the labs and I saw no African Americans. I had just come from a world where I rarely saw anyone but black people, and now, while there were international students, there were no African American students that I saw in the entire building. I went to my first class, and I was the only black in the room. We were assigned lab partners. Mine was a blond male who immediately said to me, "I bet you are here on that affirmative action stuff. Just my luck....you don't know enough to help anybody." I was dumbfounded and immediately started trying to figure out how to tell him off before going home to Tally to say – I quit. But I endured. He and I never had any kind of conversation. We just did our work together and finished the course. He dropped out of graduate school. I hung in there. Finally I heard that Dr. Kreier was back from Turkey and I went to meet him. The whole atmosphere in his lab was startling. It was crowded, seeming like there were several students and not much space. Those who were working looked up and they spoke to me! Several were obviously international, some were females, one identified herself to me a few months later as a 'lesbian and proud of it,' one said he was Jewish and that was a drawback with some of the professors in the department, and low and behold – there was Sally, a black student. Dr. Kreier was so warm and welcoming and invited me to join his lab right away. He told me I had to work with either *Babesia* or *Plasmodium* because these were the parasitic microorganisms studied in his lab. He put me to work right away....giving me readings and I could tell, this was the place for me.

While Dr. Kreier was obviously a man who loved people – all kinds of people....as a matter of fact he seemed to relish the camaraderie that comes with the sharing of thoughts and varying opinions, there were several professors that had very low perceptions of African Americans. I was so very proud of Sally, my dear friend who was the shining representation of African Americans in terms of her academic prowess. But Sally and I could cite many occurrences that demonstrated how some of the students and some of the professors seemed resentful of this. While I understood why Sally was very bothered by this, I could escape this shadow hanging over us because my world was primarily shaped by my children and my husband. Still, I had to make sure I did not let my race down by doing poorly in my classes but I also could not be derelict in my obligations to my husband and my three children. So oftentimes I was both mentally and physically exhausted but I trudged on through. After taking my comprehensive exam I came home to my husband who was anxiously waiting to see if I had passed, I told him I had but he saw no joy. He asked me why and I said I was certainly no shining star in there. His response – "You passed right? That's what's important." After all, we had heard that one grad student had failed his comprehensives and left and committed suicide. I know I owe a lot for my continuing progress to Dr. Kreier, my Ohio State guiding angel. While he never said so, I am sure he knew some of his colleagues did not think kindly of black students (to put it nicely). He chose a wonderful committee for me and I could tell each of them really wanted me to succeed. Two of them I particularly want to mention and they were Dr. Rod Sharp (I am blessed to still see Dr. Sharp from time to time) and Dr. Robert Pfister.

After all Dr. Kreier had done for me and after getting over the biggest hurdles to acquire a PhD in microbiology (finished all my required course work and passed the comprehensive exams), I now asked to see him and told him I would have to drop out for a while, because I was pregnant again. Dr. Kreier seemed to first be shocked, then disappointed in me, and maybe he was even a little angry with me. He even mentioned how disappointing it is to have female students drop out to get married or pregnant. I said, "I'll be back Dr. Kreier, I promise." I could tell he didn't believe me, but he didn't stay mad at me long and was back to his warm and 'guiding angel' self before I left. I went to Frankfort, KY. Tally had finished his PhD and had gotten a job at Kentucky State University and had taken the three children with him so I could finish up my research and graduate. Things were on hold though until this baby girl, Toni, was born. In the meantime, I applied for what was called a Ford Foundation dissertation year fellowship so I could go back to OSU. Both Drs. Carter and Kreier wrote letters of support for me and I got the fellowship. So after Toni was born and was old enough for a baby sitter, I left Tally and the boys in Kentucky and took the two girls with me to Columbus and worked my behind off to finish up my research and the writing of my dissertation. Dr. Kreier told me he never really thought I'd come back and he was proud of me. Having him say that meant a lot to me. I finished quickly and unlike how I felt about my comprehensive exams, I think I did pretty well on my thesis defense.

Kentucky State hired me. I did get some letters from some major research institutions asking me to apply for faculty positions, but having my family all together was my main goal. To be honest, I did not want to be at another major university where I would always be feeling I had to prove myself and that I was just as bright as everybody else, not just an example of 'affirmative action.' I loved being at a minority institution and they took note of my capabilities and my work ethic and I kept getting job promotions, even though I did not seek them. I started as a research associate in a research program and shortly moved up to a principle investigator and then the Program Director. I loved the people I worked with and even got along well with the 'difficult ones.' But I did not like Kentucky. The African American population was small and my children seemed to always have to prove themselves in this new environment. For example my oldest child, Charles, has always excelled academically. I remember a slip of the tongue by his math teacher who said "I am surprised at how good he is in math." Surprise was always the sentiment when my children excelled. So I convinced Tally that he should take the job he was offered at the University of Maryland Eastern Shore (UMES). I would be closer to my family in Richmond, a big bonus.

UMES also hired me and the President was so happy that they finally had an African American, and a female at that, in the sciences. For while UMES is an HBCU, there were very few blacks in biology or in the agricultural sciences. My salary came from the land grant formula funds for agricultural sciences and so I had to adjust my research program to be agricultural in nature. I worked on developing microbial insecticides (viral and bacterial) and I collaborated with a colleague on detecting the causes of subclinical mastitis in cows. I began work with the nitrogen fixing bacteria and advanced to developing plant tissue culture in my laboratory. The campus went out of its way to modernize laboratories for me and though I was expected to be a researcher, I was liked by a lot by the students who worked in my laboratories and before long I was a freshman advisor and agreed to teach courses in cell biology, environmental microbiology, agricultural microbiology, recombinant DNA technology and seminars. Many of the freshmen I advised saw me as their mentor and were always coming by to see me. I increased the enrollment in the department because several who had not chosen majors would ask me what my area was, and then say, well I want to be in agriculture too.

The President was always putting me on science related committees as the UMES representative. There again, I was faced with the old stereotyping of blacks in science. I was now older, more confident and oftentimes rather combative if anyone spoke unfavorably about HBCUs or minority students. I hated sitting around a table with representatives from the University of Maryland System and always hear things like, we cannot increase the number of minorities in our STEM programs without first putting some remedial programs in to get them better prepared. Yes, many of the students I taught or mentored at UMES came to college unprepared, but many of them were truly stellar students. I always spoke out and it seemed that everyone would bow their heads and once I would finish speaking they would move on as though I was not there. At first I would drive back to the Eastern Shore crushed and angry, but I decided I was foolishly getting overly stressed with this. I'd have to do better! Before long I did get better, speaking with more proof or with documentation and not getting emotional in my delivery. More professors and administrators at the other universities wanted to collaborate with me and began to listen to what I would say after I got a little more polished. They would ask me to collaborate with them or wanted me to send them more of these 'stellar' students for their programs. I was also often called to give speeches at a variety of conferences to talk about my opinions, my experiences, and my concerns. Well, the pressure was on again. It always seems blacks have to prove themselves and if one or some of my students did not perform well, that was justification to say – black students can't cut it no matter what Carolyn says.

So I started acquiring my own external funding for undergraduate research for UMES students. In the meantime I still had plenty of research to do because of grants I had acquired and I had become a department chair and I served on dozens of university committees.....and I had four children who deserved all of the support and attention I could give them. Also I had a very supportive husband and he deserved my support and attention too. Some of the funded grants one of my colleagues and I had required us to work 6 – 7 weeks in West African yearly – me as the microbiologist and tissue culturist and he as the plant scientist. Tally, great husband that he is, never complained about having to take care of the children while I was away.

I apparently have been a pretty good mentor. I had some guiding angels in my academic pursuits and I guess that washed over on me. My mentees were and are like family to me. They enriched my life so much because I relished seeing them achieve. My colleague Dr. Mervalin Morant and I oversaw them as they competed in oral and poster presentations and they always won (I have no examples of any of them not winning a competition and there were many.) We developed an exchange program with Michigan State and found mentees in Michigan who cared for them like family. We took them on trips, to conferences and became not only their undergraduate advisors but also their major professors for graduate degrees. Several of my students went on to earn doctorates, either with me or elsewhere and they still communicate with me regularly. I help them with their programs or initiatives still and now they help me too. My third child, Alexis, followed my life almost to the letter. Her high school days and achievements were almost identical to mine (she was even salutatorian) and she has a PhD in microbiology and is a department chair now even as she teaches and does research. The dean at the HBCU where she works says he is grooming her to succeed him as dean. I forgot to mention, I was dean of the School of Agricultural and Natural Sciences at UMES for 11 years before accepting the current job that I have as Executive Director of the Association of 1890 Research Directors (18 HBCUs). For five of those eleven years when I was serving as dean, I was also the President's Chief of Staff and Executive Assistant. Tally and I must have been pretty good parents because our children and their children seem to think so. Charles has a top level position as Global Officer for ACE, Inc., an international firm, Marcellus is in management at AT&T, Alexis is an Associate Professor and Dept. Chair at Bethune Cookman University and Toni has a county supervisory level position in Special Education in Hartford County, MD.

So, yes my life has been charmed. I married an amazing man and our marriage is still 'fun' because we enjoy each other so. I have had positions at HBCUs and that is my passion. I have been recognized with awards from the White House, from the State of Maryland, from the American Society of Microbiology, from the USDA and regionally, from the university system and from the university. I was chosen as one of *Maryland's Top 100 Women* in 2007. But what meant the most to me in terms of awards is that my family is huge.......I have so very many mentees who continuously let me know how much I meant and mean to them and the feeling is mutual. Many of them are mentors now and it is much harder for them than it was for me. It would take a whole essay for me to explain why. But, I still live a charmed life, and I still owe it to so many. I remain deeply grateful. .

CHAPTER 5.

An Academic Scientist's Life

Francis Edmund Gabriel Cox

Family Background

There was nothing in my family background to suggest that I might become an academic scientist. My paternal grandfather was a soldier and my maternal grandfather was a factory worker. My mother, Mary Elizabeth, was born in 1896 in Athlone in Westmeath in Ireland. During the eighteenth and nineteenth centuries, Ireland suffered from a series of famines the most serious of which was the Great Famine of 1845 to 1852 caused by the failure of the potato crop: over a million people died, a further million emigrated and the population fell from 8.5 million to 6.5 million. Another famine occurred in 1889 and, although it was not as serious as the earlier one, it had a similar effect on emigration as the population panicked in case history was to repeat itself and by 1891 the population had further declined to about 5 million. My grandparents were factory workers employed as weavers which gave them some stability and enabled them to raise my mother in some degree of security and comfort and I have a photograph of her at about the age of 12 with a bicycle which must have been a luxury at the time. Ireland, however, was not a particularly good place to grow up in and no place for an adventurous young woman so, like so many others, my mother migrated to Canada where she was employed in domestic service and later migrated to England where she continued with the same kind of work in and near London.

Alfred, my father, was born in 1887 in Broughty Ferry Castle, then an army garrison, in Fife in Scotland. His father, Charles Frederick, was a regular soldier who had married his wife Jane in 1875 and had five children all born in different garrison towns as my grandfather moved from posting to posting. In 1901 Charles had retired from the army and the family was living in Finchley in North London. In 1911, Charles, Jane and their children Edward, Alfred, Alice and Harriet and Harriet's son Frederick were all living in a two-bedroomed house in East Finchley. After leaving the army, Charles had worked as a drayman. Draymen were the men who drove drays, low, four-wheeled carts with no sides used for the conveyance of many different kinds of goods particularly casks used in the vintners' trade in which Charles was employed. The drays were the all-purpose vans of their day and were drawn by massive horses bred for heavy work such as ploughing. A drayman's job was very demanding and skilled and by 1911 Charles had given up this work and was employed by the Finchley council as a road sweeper. Alfred, in the meantime, had himself become a drayman.

In 1914, the First World War began and Alfred, like millions of other young men, joined the army and served with the Suffolk Regiment. The Suffolks were involved in many of the bloodiest events of the war and suffered massive losses. My father never talked about his time in the army and I discovered that this was also true for many of the others who served in the First World Way. When the war ended in 1918 Alfred returned to Finchley and resumed his career as a drayman and married his first wife Grace who died in childbirth a few years later. In 1931 my father, an Anglican, met and in 1931 married my mother, a Catholic, in one of the smartest Catholic churches in London, St James, Spanish Place and the family moved into my grandfather's house in Finchley which was then also occupied by Charles and his daughters Harriet and Alice. I was born on September 10th 1932 in a maternity hospital in Hampstead, close to Finchley and one of the smartest of the London suburbs. Living in my grandfather's crowded house was not an ideal arrangement and shortly afterwards my parents and I moved to a flat in a large Victorian house in another part of Finchley where my sister, Christina was born in 1935 and where she still lives.

The 1930s were not ideal for rearing children. Britain had just recovered from the ravages of the First World War when the effects of the Great Depression that had begun in the United States in 1929 reached the United Kingdom. With the rapid decline in American prosperity, the flow of money dried up, America and other countries stopped buying British goods and using British ships and the British economy collapsed. The greatest effect was on employment. There had been chronic unemployment throughout the 1920s but by July 1930 unemployment had reached 2 million and 2.5 million by December and was heading towards 3 million. There was a gradual recovery from early 1933 but this still meant that by 1934 16.75% of the potential working population was unemployed. At the same time there had been an increase in the use of petrol driven vehicles and the need for horse-drawn carts and carriages sharply declined. In 1934 there was an unemployment act aimed at supporting the unemployed of whom my father could have been one. Instead, he became a road sweeper, a job in which he took great pride. He also was very proud of the fact that he had managed to find work, albeit of an unskilled nature, in a time of high unemployment. He never lost his love of horses and used to take me to displays of working horses in Hyde Park in London but I was too young to sense the disappointment he must have been feeling not to be working with these marvellous animals. Things were not easy and my mother had to take up some domestic work but my sister and I never really felt deprived largely because we were part of a larger Irish and Italian Catholic community who were all living under similar circumstances.

Elementary Schooling in England in the 1930s and 1940s

During the 1930s, education in Britain was undergoing a period of upheaval and provision was massively unequal. The wealthy sent their children to fee-paying public schools (this might be confusing to readers in the United States because these schools were actually privately run). Public schools, most of which were boarding and nearly all for boys, provided the country with its university entrants, armed service officers, lawyers, clergy and politicians. In many of the big cities there were also long established grammar schools intended for bright boys whose families couldn't afford a public school education and whose fees were often subsidised by local authorities. The majority of pupils, however, had to go to local schools which provided a basic education from the age of 5 to 11 after which pupils could move to secondary schools until the age of 14. In 1931 only 20% of children received any secondary education, a situation that had concerned politicians for some time but an attempt to improve this by raising the school leaving age to 15 was rejected in 1931.

In the 1930s the school entrance age was defined as 'those who reached the age of 5 at the beginning of the school year' which was at the beginning of September. As I was born on September 10th 1932 I was not eligible to go to school until the beginning of September 1938 when I was nearly six years old and this year gap was to follow me through the rest of my schooldays. The first schools were variously known as primary or elementary schools and my local one was St Mary's attached to the Church of

England School of that name and a short walk from my home. I don't remember much about my time there but I assume that the education must have been adequate because I recall being able to read as long as I remember. Then a year later the Second World War began on September 3rd 1939. I remember this very well. Our mother had taken my sister and me for a holiday to a place called Newton Abbot near Bovey Tracey in Devonshire to where my godmother had moved and for whom a friend of my mother's was working. I remember two things, picking apples in the orchard and everyone listening to the radio in silence. My mother then had an immediate problem, how to get us back home because all the trains had been commandeered by the army. In letters to my father, my mother listed several possibilities including the three of us remaining in Devon and her taking work there. In the event, trains gradually became available to civilians and we returned home to a world of darkness, shortages and rationing. In July 1940 the Germans decided to invade Britain beginning with attacks on airfields in the southeast of the country and in September by bombing London. Finchley is less than 10 miles from the centre of London so well within the target range for the German bombers. These sporadic early bombing attacks were merely a prelude to 'the blitz' when the Germans bombed London every night from September to November 1940. Every night we heard the warning sirens and heard the sounds of anti-aircraft gunfire. These were exciting times for small boys and we spent much of our spare time collecting shrapnel. A number of bombs fell in and around Finchley and several pupils from my school were killed. Safety for children was paramount and parents were encouraged to send their children out of London as 'evacuees'. We had neither the money nor the contacts to take part in this adventure so we just lay in our beds hoping and praying.

One immediate consequence of the German bombing was that St Mary's School was converted into a first aid post and we the pupils were left to continue our education as best we or our parents could. This even involved having lessons in people's houses including those of Italians whose menfolk had been interred after Italy entered the war. My disrupted education resumed when my sister and I moved to Our Lady of Lourdes Catholic Primary School which was about half an hours walk away. This was a very good school and my English, maths, history, geography, French and Latin flourished but I learned virtually no science.

Secondary Schooling in England in the 1940s and 1950s

In 1944, an event occurred that was to change my life and that of millions of others like me. The Government returned to considering the inequalities in education. In 1930 there were 5.5 million children in primary education, but only 600,000 in secondary education thus the majority of children were still leaving the system too early. These figures were little different in the 1940s but, as a result of the war, society had changed and there was a new demand for changes in education. The 1944 Act raised the school leaving age to 15 and set up what was called the tripartite system based on ability and aptitude whereby children would go to one of three kinds of schools at the age of 11, grammar schools for the brightest and technical and modern schools for the rest. The selection was based on a single examination at the age of 11 with all fees paid by the State through local counties. The age of 11 was somewhat arbitrary, the public schools divided pupils at 13 and Scottish schools at 12. This tripartite system had many disadvantages, leading to it being abandoned in 1976 but what it did do was to allow children such as myself to enter a world of education that previously had been outside our wildest dreams. Furthermore it opened up opportunities for this kind of education to girls.

Grammar School

In the summer of 1944 I sat and passed the 11+ examination and with it admission to the Finchley Catholic Grammar School. The school had been founded in 1926 by one man, Father Clement Parsons under whose inspired headship it flourished. In 1944 it had a two stream entry each of 30 boys some of whom had not taken or had not passed the 11+ examination and were still paying their fees. The system was so democratic that I still don't know to this day who was and who was not an 11+ scholarship boy.

So in September 1944 I, together with some of my old friends from Our Lady of Lourdes School, enrolled as a new boy at the Finchley Catholic Grammar School. Everything about this new adventure fascinated me and I soon learned that the happy-go-lucky approach with which I had become familiar no longer applied. We were all expected to work and to work very hard and were expected not only to go to lessons but also to build on them with reams of homework. This meant that the carefree evenings and weekends with friends who had not been as fortunate as I was gradually declined and I formed new friendships even with boys who lived some distance away who I met on the train to school. For the first few weeks I was exhausted but I soon settled into the rhythm of study and began to enjoy myself. From the first form we studied English, Maths, History, Geography, Latin, French, Art, Religious Studies and Science plus daily sessions of Physical Education and weekly football. Looking back on it, we were not well served by some of our teachers. Most of the School's pre-war teachers had joined the armed services and what was left was a hotchpotch of those who for various reasons were ineligible for military services such as Irish citizenship, some women and some priests. It soon became clear that some of them didn't know what they were doing and at least one of the priests was a borderline psychopath. In retrospect, to quote, 'perhaps they didn't teach us much but they taught us all they knew'. The science teacher, a seminary drop-out, wasn't one of these. I hadn't done any science before but I soon became fascinated with the things that one could do with magnets, prisms, lenses, magnifying glasses and electric currents, how plants lived and the mysteries of soil. I liked the discipline of laboratory work, working with others and making and recording observations. I also learned more about drawing than I ever did in Art. I was hooked. Time passed and in the summer term I discovered that I had a talent for cricket and thereafter cricket and homework competed for the long summer evenings.

The war seemed a long way away apart from occasional mentions of yet another old boy or father who had been killed but it came closer in early 1945 when the Germans began to aim rockets rather haphazardly at London. We had little warning of these and the first thing we heard was the whine and the initial panic when it stopped and the rocket plunged to earth. We had little or no time to get to a shelter and gradually most of us, including the teachers, became blasé about the whole thing. The war ended in 1945, the men who had been fighting in the war returned, the teaching improved and we were now being taught by graduates in their specialist fields.

The School had, as I wrote earlier, a two form entry and from the very beginning an attempt was made to separate the highfliers from the rest so there were two forms called, unimaginatively, A and B. In the fifth year we were further divided into Arts, Science and Modern in order to prepare us for the next public examination, the School Certificate. The contents of the School Certificate syllabuses were much proscribed and everyone was expected to study English and Maths, a foreign language, a science subject and one other subject. We in the Science Fifth, as well as English and Maths, studied Biology, Chemistry and Physics as separate subjects, French, Latin, History, Geography and Religious Studies. The School Certificate, taken at about the age of 16 one year after the normal school leaving age, was also known as the School Leaving Certificate and this was when most pupils, even those in Grammar Schools, left in order to enter the field of work or further education. Those who obtained the requisite five subjects in either the Arts or Science streams matriculated, in other words, obtained the minimum qualification for university entrance so this exam was also known as the matric.

Although it was theoretically possible to enter a university with a simple matriculation there were very few universities that allowed this and the standard required was three cognate subjects at Advanced School Certificate level. This was also an accepted route into the Civil Service. This required a further two years at school, something that not all parents could afford or could see the advantages of. Increasing numbers of students taking Ordinary and Advanced level examinations without any intention of going into the Civil Service or going to university required something different from the stranglehold of the School Certificate examinations and in 1951 these were replaced with the more flexible Ordinary and Advanced Level General Certificates of Education (O and A level GCEs). These certificates were offered by a small number of examination boards, the most important being the university-based London, Oxford, Cambridge and Oxford and Cambridge Boards. There were also Northern and Welsh Boards and an independent Associated Examination Board. Scotland had a completely different system of Ordinary and Higher Certificates. The syllabuses offered by the various boards were often very different and schools had to choose only one board for all its examinations. My school chose the Oxford and Cambridge Board which was the one taken by most private schools and was generally considered to be one of the more demanding examination bodies but one that had more interesting syllabuses. I chose to study, Chemistry, Physics Botany and Zoology (Biology was not available as a separate subject). Those of us who were studying biological sciences were dissuaded from taking more than three subjects so I dropped Physics a subject with which I was struggling because I was not allowed to take any advanced course in mathematics. As well as our particular subjects we were also taught some Philosophy and the scientists who had not been allowed to take German at Ordinary Level were given lessons in this subject; both these extra subjects were to serve me well in the future. In 1951 I took my A levels and looked forward to saying goodbye to school for ever.

The School's A level results were dreadful and although I had not done badly many of my friends and classmates did less well than expected. Thinking and talking about it afterwards we do seem to have been inadequately prepared for the new Advanced Level GCE examination, particularly on the Arts side, and the teaching didn't appear to have advanced much from the former Higher School Certificate syllabuses. In the sixth form we were taught by three scientists, a chemist, a biologist and a physicist. When practical exercises did not go according to plan the biologist used to suggest that we wrote down the expected results in our practical books and the chemist simply asked us to ignore those particular results. The physicist, on the other hand, used to ask us to explain why we had obtained particular results. It was years later that I realised that only one of these teachers was a real scientist. For a few years I was a member of the Old Albanians, a club for old boys, but after that I never maintained any real contact with the school. I did, however, keep in touch with two particular friends, both on the Arts side, Michael Baker and John Roper. Michael served in the Royal Army Service Corps and afterwards emigrated first to Canada and then to Australia and we still keep in touch. John was, and is, one of my best friends. He served in the Intelligence Corps and afterwards studied at Oxford University where he graduated in Philosophy, Politics and Economics (PPE) and then worked for the Bank of England. After a successful career in London he was seconded to Guernsey in the Channel Islands, a 'tax haven', and then became head of financial services for the Isles of Guernsey. John played a significant role in my life and we still see each other every year, something I will return to later.

A Taste of Research

Before this, however, there had been another significant event in my life. Every summer, Mr William Cooper, a Laboratory Superintendent at the London School of Hygiene and Tropical Medicine and with boys at the school, invited two boys to work at the LSH&TM as laboratory assistants. The idea was to encourage boys to train as laboratory technicians, a career in which there was a great shortage of applicants trained up to Advanced Level. At the end of my first year in the sixth form I was selected and began to work in the Department of Medical Protozoology. The Department was headed by Professor Henry Shortt who with Dr Cyril Garnham a few years before had discovered the stages of the malaria parasite in the liver. There were people working with malaria parasites, trypanosomes, blood parasites of birds and mosquitoes all willing to share their knowledge and enthusiasm for their subject with a couple of schoolboys. The Department was also full of doctors and scientists from the tropics and we were encouraged to go to their talks. I was intrigued and fascinated: until then I had no idea of the way my career might go but now I had begun to see some glimmerings of how my life might progress. It was during these few weeks that I learned how to use a high power microscope, prepare and stain microscopical preparations, make solutions, clean glassware and surgical instruments and so I began to absorb some of the excitement of science, When I left school in 1951 I was fortunate enough to be taken on again by the LSH&TM, this time as a temporary technician, and among other duties, was given some responsibility for the maintenance of the mosquito colonies. By then I knew vaguely what kind of career I wanted to follow, a colonial medical officer or a laboratory scientist.

National Service in the 1950s

There was, however, a massive cloud on the horizon, national service. Although the war had ended in 1945, the services were desperately short of personnel and the National Service Act of 1948 amended in 1950 required all young men at the age of eighteen to serve for two years in the Army, Navy or Air Force or to work in the mines. One could defer national service in order to go on to university or further education but the policy at my school was to get it over and done with so that those who wanted to go to university could proceed to their chosen careers without interruption, in my case this was the right decision. For most of the year, the number of 18 year olds eligible for national service would be expected to trickle onto the market but in July-October recruitment took a bit of time because so many 18 and 19 year olds with A levels together with a number of graduates came available at the same time. So there I was waiting to be called up. There were a number of selection boards who were meant to direct the young conscripts to branches of the armed services suited to their experience and qualifications. The working of these boards was mysterious. After a preliminary and somewhat perfunctory interview during which the possibility of being posted to the Royal Army Medical Corps (RAMC) was raised I found myself posted to the Royal Army Ordnance Corps (RAOC). It turned out that this was not as illogical as it at first seemed. The RAMC was awash with fully trained laboratory technicians who had left school at 15 so there very few posts suitable for anybody with A levels and a few weeks laboratory experience. The RAOC's main role was to provide the army with the ammunition it needed from several ammunition depots in the UK and others scattered throughout Europe, Africa and the Far East. In addition there were vast stocks of ammunition captured from the enemy. After five years of relative neglect much of this ammunition was in a very poor state and required the attention of highly skilled ammunition examiners and inspectors. These posts required training approaching degree level and the army found itself unable to recruit enough regular soldiers for its needs and reluctantly decided that it had to train a small number of national servicemen. The training course lasted twelve months so after basic training the army could only expect a maximum of ten months useful work. Ammunition Examiners could expect rapid promotion and the trade was among the best paid in the army. The selection procedure was very rigorous and eventually I found myself with 14 others on the Ammunition Examiners Course at Bramley a rather dull village in Surrey, south of London. I was one of the least qualified members of the cohort; most of the others had degrees in chemistry, metallurgy, mathematics or physics and a small number of university drop-outs.

Life at Bramley was markedly different from that at the basic training camp at Aldershot. Drill was practically non-existent and we marched, when we had to, in a style know to the rest of the Regiment as the 'Bramley amble'. There was a new intake every month except December so at any one time there were eleven cohorts. One of the most interesting aspects of life at Bramley was that the whole cohort of 15 lived for about a year together in one hut and we were 'encouraged' to compete with other cohorts in every possible way. For track and field sports, tennis, badminton and boxing it was possible to raise a team, for association football and cricket it was less easy and rugby football required the participation of everybody. We even competed for decorating our huts and growing the best vegetables and one of my favourite army notices, of which there were very few, stated that 'The habit of throwing cabbages onto the roof of Hut 10 will cease w.e.f (with effect from) this notice'. The work was very hard with lectures and practicals every day, often into the evening, and examinations every four weeks. If anyone failed they were usually allowed to take the exam again the next month but if they failed again they were relegated to the cohort below. This is where the competitive spirit came into its own because throughout the course we all helped each other, particularly as the exams came along. In effect we had stumbled on the concept of small tutorial groups. It was while I was at Bramley that I learned how to divide my time between study and leisure, a habit that stayed with me and served me well when I eventually went to university...

At Bramley, most weekends were free. Some of us went home; some went to London but most simply lay in bed, explored the numerous public houses in the area or made friends with the local girls. In the summer I played cricket for the unit on Wednesdays and for Finchley at the weekends and even had a trial for the RAOC regiment but as I was a wicket keeper I had no chance against someone from a Stores Unit who had played cricket for his county and as, I saw in a newspaper later, was a contender for the English team.

Part of my last training month was spent on factory observation in Preston, Blackburn, Chorley and Wigan in the Northwest of England. We visited a range of factories, including car manufacturers and cotton mills where we learned about different factory procedures. These visits were eye-openers for me as I had never experienced the noise and boredom of factory work nor the abject poverty of these working class towns engulfed by clouds of choking smoke. It was all incredibly depressing and it was many years later before I could shake off my first impressions of these deprived parts of the country and half a century before I visited them again. These visits made a great impact on my political views.

My training at Bramley enhanced my scientific experience. I learned about the chemistry of explosives, the degradation of metals, ballistics, statistics and I even became a specialist on fuses and learned which batches were most likely to be dangerous. Most of all, however, I learned the importance of making rapid mental calculations, making immediate meticulous notes and the importance of having all my work countersigned. Many of the workshops at Bramley were designated 'One Man Risk' and another important lesson was that we learned how and when to take responsibility and when (i.e. never) and when not to take risks. These lessons were also to serve me well in my future career. Fortunately, we all passed every examination and qualified as Ammunition Examiners with our bright arm badges that depicted a letter A surrounded by flames, known by everyone as a 'flaming ****hole'. There was then a short break during which postings were sorted out and we learned to drive. The driving course was much more demanding than the civilian one and included driving a variety of vehicles from ordinary cars up to three ton lorries at night and without lights. The test was very demanding and included theoretical components, something that did not feature in the civilian test until many years later.

The next move was to a proper posting. The system at Bramley was for the list of postings to be put up on a notice board and we each made our choice in the order of our overall class grade. I was lucky as I was fairly near the top of the list and, together with four Bramley friends, chose the Central Ammunition Depot (CAD) at Buckingham, a small market town about 50 miles north of London. CAD Buckingham itself had no stores of ammunition and these were situated at two sites, Little Horwood and Great Horwood, a few miles away. These were essentially run as factory lines where ammunition, in this case shells and their components, came in from all over the world, was inspected for any corrosion, dampness, damage or other abnormalities, the fuses removed and checked, then replaced with new ones and the shells cleaned, repainted, labelled and repacked for distribution to serving units.

We were all very much made aware that the lives of frontline soldiers depended on our decisions and that safety took priority over speed. Each of us was responsible for one particular factory line and here the camaraderie established at Bramley paid dividends as we helped each other to make decisions and learned from those who had qualified before us. The factory lines were manned by a mixed bunch of local women ranging from a vicar's daughter to some of the most foul-mouthed individuals I had ever met. They were all fully aware of our naivety and took the micky out of us and teased us unmercifully. Soon, however, we began to give as good as we got and formed very good working relationships. We also learned that the women were very good judges of our abilities and knew who to ask for an opinion about a particular shell or fuse. One additional lesson was that I learned a lot about mutual respect. At Buckingham we worked a five day week and most weekends I went home but when I stayed in Buckingham I was sometimes invited to Sunday lunch or supper with some of the women's families or by the wives of soldiers who were on serving abroad, the latter invitations I tactfully declined which was not the case with some others who occasionally returned to camp with black eyes.

Work at Buckingham was demanding but rewarding and we were very much left on our own and very seldom saw a senior officer. Some of the work was extremely boring such as inspecting the stores of ammunition for pilfering or vandalism and each month we had to take and pass an advanced course which kept us on our toes and had to undertake some other 'educational activity' . I went on short courses as different as drama and refereeing motor cycle trials. Much of the summer was spent playing cricket and I found myself in demand for several local teams in villages where some of the senior officers lived.

My last months in the army were spent in London at Woolwich Arsenal, the headquarters of the Royal Artillery, the regiment with which my grandfather had served. Here I had a training role and instructed regular and national service soldiers about fuses, how they worked and how to identify potential hazards. Here we experienced a hard sell from recruiting officers who tried to persuade us to apply for short service commissions which was hardly surprising considering how much money and effort had gone into our training. In the local pubs we were also contacted by members of the Canadian Army and civilian mining groups. Nobody in my intake decided to stay on in the army but some others, I understand, did succumb to the temptation of large sums of Canadian dollars. As a footnote I recently saw on a television programme that Ammunition Inspectors now train for five to seven years but that also includes training in bomb disposal and activity that was not, thankfully, offered to national servicemen.

Between the Army and University 1953-1955

My national service at an end so this was the time for me to sort out my future career. This had not been uppermost in my mind while I was in the army and an opportunity to postpone the decision further came when I received a date for my discharge which was too late for me to consider university entrance in 1953. I was pretty sure about what I wanted to do but whether to study medicine or science was another matter. I had considered a career as a Colonial Medical Officer but by the beginning of the 1950s it was clear that most, if not all, of the colonies would be becoming independent and training their own doctors. The turning point came when I visited the outpatients department in a major London hospital and didn't like what I saw there. I also didn't like the idea of working in a general practice and seeing the same range of snuffly noses and aches and pains on my last day as on my first. So it would have to be a degree in science but that would have to wait nearly a year. I returned to the LSH&TM to look for some work but there were no vacancies in the Protozoology department but there was a short term vacancy for a laboratory assistant in the Department of Helminthology working with a Dr Pierre Le Roux whose interests were schistosomes. Dr LeRoux turned out to be a brilliant but cantankerous South African Boer who fell out with nearly everybody with whom he worked. He also smoked like a chimney, his cigarette ash falling onto his papers, onto animals that he was dissecting and into the tanks where he kept his snails. It was my job to maintain the colonies of snails used for the experimental transmission of schistosomes and the welfare of the mice used in his experiments. This was potentially very boring but I saw an opportunity to 'clean up Dr LeRoux's act' as it were. I ensured that the tanks were meticulously clean whereas they had formerly been covered in algae and that the mice cages were clean, regularly changed and properly labelled instead of being covered in peeling sellotaped notes. I devised better ways to feed the snails on boiled lettuce and made frequent trips to the local marked where I got fresh lettuce leaves that had been discarded for nothing. I don't know if Dr LeRoux noticed any changes but if he did he didn't say. He didn't, however, pick holes in my work which had done with his previous assistants. It was during this time that I made some of my most important contacts and formed lasting friendships with, in particular, Ralph Lainson, later to become a Fellow of the Royal Society, John Baker, Elizabeth Canning with whom I was later to run parasitology field courses and Neil Brown who later worked at May and Baker with Valerie Weston, who was to become my wife.

It was now time to seriously consider my next move, which university to go to and what to study. This was when another chance event occurred. University grants were managed by the local county education authority, in my case Middlesex, and were discretionary. The person I spoke to told me that if I waited another year I would be eligible for the maximum grant at the age of 23. This coincided with me seeing an advert for a laboratory assistant in the Microbiology Department at British Drug Houses, a major pharmaceutical company in North London. The salary was good and I could have one day a week to study for and to retake my A levels if I wished. I did not really want to do this but the opportunity was too good to miss so I applied for and got the post which involved testing antimicrobial drugs against the parasitic amoeba, *Entamoeba histolytica*. Here I learned a number of microbiological techniques including sterile techniques, culturing amoebae and bacteria and the principles of drug testing. The working hours were long but the company was congenial and on my first day I was delighted to discover that a chemist working in the next laboratory was one of my army friends from Bramley. I particularly admired the laboratory head, Janet Cartmel, who was an excellent teacher and I occasionally met the Head of the BDH laboratories, Dr Frank Hartley: our paths were later to cross when he became Vice Chancellor of the University of London. I enjoyed working for BDH going swimming and eating at local cafes on the fringe of central London but I knew that this was only to be temporary and that this was the time to get my act in order and to get myself a place at a university.

In 1955 there were about 30 universities in the UK, including six colleges in London, all offering degrees in biological subjects. Biology as such was not an option and the choice was between Botany and Zoology with some universities offering Microbiology, Biochemistry, Physiology or Pharmacology. Most universities offered courses in Zoology and the patterns of study were similar. In the first year students studied Zoology with two other subjects, usually Botany and Chemistry, in the second year Zoology was studied with one of these other two subjects and the third and final year was devoted entirely to Zoology with an opportunity to study a particular topic as a special subject in some depth. Not all universities offered Parasitology as a special subject and as this was my obvious choice there were several but limited possibilities. King's College and Imperial College London both offered Parasitology but students at other colleges could take the King's Course. As far as I was concerned, London was out of the question as I would be expected to live at home and to travel into London each day which I didn't want to do. Of all the other choices University College Exeter seemed the most attractive. Exeter offered a London degree, was a very pleasant city and was a reasonably inexpensive place in which to live. I checked with Exeter and was assured that they would be providing Parasitology as a special subject so I retook my A levels and passed which was very reassuring after quite a long gap from full-time study. I also had a verbal assurance from by then Professor Garnham at the LSHT&M that he might be willing to consider taking me on as a PhD student if I obtained a good degree. So in October 1955 with this assurance I left London to begin my full time study for a degree at University College Exeter.

Exeter University 1955-1958

Exeter is a delightful medieval city which, although it had its heart blown out during a particularly vindictive bombing campaign during the Second World War, enjoyed a spectacular situation on the River Exe, a magnificent cathedral, many Georgian terraces and was not far from the sea. The College itself consisted of rather cramped premises in the City centre, several Georgian and Victorian houses scattered throughout the city and two new sciences blocks on its outskirts. Botany, Zoology and Geology were housed in the Hatherly Laboratories and all the airy undergraduate teaching laboratories had floor to ceiling windows totally different from anything I had ever seen before. There were about 30 of us reading Botany, Zoology or Geology in various combinations, 12 were intending to read Zoology in some form or another and we joined another eight second year students. The Zoology

Department ran alternate first year courses in Invertebrate and Vertebrate Zoology and we were fortunate enough to join the invertebrate year so our course was more logical than the one for those who had joined the previous year and were rather resentful of this fact. There were only four Zoologists on the staff, Professor Leslie Harvey, an ecologist, Geoffrey Vickers, another ecologist, Ian Linn, a physiologist, and Julian Hawes, a protozoologist and the only member of staff with a PhD. Hawes was a graduate from King's College London where he had studied under the formidable Professor Doris Mackinnon, a distinguished protozoologist and one of the first women science professors in the country. There were also a few postgraduate students, mainly ecologists, and Hawes's new student, Keith Vickerman, who had just graduated with first class honours from University College London and was about to work on the protozoan parasites of crane fly (tipulid) larvae. Within my immediate group of Honours Zoologists were John Alexander who wanted to be a protozoologist, Alan Brafield, a potential marine biologist and someone who I quickly realised was going to be a serious competitor, Edward Eastwood and Clive Walters. Most of the others were intending to read Zoology as part of a General Degree. In all, it was a very nice group of people very easy to get on with. We had by then had to choose our major, three year, subsidiary, two year, and ancillary, one year, subjects and I opted for Zoology as my major, Botany as my subsidiary and Psychology as my ancillary subjects. I could, and should, have chosen Chemistry as either my subsidiary or ancillary subject but I had been taught Chemistry so badly at school that I had no interest in the subject. I had to make up for missing out on Chemistry later in order to keep up with developments in the biological sciences.

I quickly entered into the social life of the college and I joined the Dramatic Society took part in a number of plays and later became Treasurer. As my first summer approached I took part in the trials for the university cricket team and was selected for my batting and not my wicket-keeping. It was a care-free summer and we played against a number of universities and touring sides. I also played the occasional game for the Exeter city team and took part in trials for the county XI for which I played in some non-competitive games. The county team and Exeter Cricket Club played on the county ground situated below the Hatherly laboratories and, although I didn't know it at the time, Professor Harvey was a keen cricket fan and watched some of the games through his binoculars. It was on his advice that in my final year I gave up playing serious cricket although I still played in a number of minor games.

At the end of the first year all the staff, all the postgraduates and all the first and second year students attended a field course at Ladram Bay on the coast a few miles from Exeter. We lived in tents and did our own catering and studied in great detail the ecology of woodland, moorland, freshwater and marine habitats. This was very exciting and we learned a great deal. In particular I discovered a fascination for rock pools which has never left me and wherever I go, anywhere in the world where there are such pools, I spend many happy hours looking to see what I can find. Later in the summer vacation I found some well-paid work as a gardener in the Finchley Parks Department. I then spent a few weeks working as a laboratory assistant in the Protozoology Department at the LSH&TM where I was given more responsible tasks than I had had before.

Exeter was granted university status in 1956 and no longer had to offer London degrees so as we embarked on our second year we had to choose between continuing to study for the London degree or taking the Exeter one. The syllabus remained more or less the same but for the Exeter degree there was to be a Part 1 examination at the end of the second year which decided who could or could not continue to an Honours degree. All our year opted for the Exeter degree.

The second year was devoted to the vertebrates and we were joined by the 1956 intake who must have found being introduced to jawless fishes in the first lecture a bewildering experience. We also studied physiology and embryology and were fortunate enough to be taught by Dr David Newth from University College London. During the Easter break of 1957 I attended a meeting of the Parasitology Group of the Institute of Biology (later to become the British Society for Parasitology). This was very exciting for a second year student and I was thrilled to hear talks given by some of the people I had met at the LSHT&M. There was also at this meeting a final year student from Exeter, Valerie Weston, who I had only known briefly because she had missed much of her second (my first) year through asthma, an affliction that was to plague her for the rest of her life. We talked and went to a theatre together and afterwards, back at Exeter, began to see each other more and more frequently. Valerie got on very well with Keith and I particularly remember long walks along the River Exe and to the nearby Countess Wear pub with Valerie, Keith Vickerman and John Alexander.

Then at the end of the second year came the Part 1 examination but, sadly, only four of the twelve of us were allowed to proceed to Honours, the rest being encouraged to study for a two subject General Degree which meant that most could not take specialist subjects such as Parasitology or Marine Biology. This was a disappointment for some but suited those who were hoping to become teachers and wanted to keep their zoological and botanical interests open. Eventually, when we came to our final year three of us opted for Parasitology and one for Marine Biology.

My final year was a busy one and not only did we have to take a range of taught zoology courses but also an anatomical and a behavioural project. I dissected a pangolin and investigated the reactions of colourless planarians to light. We were not offered an opportunity to do any parasitological topics which was very disappointing. The parasitology course was very conventional beginning with the protozoa and working through to the trematode, nematode and cestode worms. Julian Hawes taught us about the protozoa but from a very eclectic viewpoint. We studied in some detail obscure parasites of marine worms but nothing about malaria or trypanosomes. The highlight of this part of the course was having Keith Vickerman as a demonstrator as his enthusiasm for the subject was infectious. The part of the course devoted to helminth worms was very different and Leslie Harvey provided us with an endless supply of hosts to dissect and parasites to find. We learned a lot about the parasites of small and large mammals, frogs and birds but nothing about important parasites such as schistosomes or filarial worms. Looking back on my final examination papers I realise how unimaginative they were. However, I passed with an Upper Second Class degree and was ready for the next stage in my career.

Before this I must fast forward. The friends I made at Exeter were good ones and our lives were later to become interconnected. Keith Vickerman became a lecturer at University College London and later Professor of Zoology at the University of Glasgow and a Fellow of the Royal Society, Alan Brafield became a lecturer at Queen Elizabeth College and later a colleague at King's College when the two institutions merged. Keith was my best man at my wedding and I was Alan's at his. John Alexander became a research assistant in the Biophysics Department also at King's College and later a lecturer at the University of Salford.

London School of Hygiene and Tropical Medicine 1958-1959

It had always been my intention to study for a PhD and after an abortive attempt by Julian Hawes to find me a postgraduate place at Cambridge and a suggestion from Professor Harvey that

I should remain at Exeter I decided that the LSH&TM was the place for me. Professor Garnham had always assumed this but after some discussion during which I told him that I fancied an academic career rather than a purely research or field-orientated one, he suggested that I should take the Academic Postgraduate Diploma in Parasitology and Entomology (DAP&E) now an MSc. I was a bit disappointed at not being able to get down working for a PhD straight away but as it later turned out this was a very good decision. The DAP&E was a very high powered and selective course and getting financial support was a challenge but Professor Garnham managed to persuade my local authority, who had financed my undergraduate studies, to pay my fees. This was something that was almost unheard of but, as I was later to discover, Professor Garnham was a man who usually got his way. The DAP&E course was run in conjunction with the Diploma in Tropical Medicine and Hygiene (DTM&H) a medical postgraduate qualification largely aimed at those who had spent a number of years in the tropics so I found myself in a class with much older men and women many of whom were already authorities in their fields. One in particular, Ken Warren, an American graduate from Harvard who was then working for the United States Institutes of Health, was a world authority on schistosomiasis and other students had made major contributions to medical and veterinary parasitology. There was only one other recent graduate, C.P. Ramachandran, a shy Indian Malay and we immediately formed a bond. The work was demanding and intensive with lectures, seminars and practicals every day but we still found time to socialise and the many discussions over coffee in a nearby café with Ken Warren and some of his American friends played a large part in my life at the LSH&TM. There was one disconcerting aspect about being a student at the LSH&TM and this was the hierarchy and the technicians who had been my former friends and colleagues now kept their distance and addressed me as Mr Cox. There was one intermediate examination which we all passed and said goodbye to our DTM&H friends whose course lasted only six months. Then we were in at the deep end with more intensive instruction, more demanding practicals and a research project to complete. I was required to cultivate a recently isolated strain of the ciliate *Balantidium coli* throughout its temperature range. I didn't realise at first that I was in sole charge of keeping this ciliate alive and that if I lost it the LSHT&M would have lost the only strain it had in culture for ever. It was no longer a five day week because the beast had to be fed every day. I was, however, very lucky because the skills I had learned while culturing *Entamoeba histolytica* at BDH were the same as those required for the cultivation of *B. coli* although I had to compete with real research workers for the use of the only departmental autoclave. One interesting aspect of our research projects was that we had to share our expertise with each other so we all learned a lot about a number of different techniques. Of particular importance to me was learning how to freeze blood parasites.

Again, fast forwarding. I kept in touch with Ken Warren and visited him later in New York where he had become Director of Health at the Rockefeller Foundation but sadly he died in 1996 at the relatively early age of 67. C P Ramachandran and I became great friends and when his daughter came to school in England Valeria and I acted as her guardians. CP by this time had joined the World Health Organization and was to play a major role in the global programme to eradicate lymphatic filariasis and became an Honorary Fellow of the Royal Society of Tropical Medicine and Hygiene and was later elected to a Fellowship of the Malaysian Academy of Sciences. Our paths kept crossing and I still remember a breakfast meeting with Ken and CP when my family and I were staying with him while I was interested in a post with the World Health Organization.

Although my grant was fairly generous I needed to earn some more money and Professor Harvey introduced me to a former student of his who was working at Sir John Cass College, a college of further education, in the East End of London. She needed a demonstrator to assist her with A level Zoology practicals so for one evening a week I travelled the short distance to Aldgate and taught a batch of very bright students who were studying A level Zoology in the evenings as well as holding down full-time jobs. I realised how lucky I had been to be able to study full-time at school. This experience was very valuable for me as I was later able to claim some teaching experience which enabled me to obtain a part-time teaching post and to become an A level examiner.

In the meantime, Keith Vickerman had returned to University College as a lecturer I n Zoology. University College was only a short walk away from the LSH&TM so we saw each other frequently and I had the opportunity to meet his colleagues, most of whom were very distinguished, including the Nobel Prize winner, Professor Peter Medaway, David Newth, who had taught us at Exeter, John Maynard Smith the theoretical evolutionist and a couple of vertebrate palaeontologists who didn't seem to have a good word for anybody. Coffee- and tea-time discussions were always very stimulating and I was privileged to take part of them. Keith also benefited from our friendship and he frequently came to seminars at the LSH&TM and met some of the students and members of staff. On one occasion he introduced me to Dr Ben Dawes who had taught him parasitology on the inter-collegiate parasitology course at King's College. Keith later told me that I had created a good impression.

King's College 1959-1997

In June 1959 we all took and passed the DAP&E examination and most of the students returned to their own countries or to study for PhDs elsewhere in the UK or abroad. I was just beginning to think about my own PhD when another life-changing event occurred. I don't know who drew it to my attention but there was an advert in the scientific magazine, *Nature,* for an Assistant Lecturer in Protozoology at King's College London. Most UK universities appointed assistant lecturers which were usually three year appointments with the possibility of reappointment onto the permanent lecturer scale. These posts were intended for recent PhD graduates who wanted to get a foothold into the academic world. I felt that I was totally unqualified for such a post but this was an opportunity not to be missed so I discussed this with Professor Harvey, Professor Garnham and Keith who all felt that although my chances of getting this post were slim I had nothing to lose by applying. In the event I did apply and was selected for interview and interviewed by Professor James Danielli, the head of the Department, a distinguished chemist who had accurately postulated the nature of the cell membrane on theoretical grounds, the Principal of King's College and Dr Ben Dawes. I had given a lot of thought to how I might structure a course in parasitic protozoology and obviously impressed the panel and was not only appointed but appointed on an advanced salary scale. I don't know if my previous encounter with Ben Dawes had anything to do with this and he never said.

So in June 1959 I had a job at King's College starting in October with a salary that would only be paid at the end of that month. I needed some money and had to find some work so I registered as a supply teacher. Supply teachers filled in to replace a permanent teacher for a short time or to fill a post that was temporarily vacant. My first school was a grammar school in Romford, a town in Essex east of London. For four weeks I was to teach biology to A level pupils during the temporary absence of the permanent teacher and also run practical classes in physics. This was a splendid school. All the staff wore suits and gowns and the boys wore smart uniforms, it was very much like my own grammar school.

The teacher I was replacing had prepared meticulous notes and a group of senior boys knew more about physics practical classes than I did and we worked together as a team. I had never been so exhausted in my life and just crashed out when I got home before rousing myself to mark essays, homework and practical books activities that also occupied much of the weekends. The weeks passed very quickly but came to an end and I was offered a post at a secondary modern school in Wood Green, a pleasant suburb in north east London. Here my experience was so very different from Romford. I was asked if I would teach English, Maths and some sport but I was given no guidance at all. All the school seemed to want was someone to sit in front of a class and to keep the boys and girls reasonably occupied and quiet. The pupils I taught were aged 13-14. They were very nice kids but all they seemed to want was to leave school and to get jobs. They were not unintelligent but had been badly served by the education system. I tried to introduce them to algebra and geometry based on my practical experience in the army. I soon discovered that this was useless because most of them could not do any simple addition, subtraction, multiplication or division. I tried to get them to understand the mathematics involved in football league tables and the probabilities of their teams beating their next opponents but it was all too late and I was not qualified to teach the kind of mathematics I had mastered well before the age of eleven. English was even worse. Very few of them had read anything other than popular newspapers and magazines and their knowledge of spelling and grammar was minimal. The really sad thing was that among this mass of underachievement there were some really bright kids but there was no way in which they could be extracted from the morass of mediocrity. Until then I had thought that the system whereby children could be allocated to grammar schools for the most able and secondary modern schools for the less able was a very good one. I was later to discover that my experience was not unique and although the grammar schools catered very well for the brightest students insufficient attention had been paid to the education of the bulk of the population, a situation that eventually led to the partial elimination of the grammar schools. I know that there were some poor grammar schools and some excellent secondary schools but I have never understood why so little effort was made to raise the standards in the majority to those in the very best. It took me some time to realise that in the name of equality left wing government policies always disadvantaged those whose needs were greatest.

In October 1959 I began my career as a university lecturer at King's College London. Kings College occupied a site in the Strand on the River Thames in the centre of London. The Zoology Department consisted of ten members of academic staff including James Danielli, a biochemist and Fellow of the Royal Society, Ben Dawes, a helminthologist who had written two major books on this subject, Don Arthur, a world authority on ticks, Eric Barnard and Gerry Bell both chemists and cell biologists, John Cloudsley-Thompson, an ecologist and Barry Cox, a vertebrate palaeontologist. There were three assistant lecturers, Shirley Hawkins, a cell biologist; Lewis Wolpert, a developmental biologist who had already began to establish a name for himself, and myself. Six of us, Ben Dawes, Don Arthur, Barry Cox Lewis Wolpert, Eric Barnard and I were later to become professors and Lewis Wolpert and Eric Barnard were also to become Fellows of the Royal Society. It was a very unusual Zoology Department in that only seven of the eleven staff had degrees in Zoology. The Zoology Department occupied rooms on the second floor and basement of the Strand building and I was allocated to one of the basement rooms next to those of Wolpert and Barnard. Also in the same basement was the Biophysics Unit in the Department of Physics. This was a pioneering unit headed by Professor Sir John Randall, a physicist who had been involved in the development of radar. The aim of the unit was to apply the tools and theories of physics to the solution of fundamental biological problems. Also in the unit were Maurice Wilkins, a Fellow of the Royal Society, who was later to receive the Nobel Prize for Physiology in 1962 for his work, with Rosalind Franklin on unravelling the structure of DNA, and Jean Hanson, also a Fellow of the Royal Society, who had elucidated the mechanism of muscle contraction. This was a congenial group and Lewis Wolpert, Eric Barnard and I tended to enjoy their company in preference to our own department. There was quite a lot of interest in the protozoa at King's, Shirley Hawkins and John Randall were working on amoebae and ciliates as models for problems in cell biology and there was a constant flow of visitors with protozoological interests passing through the Biophysics Unit.

All of this was, however, something for the future and my immediate problem was how to prepare twenty lectures and ten practicals on parasitic protozoa for students from several London Colleges who were coming to King's to take the special parasitology course as well as ten lectures and five practicals on the protozoa for the combined King's College first and second year course. On top of all this I had been allocated a number of tutorial students and was expected to give a series of lectures on the sponges after my protozoa lectures. Fortunately I had kept meticulous notes from my DAP&E course which I was able to adapt for lectures and practicals for the King's course. I was also able to borrow material from the LSH&TM and Keith Vickerman for my practical classes. At the risk of being big-headed the course I ran was much better than the one at Exeter and included a lot of material of medical and veterinary importance. The special parasitology students were incredibly bright and desperately keen to learn and most went on to academic careers. I had my work cut out to keep up with the literature and not for the last time the library at the LSH&TM came to my rescue. I also discovered how generous the people I had met at the LSH&M and through the Parasitology Section of the Institute of Biology were. Parasitologists are very nice people and always keen to help each other out but on the other hand I could not believe the acrimony in other fields of zoology particularly palaeontology.

The first term over, I now had to decide on what I was to study for my PhD and Professor Garnham suggested that I should work on some newly discovered malaria parasites of rodents about which practically nothing apart from their morphology was known. The immunological aspects of these parasites turned out to be very rewarding areas of research and I soon began to obtain publishable results some of which were presented at laboratory meetings of the Royal Society of Tropical Medicine and Hygiene, a society that I had joined and where I began to meet a lot of people who had made important discoveries in the field of tropical medicine. These laboratory meetings were very exciting because they offered me opportunities to discuss my discoveries informally with some very senior people. My connections grew and I gradually became accepted as a member of the worldwide realms of parasitology and tropical diseases and began to frequent local and national meetings.

My second year, 1960-1961, as a lecturer was slightly easier than the first mainly because I had gained experience and confidence and also because it was the vertebrate zoology year in which I was not involved. I was, however, expected to supervise the departmental finances which was quite a challenge partly because the Chief Technician, a man of the old school, was very reluctant to provide me with all the information I needed. At the LSH&M the chief technicians were regarded as gods and I had accepted some of these attitudes when I moved to King's. The disciplines I had acquired in the army served me well and I held my ground and discovered that there were no financial irregularities, something that I had feared, merely incompetent bookkeeping. By this time

I had had good dealings with people in the accounts departments and they advised me on good procedures which I implemented to the benefit of everybody and something that Professor Danielli had not failed to note. Early in 1961, nearing the mid-point of my probationary period, he took me aside and intimated that I could expect to be given a permanent appointment from October 1961. In 1962, Danielli and most of his team departed to the University of Buffalo in the United States which was a great loss not only to the department but also to the college and British science.

Settling Down

For my first year in London I lived at home in Finchley with my mother and sister but as soon as I could afford it I moved into a flat in Kentish Town close to Hampstead Heath which was near to where I had been born. The flat had been found by David Garrood, an Oxford University contemporary of my school friend John Roper, and David invited me and John Boulter, another Oxford graduate, who also worked at the Bank of England, to share the flat with him. Valerie Weston, my friend from Exeter, had also returned to London and had been working on anti-schistosome drugs in the parasitology laboratories of the pharmaceutical company, May and Baker, later to become part of the massive international Rhone-Poulanc. Valerie was working of the action of anti-schistosomaL and had published a paper on this subject. Unfortunately the asthma that had disrupted her studied at Exeter returned and was aggravated by the chemical fumes at May and Baker so she decided to abandon parasitological research and to become a schoolteacher. Without any teaching qualifications she was fortunate enough to be appointed as a biology teacher at Sutton High School, an excellent girls' school not far from where she lived in Surrey, and was able to live at home. At the same time I had moved into a flat with two friends in Kentish Town in north London. We the decided that it was a good time to get married which we were on April 15th 1961 at St Patricks Church in Soho in the centre of London. We then rented a flat in Sutton, a south London suburb, and commuted to London while she drove to work at the nearby Sutton High School. Her asthma improved and we settled down to married life often with the evenings spent with her marking and preparing lessons while I prepared my lectures. Valerie later obtained a post at one of the best girls' in England, the North London Collegiate School, and we moved to Boreham Wood in Hertfordshire twelve miles north of London. We later moved back to Surrey, to the pleasant market town of Epsom, where she became a part-tine lecturer in biology at the local College of Further Education, the North East Surrey College of technology (NESCOT). In January 1967 our daughter, Joanna, was born, and three years later we moved to a large house in Ewell a historic village near Epsom. In the meantime, David Garrood had married a Swiss girl, Frida Baumgartner, who later became one of Valerie's best friends when they were also living in Epsom.

Now back to 1961 which was a very eventful year. In that year Professor John Corliss of the University of Maryland, a distinguished protozoologist and expert on ciliates, arranged an exchange with Julian Hawes at Exeter and, through Keith Vickerman, we got to know him very well. Largely due to his influence I was invited to present a paper at the First International Congress of Protozoology in Prague. Britain was still in the grip of austerity and one was only allowed to have £50 of foreign currency. Scientists going to international conferences were exempt from this restriction so Valerie and I set off to Prague by train, a tedious and filthy journey particularly as the Czech trains burned lignite. We travelled with Keith Vickerman and John Alexander and also a botanist from Leeds, Gordon Leedale. In Prague we met up with a number of other British protozoologists including Elizabeth Canning. This was also a wonderful opportunity to meet other protozoologists especially those from behind what was then called the 'iron curtain'. The congress was a great success and our hosts arranged a number of tours for us. Valerie and I travelled back from Prague by boat through Germany to Bonn. I had never been abroad before so this was a memorable experience. The most important part was, however, making so many useful contacts and friendships and being accepted as part of the international community of scientists. I could hardly believe that only three years before I had been a student at a minor British university and was now on Christian name terms with some of the most important figures in the field of protozoology.

Research

My research was progressing well, albeit slowly, and my teaching improved with experience. In 1963 I began to work with two clinicians, Douglas Bilbey and Tom Nichol in the Department of Anatomy on the floor above the main part of the Zoology Department to which by this time I had moved. They were interested in the activity of the reticulo-endothelial system in disease conditions and asked if I would cooperate by examining what happens in mice infected with malaria parasites. We discovered that at the height of the parasitaemia the reticulo-endothelial system completely shut down and was unable to remove any particles, presumably including bacteria, from the blood. In other words we were the first to describe and explain immundepression in a rodent malaria model and published our results in *Nature*. This fruitful line of research might have continuds had not Douglas Bilbey departed for the United States and I returned to my own studies but it did set my thoughts in a new direction.

In July 1993, Julian Hawes died suddenly from an overdose of alcohol and drugs with no indication that this was anything other than a tragic accident. This left Exeter with a problem and Leslie Harvey asked Keith Vickerman and me if we would give his protozoological lectures and practicals in the autumn term which we did. Julian Hawes needed to be replaced and although I was tempted to apply I had become too deeply involved with front line science in London to move to a quiet provincial city attractive as this might have seemed. In any case, I had my PhD to complete and I could not envisage doing so in Exeter which was at that time rather a backwater. The post was advertised and given to a friend of mine, Peter Walker, an Oxford graduate who had just finished his PhD working on trypanosomes at the National Institute of Medical Research at Mill Hill. My forebodings about Exeter proved to be justified and Peter never did any more significant research and a few years later abandoned academia altogether, studied for a medical degree and ended his career as a country doctor in a one-man practice. I gave a seminar at Exeter in January 1965 and didn't return for over 40 years.

In the early summer of 1964 I eventually finished and presented my PhD. My examiners were Cyril Garnham and Ben Dawes, something that would not be permitted today. The viva went very well and my thesis was accepted without any corrections or amendments. After the viva, Cyril Garnham took me aside and suggested that I should choose a field of research other than malaria because, in his words 'it will be all gone in ten years'. It is worth commenting on this. In 1955, the WHO was presented with a document stating, *inter alia*, that it was '...not unreasonable to begin planning for world-wide eradication of malaria'. This suggestion was taken up enthusiastically by the WHO and other health organizations, adopted by the World Health Assembly in 1956 and implemented as the Global Eradication Programme in 1957. 1964 was, therefore, a period of optimism but things were to change very fast and after some initial successes, the global eradication programme became unsustainable and was abandoned in 1969, the WHO quietly changed the name of the Malaria Eradication Division to the Division of Malaria and Other Parasitic Diseases and the word 'eradication' gradually drifted into abeyance. My resolution to retain an interest in malaria was not based on the possibilities or otherwise of eradication but on the assumption that there were fundamental problems in biology, particularly immunology that studies on malaria might help to resolve. Nearly half a century later my assumptions appear to have been well founded and malaria research has become a growth industry.

There are two advantages to having a PhD, the title itself and being able to supervise PhD students. By the 1960s individual research was no longer feasible or desirable and what were required were research teams. Space at King's was very limited and I could only plan for a maximum team of myself, a research assistant, a technician and two or three PhD students preferably one each year. This turned out to be unrealistic but I did manage to get a technician and permission for one PhD student. The next problem was how to finance my research. By the mid-1960s universities were no longer able to support increasingly expensive research so it was necessary to obtain funding from one of the Research Councils the most appropriate being the Medical Research Council. Here I discovered a problem. Zoology departments were not favoured by the Medical Research Council, who were already pumping vast amounts of money into the Biophysics Department, but were well supported by the National Environment Research Council. I therefore sought and obtained a PhD studentship for someone to work on the blood parasites of British small mammals. Fortunately one of our own students, Frank Newell who later became a lecturer at the University of Reading, was very interested in this project. He was followed by Alan Young, a graduate from Imperial College London, who later became Head of the Tick Unit at the International Laboratories for Research into Animal Diseases (ILRAD) in Kenya. Alan came to King's on the recommendation of Elizabeth Canning because his interests in blood parasitic protozoa and their tick vectors were compatible with my interests and those of Don Arthur, the tick expert, who by this time had become Head of Department. It would be invidious to pick out particular PhD students but two who are relevant to this essay are Mike Turner, who later joined Keith Vickerman in Glasgow and became a Professor in his own right, and Sara Brett, who later joined Eddy Lie's team at the Welcome Laboraories as did my Research Assistant Stephanie Millott. I kept a close eye on these students but meanwhile pursued an agenda of my own.

My own interests had now turned to the burgeoning field of the immunology of parasite immunology and I had become intrigued by the immunological differences exhibited by four rodent malaria parasites, the virulent *Plasmodium vinckei* and its morphologically identical but avirulent counterpart, *Plasmodium chabaudi*, and the virulent *Plasmodium berghei* and the less virulent *Plasmodium yoelii*. Nearly everyone working with rodent malaria parasites was using *P. berghei* so my interests became focused on the less virulent *P. chabaudi*. In the meantime, Frank Nowell had isolated a strain of *Babesia microti* from wild voles and I had cleaned it up and adapted it to laboratory mice. *Babesia* parasites are intraerythrocytic protozoa related to the malaria parasites but transmitted by ticks and some species such as *Babesia bovis* and *B. bigemina* are responsible for major and calamitous diseases of cattle. What was interesting about *B. microti* was that it caused what appeared to be a long low level chronic infection. In contrast another rodent parasite, *B. rodhaini,* caused a short fulminating and invariable fatal infection resembling the infection seen in cattle. What interested me was what happened in *B. microti* infections. The initial infection quickly reached a peak before declining and persisting at a low level. This meant that there was an effective immune response which brought the initial infection under control but was unable to eradicate it completely and this raised a number of interesting questions. I was also intrigued by the fact that this seemed to be what happened in *P. chabaudi* infections. The question I asked myself was would an infection with one of the avirulent parasites protect mice against subsequent infection with its virulent counterpart. Alasdair Voller at the Zoological Society of London was asking the same kind of question but using rats so we collaborated and found that animals that had recovered from infections with the avirulent P. *chabaudi* were resistant to subsequent challenge with the virulent P. *vinckei* but not to challenge with P. *berghei*. Alan Young and I then found that the avirulent B. *microti* protected mice against the virulent B. *rodhaini*. A pattern was beginning to emerge in which avirulent species of blood parasites were able to protect against different species of virulent parasites. This was beginning to undermine immunological dogma that immunity was species and strain specific. What happened next was totally unexpected, I fund that B. *microti* also protected mice against the rodent malaria P. *vinckei* and that the avirulent malaria parasite P. *chabaudi* protected mice against the virulent babesia B. *rodhaini*. These findings were greeted with a great deal of scepticism as my results ran counter to everything that was known about specific immunity. By this time I had managed to obtain research grants from the World Health Organization and the Wellcome Trust and had a graduate research assistant, Sheila Turner and had been joined by an American immunologist Catherine Crandall. We extended our range of examples of what we called heterologous immunity and ruled out any possibility of it being due to cross-reacting antigens but we still didn't understand the mechanism.

In 1975, I examined a PhD submitted by an Australian veterinarian who had been studying at the Royal Veterinary College close to King's College and who was then working with Tony Allison at the Clinical Research Centre at Harrow. Both Ian and Tony were interested in my work and we began to speculate about the possibility that the protection that I was observing might be another example of the phenomenon of non-specific immunity that was being considered as an adjunct to anti-cancer therapy. *Mycobacterium bovis* BCG (Bacillus Calmette Guerin) was one known stimulator of such non-specific immunity and had been tried in conjunction with conventional antigens in attempts to vaccinate animals against malaria and babesiosis. Ian therefore suggested that we should investigate the effects of prior exposure to BCG on subsequent infections with *B. microti* and *Plasmodium yoelii, P. vinckei, B. microti* and *B rodhaini*. The results were staggering, intravenous and intraperitoneal (but not subcutaneous) administration of BCG completely protected mice against *B. microti, B. rodhaini* and P. yoelii, considerably reduced infections with *P. yoelii* and had a lesser but significant effect on infections with *P. berghei*. BCG is a living agent so Ian and I began to look for a possible non-living agent capable of inducing non-specific immunity and independently hit upon *Corynebacterium parvum* and pooled our results. In summary, we discovered that *C. parvum* given intravenously or intraperitoneally, but not subcutaneously, protected mice against subsequent infections with *Babesia rodhaini, B. microti, P. vinckei* and *P. chabaudi* but not *P. berghei,* that the parasites died inside the red blood cells and that this was due to the action of some unidentified soluble mediator. We then examined a whole range of cellular and sub-cellular mediators of non-specific immunity with essentially the same results. Unfortunately at that time I had neither the facilities nor skills to pursue this problem further. Ian went on to incriminate γ-interferon and other pro-inflammatory cytokines in a number of disease processes and much later we and others showed that Nitric Oxide (NO) was involved in parasite killing but the phenomenon of heterologous immunity which subsequently spread to infections with many other microorganisms and parasites has never been satisfactorily resolved.

While I was working with Ian I was also collaborating with Nina Wedderburn (who had initially introduced me to Ian) on the interactions between oncogenic viruses and blood parasites, how the viruses depressed immune responses to the parasites and how the parasites exacerbated the effects of the viruses. Unfortunately I had to abandon this line of research due to lack of suitable facilities which was a great mistake given the subsequent discovery that HIV and other retroviruses have adverse effects on infections with a number of agents.

Teaching Parasitology at King's

Now back to what was happening to teaching at King's where, as in and other London Colleges, everything was undergoing major changes. As I mentioned before when I was describing my time at Exeter, Zoology Honours like other science courses were constructed from three subjects in the first year, usually Honours Zoology plus Subsidiary Botany and Ancillary Chemistry (or vice versa), Zoology, Zoology and Botany (or Chemistry) in the second year and Zoology, Zoology and Zoology (special subject) in the third. King's offered Parasitology and Cell Biology as special subjects in the final year and these courses were attended by students from other colleges. In return, Queen Mary College offered Marine Biology which was taken up by some of the King's students. In addition to the BSc students there were also 'intercalated' medical students who were allowed to take the final year of the science degree course between the completion of their pre-clinical studies and the beginning of their clinical studies. This arrangement was funded by the Medical Research Council and was aimed those medical students who wanted to study for PhDs and other postgraduate qualifications for which the pre-clinical training was in itself inadequate. Parasitology was an obvious option for these students and as only the top twenty five per cent of medical students were eligible to tale the intercalated degree we had some of the brightest students in the college. Because of this arrangement and the presence of students from other colleges I was privileged to be able to teach some of the most able students from the whole of the University. I have watched their careers as parasitologists develop and I still get e-mails from top ranking medical consultants and professors thanking me for introducing me to the field of parasitology. One particularly interesting aspect of parasitology teaching at this time was a joint King's/ Imperial parasitology field course at Slapton Ley in Devon in the West Country. The course, the first of its kind in Britain, arose after discussions between Elizabeth Canning and me because we were concerned that the students we were teaching did not have any field experience. Slapton Ley, run by the Field Studies Council for this very purpose was an ideal place for such a course. The situation was perfect with the sea and a freshwater lake only a short walk away and numerous nearby terrestrial habitats to sample. The laboratories were adequate and we brought nearly all of our own equipment and a technician. The course tutors as well as Elizabeth and myself were two helmintologists, June Mahon and Neil Croll both of whom died from natural causes tragically young. Kaye Lyons replaced June Mahon and the course was so successful that that the four of us published a paper on the parasites of Slapton Ley which became a classic of its kind.

This traditional way of teaching, which had survived for over half a century, came to an end when the University decided to organise the teaching into a modular pattern. Each degree course was to consist of a minimum of three and a maximum of four course units each year with a wide choice of options. Course units were specifically designed for years 1, 2 and 3 and this completely overthrew the traditional invertebrate-vertebrate years pattern. Don Arthur had no interest in any change and I had responsibility for reorganising the Zoology teaching in order to fit it into the new university and subject-wide framework. It soon became clear that the 9-12 unit system would not be universally workable or acceptable and at King's and elsewhere, except Imperial College, some of the units became half units so students were expected to take the equivalent of eight half units each year. The immediate result was that students who wanted to take a degree in Zoology could, within certain limits, dispense with units that did not particularly interest them. This resulted in a proliferation of units particularly in the second and third years. As far as parasitology was concerned there was a new half unit on parasitism in the second year and a revamped parasitology half unit plus new half units in parasite ecology and parasite immunology in the third year. The revamped parasitology course was developed after discussions with Dr Bridget Ogilvie at the Wellcome Trust and was replaced by one called the Parasitology of Tropical Diseases based on the five diseases listed by the WHO as the most important, malaria, African trypanosomiasis, leishmaniasis, Chagas disease, schistosomiasis and filariasis. There were also new units in Parasite Immunology and Parasite Ecology. The Parasite Ecology course was based around a field course run at the King's College field station at Rogate in Hampshire that had replaced the one run in conjunction with Imperial College. This course was also taken by students from Royal Holloway College and one of their staff, John Lewis a heminthologist, joined the teaching team as did Phil Whitfield from King's who had replaced Kaye Lyons. In addition to the formal teaching, all students had to undertake a full unit research project. The net result was that students interested in parasitology could take a total of 3.5 units, 3 of which, three of which were in the final year. Similar programmes applied elsewhere in the College and brought the Biophysics Department into mainstream biology teaching. Of particular interest there was a half unit on ethical issues run by the Nobel Prize winner, Maurice Wilkins and distinguished visiting lecturers. As I taught the whole of the parasite immunology course, much of the parasitology of tropical diseases, took part in the field course, supervised projects and also ran a very popular immunology course my teaching load was bordering on the excessive.

In the meantime, however, my academic career had been progressing satisfactorily and I was promoted from Lecturer to Reader in Parasitology in1969 and to Professor of Parasitology in 1974 and had been awarded the degree of Doctor of Science (DSc) in 1973. I had also been acting head of department for the academic year 1973-4.

Changes at King's College

The nature of the Zoology Department at King's began to change after the departure of James Danielli and most of his team in 1962. Don Arthur, who took over as Head of Department, had made his considerable reputation in the world of ticks but he was largely interested in their taxonomy and not in their importance as vectors of disease and so began to become side-lined in this field despite urgings from Alan Young, who was later to become a world authority on tick-borne diseases of veterinary importance, and myself who had hoped that the Zoology Department might establish itself as a centre of excellence in parasitology which had gradually slipped towards Imperial College. There had been a strong helminthological resurgence in the Department after the retirement of Ben Dawes, firstly Kaye Lyons, a talented electron-microscopist who was instrumental in recruiting Phil Whitfield, from Cambridge, who had a broad interest in helminth worms in-cluding schistosomes. Kaye Lyons left to become a schoolteacher and was followed by Roy Anderson, one of the most significant mathematical biologists of his generation (who was subsequently elected a Fellow of the Royal Society, became a Government sci-entist, was knighted and became Rector of Imperial College). Roy was the succeeded by Nigel Evans who soon left in order to take up a post with a major pharmacological company and was never replaced. Don Arthur then turned his attention to pollution in the River Thames which had the effect of filling the department with freshwater ecologists.

Over the next few years the department recruited a number of scientists including Michael Stoddart, a small mammal ecologist who later became Professor of Zoology at the University of Tasmania and David Pye, an expert on bat sonar, who moved to a professorship at Queen Mary College. When Don Arthur retired he was replaced as head of department by a vertebrate palaeontologist. I saw no future in the Department of Zoology at King's as a centre of parasitological excellence or so I began to look elsewhere for a more congenial atmosphere in which to work and in 1985 I moved to the Biophysics Department at King's College but this was something for the future.

Reorganization of the University of London

Beginning in the he 1980s there began a massive reorganization within the University of London. The University was essentially the administrative hub of an organization that consisted of a number of independent colleges. Originally in the 1830s the University consisted of only two colleges, King's College and University College both world class institutions whose well-publicised rivalry masked mutual admiration. Gradually the number of single- and multi-faculty colleges increased and by the 1960s the federal organizations, were 18 non-medical and 16 medical units. The University had become unwieldy and included at least 12 institutions large enough be universities in their own right These included, Imperial, King's, University, Queen Mary, Bedford, Royal Holloway, Westfield, Queen Elizabeth and Chelsea Colleges. In 1985 it was proposed to merge Bedford with Royal Holloway King's with Queen Elizabeth and Chelsea and Queen Mary with the science departments from Westfield. There followed a redistribution of academic subjects and staff and science teaching and research was to be concentrated on five centres, Imperial (which had always stood apart from the federal university), University (which had hardly been affected by the mergers),

King's, Queen Mary and Bedford Royal Holloway. One of the greatest changes was the reorganization of the medical schools that had been largely autonomous so that they all became part of one or other of the multi-faculty universities. King's merged with Guy's and St Thomas' Hospitals.

I have rather laboured this subject because I was intimately concerned with the events that led up to the mergers and their consequences. By the late 1970 the governance of the University needed to be brought into line with the events that were happening in the university as a whole. The overall governing body of the University was the Court but the academic body was the 120-strong Senate which included the heads of constituent institutions and 40 elected teaching staff, 15 representing their colleges and 25 representing academic boards of studies. In 1976 only one year after becoming a professor, I was elected by my colleagues to represent King's College on the Senate and its constituent body the Academic Board.

In 1986, after I had moved to the Biophysics Department, I was elected as Vice-Dean of Science and in 1988 Dean of Science of the University of London and I obtained membership and chairmanship of the of the Faculty of Science Board of Studies and an *ex officio* member of the Board of Studies in Medicine. I was, therefore, very much involved in the planning and discussions leading up to the 1984 mergers and changes in the provision of medical teaching. This was a very interesting time that involved membership of various committees and working parties and lots of discussions with heads of schools and faculties. Being Dean of Science had its up sides particularly participation in the degree ceremonies held in St Paul's Cathedral, Westminster Cathedral and Westminster Abbey. On these occasions I had an opportunity to meet and talk to our charming Chancellor, Princess Anne, the Queen's daughter. On these occasions my wife and I had dinner with the honorary graduands, a small number of people who had reached the top of their various professions. I particularly remember, the philosopher Carl Popper, the heart surgeon, Michael DeBakey and the singer Bob Geldoff.

This was also a difficult time because I had to make or agree to decisions that might not be in the long term interests of my department or myself. In 1984 King's merged with Queen Elizabeth College (QEC) founded in the 1950s and with relatively modern buildings in Kensington, one of the most pleasant and most expensive (of which more later) parts of London. QEC did not have separate departments of Zoology and Botany but a single department of Biology and also departments of Microbiology and Biochemistry within which there was some immunology. The other partner in the merger was Chelsea College, a former college of further education also sited in one of the most desirable parts of London, with departments of Botany and Zoology and also Immunology but with rather old and rundown buildings. The mergers created excess of space and staff so there was an urgent need to rationalise both the physical and human provision. My interests were then much more allied to those in the Immunology Department at Chelsea College which had joined the Biophysics Department at King's so I moved to this department which occupied its own building in Drury Lane, a short walk from the main building in the Strand. I did not want to move to Chelsea so Professor Bob Simmons, the head of the department and who was soon to be elected a Fellow of the Royal Society, arranged for me to have an office and laboratory in Drury Lane. The Biophysics department had been, and was, doing world class research mainly in cell and molecular biology and was a very stimulating and exciting place to work. There was a real problem, however, I was now 53 and having spent the past four years mainly in university administration I was going to find it very hard to get any significant research funding. In the event, this didn't really matter.

There was no possibility of the biological sciences continuing to operate from three sites and it was reluctantly agreed that everyone should move to the Kensington site but this would inevitably incur a vast amount of friction. It was then that the Principal of King's, Sir Stewart (later Lord) Sutherland, with whom I had worked closely in the University of London and was later to become Vice-Chancellor, asked me to form a working party to rationalise the provision of biological sciences in the new College. He had already invited Sir John Maddox, the editor of *Nature* and a member of the Council of King's College to serve on my working party and he and I began to consider who we didn't want to serve with us and to set out our terms of reference. For a number of years I had been chairman of the college's technical committee and had overseen a complete restructuring of the technical staff so I thought that I knew how to deal with heads of departments. What I hadn't appreciated was the enmity and distrust that existed between certain of the biologists from the three colleges. We soon realised that compromise was not likely to succeed so we undertook a detailed look at the teaching provision, the staffing and the research success of the disparate groups and decided that the most logical approach would be to merge non-medical biochemistry, immunology and biophysics (which was to become cell and molecular biology) as the Department of Biomolecular Sciences and the remaining biologists from Zoology, Plant Sciences and Microbiology would form a second department of Life Sciences focussing on whole organism biology and to bring the two departments together in a School of Life and Biomolecular Sciences. We also proposed that individuals should be free to move between departments but very few took up this offer and a number of members of staff decided to take early retirement. In the event, more by luck than judgement, we arrived at a situation where we had an equitable provision of courses and staff to run them although it was obvious that further pruning of staff would soon necessary. It was suggested that I should be Dean of the School of Life and Biomolecular Sciences but I was still Dean of Science at the University of London so I declined this offer and control of the School passed to the very capable and energetic Professor of Biochemistry from Chelsea, Harold Baum, and I became his deputy. Shortly afterwards, when the dust had settled, I moved to the Kensington Campus situated in one of the most beautiful (and expensive) parts of London surrounded by Embassies and within walking distance of some of the nicest parks and the best shops and pubs in London. The bonus for me was that I did not have to travel to the centre of London but could take a train to Kensington Olympia and then a pleasant 15 minute walk through Holland Park with peacocks, Japanese gardens and open air concerts on the way to the College. The Campus, formerly Queen Elizabeth College comprised delightful old buildings and newer buildings that embraced the worst of 1950s architecture and which, less than 40 years later, had become very run down and shabby. The one advantage, however, was that I and the other four immunologists had adjacent offices and laboratories in a newly refurbished corridor.

On a tangential subject relating to my time on the Senate of the University of London, the university had representatives on the governing bodies of a number of schools and Senate members were expected to serve on these bodies. I was very fortunate to be appointed to the governing body of the North London Collegiate School, one of the best girls' independent schools in the country and where my wife had been a teacher. This again was a new experience which I thoroughly enjoyed and I learned a lot about private education I was also on the governing body of a an excellent state girls' school, Camden School for Girls, which made me realise that state run schools could be very good. Later I was to become a governor of one of the best boy's schools in England, King's College Wimbledon. My experience as a governor came in very useful when I was appointed to governing body of the North East Surrey College of Technology (NESCOT). NESCOT is a college of further education a short walk from my home and I was privileged to be involved in the guiding the college through troubled times to its present status as 'excellent'

Early "Retirement" and Change of Direction

Ex-Deans, especially those who had ruffled people's feathers, find it difficult to return to the day to day routine of an academic department and I was no exception and never settled down to life at Kensington which had none of the excitement of life at Drury Lane. There was also the problem alluded to above that there were too many staff and that some had to go. A logical, but difficult, solution would have been to get rid of the non-productive staff, most of which had secure contracts of employment, but instead, the College decided to offer early retirement on very attractive terms to the highest paid staff with the inevitable result that many of the brightest and best who had many alternative opportunities took up this offer and I was one of these. The bargain that I was offered was effectively a half time appointment on full salary. No doubt the College expected me to work full time but I had other ideas.

As a result of a casual conversation with Professor David Bradley at the London School of Hygiene and Tropical Medicine, I was invited to be a Senior Visiting Research Fellow at the School. There was no salary attached but there I could have office space, a computer and access to all the School's facilities. I accepted with alacrity and have been working there on a part time basis in the Department of Disease Control in the Faculty of Infectious and Tropical Diseases now headed by Professor Simon Croft. When I first started working at the LSH&TM, I did quite a lot of teaching but this has gradually declined although I still enjoy giving the occasional lecture. There was one unexpected bonus that awaited me at the LSH&TM. The MSc Parasitology students each year went on a field course to familiarise themselves with as wide a range as possible of parasites and vectors. The course had been run annually in Yorkshire in the north of England but it was decided to move it to Slapton Ley in Devon with which I was very familiar from my days at King's and I was invited to join the teaching staff which I did for several years until my wife's illness prevented me from being away from home for too long. These courses were very rewarding and enjoyable and provided a marvellous opportunity to get to know the students very well. I think that my familiarity with the area must have been an asset. I also enjoyed the company of the other staff, Bob Sturrock, Jo Lines and Chris Curtis with whom I shared an office in London.

I have one important role at the London School of Hygiene and Tropical Medicine and this is that I am Module Organiser for a module entitled Biology of Infectious Diseases, one of four compulsory core units in the Distance Learning MSc Infectious Diseases course. These distance learning courses offer students courses directly comparable with those offered to internal students. My duties include writing and revising the course material and dealing with student's queries via e-mail. I head a team of five and we are also responsible for setting and marking the examination. It is always a pleasure to deal with student's queries and problems. This is very demanding but well worthwhile and I thoroughly enjoy the work and it is one way that I can repay the LSH&M for its hospitality.

Fortunately, as well as my research and teaching skills, I have developed a number of writing and editorial skills, for example I had been Editor of the journal of *Parasitology* for over 15 years and had been on the Editorial Boards of several other journals and had written a number of popular articles as well as editing a very popular text book, *Modern Parasitology*. I let it be known that I was looking for something interesting to do and was immediately asked to edit the monthly journal Parasitology *Today* (now *Trends in Parasitology*) while the editor, Caroline Ash, was on maternity leave in 1991. Caroline was married to Simon Croft who was later to become Head of the Faculty of Infectious and Tropical Diseases at the London School of Hygiene and Tropical Medicine when I was working there. Shortly afterwards I was approached by the Wellcome Trust to edit a major work to commemorate the 60th anniversary of the death of Sir Henry Wellcome. A steering group had decided that what was required was a high quality illustrated book and had elicited chapters on particular diseases from a number of eminent scientists and clinicians who were not historians but who had had first-hand experience of the conditions they were to write about. A vast amount of material of varying quality had been accumulated but no real thought had been given as to how this should hang together so the Trust decided to appoint an editor and this is how I became involved. The task was much bigger than either I or the Trust had imagined. Not all the eminent invited authors had produced anything and those that had had produced manuscripts some of which were near perfect and some piles of handwritten notes on paper. This I only discovered when I was given a file on the project in a luxurious office in the Wellcome Trust building in Park Square West close to Regents Park and the medical centre of excellence on Harley Street. I was also given the assistance of a secretary, Rosemary Tilden, who had semi-retired after a long service with the Trust and who was familiar with the world of medical literature and the Trust's collections of illustrations. Rosemary and I had to find the majority of illustrations and to put each chapter into a uniform format. Nothing was too much trouble for Rosemary and she spent a lot of her time visiting libraries and cycling around London to bring me photocopies of articles that I required. Rosemary had a unique second sense and often found articles that I hadn't even thought about. Some authors remained uncooperative and we had to abandon a few of the planned chapters and others I had to completely or partially rewrite. Eventually we had gathered all the material that we needed but, to my surprise, no thought had been given to a possible publisher. I then contacted a number of major publishing houses but all of them would have priced the book at a level above that which the Trust had expected to be able to sell it. Remember that this was to be a high quality copiously illustrated book and that publishing houses at that time were not in the market for this kind of publication. Eventually, I persuaded the Trust to publish it itself and in 1996 *The Wellcome Trust Illustrated History of Tropical Diseases*, a volume of which both the Trust and I could be very proud, was published.

It was while working on *The Wellcome Trust Illustrated History of Tropical Diseases*, that I met Leslie Collier, a microbiologist, who had written a chapter on Trachoma. Leslie was one of the editors of the multi-volume, multi-author volume *Topley and Wilson's Microbiology and Microbial Infections* then in its eighth edition. The editors had been considering including Parasitology in the ninth edition and Leslie set up a meeting where we discussed this possibility and as a result it was agreed that there should be a Parasitology volume co-edited by Derek Wakelin, a former student of mine then a Professor at Nottingham University, Julius Kreier, a retired professor from the Ohio State University (also a contributor to this volume), and myself. Derek would be responsible for the helminth worms and Julius and I would deal with the protozoans. The next stage was to recruit authors for the individual chapters and, because of my experience with the illustrated history, we carefully selected authors that we knew and could trust to deliver good manuscripts on time. In the event, everything went without a hitch and Volume 5, Parasitolology, in the ninth edition of Topley and Wilson was published in 1999. It was a great success and was followed by Volume 6 in the tenth edition in 2005.

I had written chapters on the history of parasitology and the classification of the protozoa In Topley and Wilson as a result of which I was asked to write a chapter on the classification of parasites in the tenth edition of *Manual of Clinical Microbiology* and also the tenth and eleventh. My chapter on the history of parasitology also attracted a great deal of attention and I was asked to write articles and chapters on this subject in a number of publications. I gradually moved away from laboratory-based science to become a historian as a result my career instead of petering out gradually took on a new momentum.

2009 marked the centenary of the Royal Society of Tropical Medicine and Hygiene (RSTMH) to be marked by an international conference in London and the publication of a book to celebrate the occasion. The Centenary Committee was divided about the nature of the book but advisers from the Wellcome Trust suggested that we should do something very different and produce an interactive CD-ROM. I was invited to edit this and thus began probably the most demanding and exciting phase of my career. We decided on a disc covering 6000 years of tropical medicine with three time lines, world events, medical and disease events and tropical medicine events. I had the task of writing the script and my Wellcome colleagues and I searched the Trust databases for items to illustrate every event. The disc was fully interactive and after a few false starts the final version was published in time for the International Congress and sent to all Fellows of the Royal Society of Tropical Medicine and Hygiene.

I had been on the Council of the RSTMH several times, had been Meetings Secretary and had edited the Society's journal. *Transactions of the Royal Society of Tropical Medicine and Hygiene* but I was very surprised when in 2008 I was asked if I would be Acting Chief Executive Officer of the Society pending a permanent appointment. This was something completely new for me but the President, Professor David Molyneux, was very persuasive and I agreed to take on this task for a few months. The Society was then undergoing a number of changes. A few years earlier the Society had moved from its original home, Manson House, to offices rented by the London School of Hygiene and Tropical Medicine. Although my appointment was intended to be a temporary arrangement I took the opportunity to try to bring the organization of the Society into the21st century and spent a lot of time with the Society' extremely competent administrator, Caryl Guest, delving through the Society's archives and pruning where necessary, revising the Society's handbook and working with my IT colleagues from the Wellcome Trust, who by this time had set up as an independent company Mantaray, to make the Society more web based. After the appointment of a new CEO and administrative team the services of Mantaray were dispensed with, the plans we had made were abandoned and the Society moved to new offices, since them I have had no real contact with the Society.

The International Stage

This seems to be a good time to reflect on one of the bonuses shared by scholars, particularly scientists, not open to members of most other professions. Science is international and provides opportunities to travel and to meet scientists in other countries. I once heard a famous actor say that an actor's life is difficult and uncertain but the compensation comes when one is greeted on familiar terms by one of the big names in the profession. Science is very much like this. As my research progressed I presented papers first at national and international meetings and then to my surprise and delight began to be invited to present keynote lectures and to chair sessions at international meetings, congresses and conferences. These invitations took me to the United States, Canada, Colombia, Kenya, Nigeria, Malaysia, Singapore, China and Australia and most countries in Europe. I also began to be invited to examine PhDs in the UK and elsewhere, in all I have now examined over 120 PhD theses and 23 Higher Doctorates and have advised on the appointment of professorships and memberships and fellowships of Learned Societies. It always gave me a thrill to know that somewhere, often many miles away, a committee or advisory group had considered me suitable for some task or other. Being an editor of an international journal brings with it a great responsibility to the journal and to the subject itself as well as many trials and tribulations but also opportunities to interact, albeit at a distance, with other scientists. It has always been a pleasure to follow the progress from the submission of a manuscript to its final publication and to observe the change of tone in any correspondence, now e-mails, from formal to friendly Christian name terms and to meet contributors face to face. It is also gratifying when an author says how much the editorial process has improved not only the manuscript under consideration but also subsequent ones. There have been many occasions at conferences when someone has come up to me and said "You don't know me but you published one of my papers and I wanted to meet you to thank you personally". I still get letters and emails from young scientists asking me for my advice and I always reply as fully and encouragingly as I can, sometimes at length, but, sadly, whereas a few years ago I would always get some acknowledgement or note of thanks the anonymous nature of the e-mail has virtually destroyed this courtesy.

At various times over this period I found time to serve as Meetings Secretary of the Royal Society of Tropical Medicine and Hygiene, Secretary of the British Society for Parasitology and Treasurer of the British Section of the Society of Protozoologists. All these positions brought me into contact with numerous other parasitologists, both in the UK and abroad, and new sets of contacts. The meetings of these societies were always great fun and it was a pleasure to be able to contribute to their success.

Gresham College

My early retirement provided me with opportunities to look for new challenges, some of which have been discussed above. The most important event occurred in 1993 when I saw an advertisement for the Chair of Physic at Gresham College. Gresham College is a venerable institution established in 1597 under the will of Sir Thomas Gresham to provide free lectures for those who live or work in the City of London. In its early days, Gresham College, was one of the foremost institutions of learning in the United Kingdom and its seven professors, each with responsibility for one of the following fields: Astronomy, Divinity, Geometry, Law, Music, Physic and Rhetoric were drawn from the elite of the Oxford and Cambridge Colleges and other eminent people including Sir Christopher Wren and Robert Hooke. The College had a chequered history and for a while lost its way when the Royal Society withdrew from the College in 1666 and re-joined again and nearly disappeared during and after the Second Would War. In 1991 the present College was established in a medieval inn, Barnard's Inn, in the heart of the City since then it has gone from strength to strength. When I looked at the names of the Professors I saw that they included some of the most eminent scholars in the UK including the Nobel Prize winner, Lord Porter. Nevertheless I applied and, to my surprise, was appointed Professor of Physics for three years during which time I would have to deliver 36 lectures, twelve each year. The topic I chose for my first series was 'Tropical diseases; their problem or ours?' Over the next two years I gave lectures on several aspects of public health, overpopulation, the AIDS epidemic, fake drugs and the potential of the techniques of molecular biology. One, which I could repeat today, was titled 'Malaria: the battle we cannot win'. The wonderful thing about Gresham College is that all the lectures are free and open to anyone who wants to come to them.

In 1995, my term of office came to an end but I had no idea then how much my life was to become involved with the College. Gresham College is a small organization with very few staff and is headed by a Provost, a part-time appointment. The first Provost of the revived College, Peter Naylor, had sadly died in office and had been replaced by a banker, Andy Prindle, who decided to step down after a short period in office and in 1999 I was asked if I might be willing to apply for the post. This was a very attractive proposition but in 1998 Valerie, my wife of then 38 years, was diagnosed with lung cancer and I felt that I could not undertake any long term commitment. Nevertheless I agreed to be Acting Provost until a permanent appointment could be made. Gresham College receives its funds in equal part from the Mercers' Company and the City and Corporation of London and I began to meet and negotiate with senior figures in the worlds of finance, law and politics. This was a very exciting and very enjoyable period of my life and Valerie was well enough to enjoy some of the social activities with me. This lasted only a few months but opened up a new world to me but this was not to be the end of my association with the College because shortly afterwards I was made a Life Fellow of the College, only the second such honour, which meant that I could continue to serve on the Academic Board. Valerie died in 2006 and I found myself more and more immersed with the activities of the College which has grown out of all recognition over the past decade or so under the stewardship of Provosts Sir Tim O'Shea, Lord Sutherland and Sir Roderick Floud and the Academic Registrar, Barbara Anderson. The College now puts on over 140 events each year, mainly lectures and recitals, at Barnard's Inn, the Museum of London and Christ Church in the City of London. All the lectures are videoed and put on the web and the number of hits each year is over a million. I am very much involved in formulating the programme and attend as many lectures and recitals as I can often accompanied by an old friend, David Garrood's widow, Frida.

So as I begin the ninth decade of my life, things are very good. My health is good, I travel a lot, I do lots of things other than Gresham College, and I have a lot of friends and a daughter who is geriatrician living in Newcastle in the North East of England and whose company I enjoy. A group of ex-King's biological colleagues meet up in a pub in Kensington every month and it good to keep up with all the news and gossip. For a number of years I have served on the Open University's Research Degrees Management Group that oversees all aspects of PhD and Research MSc from application to nomination of external examiners and still enjoy taking part in decisions of great importance to the students. I am also a member of a trust, the Gladstone Trust that awards small traveling grants to undergraduates. As I am writing this biography I am aware that I have a chapter for the 11th edition of the *Manual of Clinical Microbiology* ready to be sent off and a book on parasites nearing completion. I have just come back from trips to Iceland and Jordan. I don't have any other major projects lined up but one never knows. One thing is certain; I don't want to be idle.

Envoi

The Editors of this volume of Reflections and Connections, and Otto J. Crocomo, Julius P. Kreier and William R. Sharp, asked me and the other contributors to give some advice to young scientists. In another essay in this volume, Julius Kreier writes about the role of chance in one's life and the need to seize opportunities and to act on them when they arise. I absolutely agree with this sentiment. The analogy that comes to mind is that of waiting for a bus. There is no point in grieving about missing a particular bus because another one will come along but, this is important, it must be one going in the right direction. If the bus is not going in the right direction one can always get off and either go back to the starting point or catch another one that is going in the right direction. I have thought long and hard about what original advice to give and have reached the conclusion that the sort of advice that is usually given is often too trite and of little real use. Looking back on my career the best advice I can give is to cultivate clever people and to avoid stupid ones. I see now that the people that have helped me most have been very successful and have achieved honours such as fellowships of learned societies, high academic honours and civic honours. Clever, people, I have discovered, are outgoing and willing to share their time and experience with others but expect, in return, some kind of response and feedback. In this chapter I have named a number of those who have helped me and I apologise to those whose names I have omitted. When I say 'stupid' I don't mean uneducated because some of the most stupid people I have met have been highly educated. In contrast to clever people, stupid people are inward looking, unwilling to share anything and suspicious of everybody and everything. They are also dishonest and devious and cannot be trusted. I have only once worked closely with one such person and, believe me, they sap the very soul and should be avoided at all costs. Sadly, such people do have their acolytes and they too cannot be trusted and are best avoided.

Finally, the most important people in one's life are relatives and friends. I have been very lucky with my family and friends and here I can list only a few, some of whom, I have already mentioned. My daughter, Joanna, a geriatrician who devotes her life to the care of the elderly, my sister Christina, a serious world traveller now retired after a career teaching youngsters sport, my school friend, John Roper and his wife June, my university friends Keith Vickerman, his wife Moira and Alan Brafield, my postgraduate friend, C. P. Ramachandran, my research colleagues, Ian Clark and Eddy Liew, and my constant companion, Frida Garrood. I have already mentioned how these people have impacted on my life. There are also so many others, including students and colleagues and my late wife's friends, and others who are now dead, whose names would fill so many pages that it would be invidious to list any of them. In concluding this essay, I hope that my recollections fulfil the Editor's aim of bringing together the reflections and connections of a very ordinary scientist.

Francis Edmund Gabriel Cox Photo

CHAPTER 6.

An Ethnic Woman Scientist's Life

Geetha Ghai

Where does one begin is a question I have asked myself so many times. At this stage in my life the opportunity that William (Rod) Sharp has provided is a gift. I sincerely hope some value is found for those that may choose to read this story. I am a scientist trained as a plant biochemist from a university in the Indian subcontinent. My career took me through different phases in academia, industry and then back to academia. None of this, by the way, was planned other than the move to the United States to conduct research, which was a dream come true. A move that was supposed to be for 5 years turned into a lifetime. These experiences were filled with joys, challenges, heart aches and growth. A naive ethnic woman navigating her career and family through life in a foreign culture to achieve those scientific dreams stuck in my mind is the story I want to tell. Actually, for me coming of age emotionally and career wise really happened here in the United State. I would like to dedicate this story to all my family specially my husband and children, friends, mentors and acquaintances that have touched my life and made this dream a possibility.

My Beginning

I was born in Madras, now called Chennai, in India into an influential affluent "Hindu Brahmin' family on January 7th 1946. I will come back to religion and beliefs later on. Madras is and was the fourth largest metropolis, a coastal town with beautiful beaches in the southern part of India. This was the year that India had obtained Independence from the British rule. I was told and have read of the great expectations, merriment as well as the sad assassination of Mahatma Gandhi in the country. I need to explain a bit about my parents' families for you to comprehend my life.

My mother Meenakshi was from a town called Coimbatore, now called Kovai, a city with cotton mills about 300 miles from Madras. In her own right, she had a degree in Chemistry and was the second child out of four living children, with an older sister younger brother and sister. Her father, Ramaswamy, was a high school English teacher, while her mother Seethalakshmi was a home maker. This grandfather of mine was an avid bicyclist, swimmer and regimented exercise fanatic! As long as I can remember, he exercised for two hours each day. I have heard many stories like him jumping into the river Ganges and swimming across and riding the bicycle to Kodaikanal 6800 feet above sea level, on the Western Ghats of India. My maternal grandmother Seethalakmi was a loving homemaker with a lot of common sense.

My father, Subramanian, was a chemist as well and also the second child out of four children with an older sister, younger brother and sister. He obtained his Ph.D. from Liverpool, England and had returned to India in 1945 after the war. Yes, and my parents' marriage was arranged and a happy one at that. My paternal grandfather, Seethapathy, was a doctor; orphaned at age 4 and a self-made man a doctor by training. My paternal grandmother, Kanaka, had a middle school education and was a wizard of many trades. Many stories of how my paternal grandfather would serve those families that helped him grow up once he became a doctor anytime of the day are still talked about in the community. My grandparents, both paternal and maternal, along with my uncles, aunts, siblings and cousins were instrumental in shaping my early childhood.

My birth was in my paternal grandparent's house in a small room. This huge ancestral home was built by my grandparents with many rooms and a beautiful garden and they named it "Gokulum". A lady doctor, also a distant relative and family friend, came to Gokulum to deliver me. I have been told that the day I was born, the first child of the eldest son in the family, was a great day of joy and celebration. This house was always full with people family, friends and artists, mainly classical musicians, who had no place to go. Apparently, it was also the case on the day I was born. One famous homeless musician Tiger Varadachari, who was given refuge in my grandparent's house, wanted to see if this new born child (i.e. me) had any musical aptitude. And so, he decided to sing a special song from the Karnatic music style, a form of classical music in southern India, titled "Geetharthamu" meaning enchanted music in Tamil, So, this great music loving family named me Geetha meaning music.

Early Childhood

Till I was nine years old I was with my parents in a town called Kanpur in the state of Uttar Pradesh, UP for short, in the north of India. This town is about 300 miles from Delhi the capital of the country. In today's India, the state UP is considered one of the least developed areas and there are many jokes about the people and policies of this state. During my childhood I had a blast in Kanpur and had no clue about the general problems of the state or the country. My father used to bring home guinea pigs and rabbits for us to play with and take care of from his work.

My father, Dr. T.S. Subramanian, had a powerful position in the post British raj as the director of an army ordnance laboratory in Kanpur, where he was supervising more than 1000 people consisting of scientists to army truck drivers. He was 34 years old when he took the job in 1945. He supervised work on many pest control projects especially grains and fruits. This gave him the ability to expose my twin brothers and me to science. He was also an avid gardener and thought our gardener how to graft roses. I used to walk with him in the garden and feel on top of the world with pride while watching and learning the techniques. We had a beautiful garden with many varieties of roses all grafted by my father. He never abused his power, always respected human dignity and helped whoever was in need. He was not a religious person and always said his work and serving people was his religion. Although not religious he respected people with religious beliefs and always followed their wishes and traditions. I saw how people worshiped the ground he walked on and this had a lasting impression on me.

My mother on the other hand was a religious woman. She was absolutely devoted to her family. We always had delicious meals breakfast lunch and dinner. She attended to every detail from getting us ready to packing school lunches and being there when we came back from school. I could talk without inhibition about anything with her. Her ways were more eastern, mingled with some superstition and my father's more western with facts and evidence. She would say for as long as I remember that east has met west about my father and her relationship. She knew the periodic table by-heart, which amazed my two children when they visited her in India in the 1980s. I will come back to this later. Both my mother and father were progressive thinkers in their own ways. I do miss them both very much and wish they were around for me to bounce ideas especially when life gets challenging and confusing.

I attended a catholic school St. Mary's Convent in Kanpur. This was an English medium school as opposed to a vernacular language school. Even today, there are many vernacular language schools in India which is a strong cultural asset to society. During summer vacation we always travelled to south India where my parents were from. First to Madras then to Coimbatore and visited the joint family summer vacation home to a hill station 6800 feet above sea level called Kodaikanal. My father would take me on long walks in Kodaikanal, teaching me how not to lose altitude while hiking, to enjoy nature without polluting the environment while sharing information about many high altitude plants wild and cultivated. These trips were a joy especially Kodaikanal and Coimbatore.

I was totally pampered in Coimbatore, my maternal grandparents place as my brothers and I were the only grandchildren in that side of the family. None of my mother's siblings had any children. My aunt Saraswathi, my mother's younger sister, was a mathematics teacher in a Convent in Coimbatore. My visits to Coimbatore always included one or two visits with my aunt to her school where I got royal treatment by her colleagues and students. I loved all the attention given to me. I was free with no obligations or chores and never wanted this part of the vacation to end. Madras on the other hand was a bit dicey I'll talk about this in a bit.

When I was nine I developed some illness with a low grade temperature. No doctor in Kanpur, even those highly specialized in the army hospital could diagnose the cause for my low grade temperature. By process of elimination it was suggested that I had tuberculosis. It was likely as I was a very skinny child. My paternal grandfather did not agree with this diagnosis and he insisted that I be sent to live with him and my grandmother in Bangalore where we had a house and the weather was really good. My aunts, father's sisters, felt this move would also teach me Tamilian culture and I could learn the language Tamil. They believed I was not exposed enough to these ways living in northern India. For the life of me I do not understand even today why my father agreed to this plan. I regret not having asked him the reason for such a drastic measure. My mother and I cried our heart out when my parents left me in Bangalore with my paternal grandparents and the extended family and travelled back to Kanpur which was 2000 miles away. I remember that day so vividly like it was just yesterday. Incidentally, the fevers and sickness just disappeared in a while, no medication no treatment except a teaspoon of yucky cod liver oil! Thinking about that time away from my parents makes me sad. In retrospect this phase of my life did shape me to be who I am today.

The house in Bangalore was a ranch style one with a nice big yard. There were many tropical fruit and flower trees. I used to climb a fig and mango tree. We used to pluck the fruits right from the tree and eat them. There were very fragrant flowers that belonged to the magnolia family. There was a swing and a slide both custom built as there were no readymade sets in the market then. When my cousins and I were not in school or studying we would be playing in the yard. It actually was a fun time.

I was my grandfather's pet or so the rest of my cousins thought. He did favor me. He would wait to eat and would not eat if I was not there. This was his way of bringing me back to health. I must say I did feel special as here was the oldest patriarch of the family waiting for and on me. School was altogether another story. My aunts insisted that I be admitted to the vernacular (Tamil) medium school where English was a second language. So here I was in fifth grade learning every subject in Tamil. I had to start from the word go, alphabet to grammar like in kindergarten or first grade. It was a traumatic experience. I had no choice so I did it. In six months I was quiet fluent in reading writing and understanding the language which taught me a lesson that hard work and nose to the grinding stone does pay off.

However the euphoria for this achievement was short lived as my, protector, hero and sustainer, grandfather suddenly took ill. My father and mother were asked to come immediately to Bangalore. When my father walked into the room where my grandfather was lying he said, "I was waiting for you go get something to eat" I still remember before my father left the room my grandfather died. That was the time I learnt about death. I started wondering and thinking about death but never had a chance to talk about it with the adults. For some reason I felt it was a taboo subject and children are not supposed to ask questions about life and death. Self-imposed taboo why I thought that I still do not know, perhaps the culture made me think such subjects were off limits as forbidden topic. I felt very sad but could not comprehend the graveness of the situation.

Secretly, I thought my parents will take me back to Kanpur now that my thatha, is what I called my paternal grandfather, was dead. It was decided that we will stay in Bangalore to complete the school year and then move back to Madras. The rest of the school year was uneventful except for the girl's guide competition in which I won a second prize for identify and naming the flowers. For me it was an achievement but no one else in the family said anything. All these experiences with nature I am sure influenced my decision to study subjects related to biology. It was too early to decide or know what one wanted to do other than the fact that many of my older cousins were considering medical school. I felt intense pressure that I had to get an advanced degree and not let the family down.

After the school year I, with the rest of the family, moved back to Madras. My aunts convinced my father and mother that my education should not be disrupted and I should finish through my high school in a Tamil medium school in Madras. So my secret yearning to go back to Kanpur was squashed. I was enrolled in a catholic school St. Rafael's the Tamil medium section of St. Thomas Convent, in Madras. This sea city, a bigger one than Bangalore a table land about 3000 feet above sea level with pleasant weather, was hot and humid. Initially it was traumatic but I really learnt to like the school. I made some very close friends with whom I was in touch till I left India. My teachers were absolutely wonderful. I loved mathematics and was not sure what I will do after graduating from High School.

Around that time my father traveled to Australia and New Zealand for business and came back with a lot of stories. He also brought me coins and stamps from various countries along with a name of a friend's daughter Kelly Cox. This started me on serious stamp and coin collection. Kelly and I corresponded as pen pals for a year or two and then stopped. My coin and stamp collection was pilfered by my uncle in Madras. When I complained to my paternal grandmother, Kanaka, she said he has a problem just ignore it. I guess she was saying that he is a kleptomaniac; that made me very frustrated but I had no other recourse so I kept shut. I believe this experience made me keep my mouth shut when it concerned authority and scared of whistle blowing. It took a long time to overcome that fear. Till today no one in the family knows this except my long gone grandmother.

Going back to my schooling, the educational system in India was and still is very similar to the British system and students had to choose between mathematically driven or biologically driven subjects very early on. As most of my older cousins were gravitating towards medicine or biological sciences I too gravitated towards those subjects. Leadership wise I was popular in school and was elected as the school pupil leader. We had to attend chapel whether we were Christians or not. As a catholic school this was the norm, each student's religion or belief did not matter. If we were perceived as not obeying rules we got a hit with the ruler on our knuckles, all the things that went with a strict authoritarian educational system.

St. Rafael's a vernacular medium school was considered second tier in the St. Thomas convent system. It also cost way less than the English medium schools. None of my other cousins attended that school. All of them went to English medium schools and I felt like the step child and developed the Cinderella complex. I never shared this feeling with my father but did share it with my mother. My mother was really sad but her hands were tied due to her position in the family hierarchy. So, we both just cried.

Later on in life I found out that my father paid for my stay and schooling while the treatment given to me was that of a step child. More recently in talking with my cousins whose parents were with them in Gokulum, Madras, I was surprised to hear that all of us felt the same. That meant the orphan feeling was universal for Gokulite children. I remember the house having many extended family member's second and third cousins. It was customary in the family culture to help everyone in need so my grandparents would take them all under their roof. This meant feeding over 25 people breakfast lunch and dinner on the wages of my grandfather. I speculate the family supporting nature of my grandfather originated from having been orphaned at age four thus leading to this behavior of taking in all the under privileged family members, especially, his and my grandmothers siblings and their families. I marvel at my grandmother who supported his wishes with open arms. She maintained her identity by being a midwife even delivering babies in the nearby slum. I cannot imagine the pressure on my grandmother and the other women in the family.

This extended joint family living and my families giving nature had a lasting impact on me. Even in the US our house was always open to friends and family in need. I remember once my son, Pranav, commenting that our house was growing exponentially. That is when I realized how I was influenced by my childhood experiences. This realization was reinforced by the comment my father made on one of his trips to my house in the US that he did not realize that taking in needy people was a genetic trait in the family. My growing conditions along with the family environment not only emphasized education as being one of the most important assets but included providing a helping hand to those in need being of equal if not more important.

The emotional impact and challenges of my childhood was very difficult to write about. I have done so to reiterate the fact that every one's life, be it from an economically privileged or less privileged circumstances has challenges. These challenges can be overcome with persistence, mere gut and common sense. Of course the surrounding environment has to be conducive with supportive mentors. All it takes is one person to believe in your ability and skills. My mother and father were certainly among them along with some of my extended family, teachers in high school, and close friends. I do feel a sense of dismay that I did not have the foresight to thank them all.

Another thing my childhood did was to make me stay in the shadows behind the team and shy away from limelight, be assertive but not aggressive while sharing the rewards equally in a fair way. My observation is that some of these behavior patterns are not conducive to progress in our society today in the US. Many a times it is misunderstood for being meek and timid. Quiet on the contrary when push comes to shove the tiger in me will come out. I would like to say that withholding judgment before jumping to conclusion is always a good strategy. However if one makes a snap judgment then that person should be willing to change that opinion if it turns out to be otherwise. I did feel the need to digress with the hope that this may help the reader understand my moves and behavior in the future.

My stay in Madras was from 1956 to 1961. The culture in my home was centered on religion, music, and dance. I was woken up at five in the morning by my grandmother to practice music and my dance teacher came home after school. We were allowed to go to religious movies may be 2 times a year if they were produced. That required a great deal of persuasion. No popular culture equivalent of Bollywood/Hollywood was allowed what so ever! My cousins and I were never allowed to visit any friends but friends could visit us. This way the adults could keep an eye on who we were associating with. Once I remember wanting to visit a friend's house. Secretly I had planned to go but got caught in the act and that was the end of it. There was no pocket money. I used to be given an exact amount needed to reach school by rickshaw and come back home. Many a times I used to walk and use the money to by some pastries from a bakery as I was not allowed to eat outside. Every move was monitored except when I was in school. Sometimes my younger cousins were asked to spy on me this made my cousins closer to me. We developed such a bond that we used to watch out for one another. We had nicknames for all the adults and had our own secret way of communicating. It felt oppressive but in retrospect it did build a great deal of character. It thought me how to be tolerant, patient, disciplined and self-reliant.

I need to describe one incidence in particular that made a great impact which shaped some of my behavior and character. I was about 13 when my aunt accused me of taking money that her daughter had collected in school for a fundraiser. Essentially, she said I was a thief. My aunt had found some money in my cupboard and thought that I pilfered her daughter's school funds while her daughter had misplaced it. When her daughter found the money there was an apology and I was requested not to complain to my grandmother. The money in my cupboard was what my father used to give me when he visited which I saved. I describe this incidence in great detail because it has had a very lasting impression on my behavior. I try not to jump to any judgment till I have all the facts.

My grandparents (both paternal and maternal) were religious people. Paternal grandparents in Madras were particularly tradition bound with a lot of dos and don'ts as dictated in the Brahmin tradition. Totally vegetarian food no meat, fish or eggs were allowed in the house. Non Brahmin's were not allowed into the Kitchen and many rituals during religious ceremonies that had to be followed. Many of these traditional religious activities were similar to orthodox Christianity or Jewish faith. Maternal grandparent's house was a bit more relaxed. They did follow some of the religious tradition of a Brahmin house hold along with vegetarian food habits but did not have many taboos like non-Brahmins cannot enter the kitchen. These topics are so complex that I can write volumes on this culture. However, my vegetarian believes and lifestyle stem from this upbringing. Later on in life when I was forced to eat non-vegetarian food like fish or meat my brain would reject it and I was unable to swallow the food. I just gave up and decided food should be enjoyed and not force fed. Till today I am a vegetarian as a matter of fact "eggetarian". I am so glad for being an eggetarian especially in the US, since it makes my life a bit easier. I can fix an omelet for dinner and not think what will I eat or feed my family! Being a vegetarian is easier today in the US that in the 70's when I was looked upon as a freak of nature.

Religion on the other hand was more complex in nature and something I always questioned. I saw the world that I was exposed to then always rationalized religious rituals however, when it came to discrimination on the basis of caste, economic or educational status, I rebelled just as my father and mother did. I would try to trick my family in Madras and sneak people of all castes, religion and background into the kitchen. Something I learned later on that my father used to do as well when he was growing. I rebel against any form of caste system be it religious or economic. Fundamentally, this has been my social and cultural belief. I have always been an agnostic. In today's terminology I would classify myself as a Humanist. I believe my religion has always been hard work, ethics and fairness. I do respect all faith but I do not belong to any and if I am forced to conform I kick hard. I do think these beliefs come partly from my own growth and partly my upbringing influenced by observing my father's behavior. Although, I loved my mother intensely, her religious beliefs and pious nature did not rub off on me too much.

I finished my high school in 1961. Unlike in the US graduating from high school or for that matter from college or university is not celebrated too much in India. It is taken as completion of schooling and ready to enter college or the workforce. That is about it and I was more than ready to start college. Especially since it meant that I would be with my parents again. By then they had moved to a town called Ahmadabad in the state of Gujarat in India. This town is a night's ride by train about 250 miles from Bombay the financial capital of India and took about 8 hours. In the changing environment today Bombay has been renamed as Mumbai. In those days, steam engine trains used to take 8 hours to travel 250 miles. To my knowledge things have not changed much.

I feel the need to talk about the name changes. I wondered if all the name change of cities, towns and streets taking place in India today is a backlash and rebellion against British colonization. In talking to various people it certainly appears that way. The names that are being given today are pre British Raj names like Kovai for Coimbatore, Chennai for Madras, and Mumbai for Bombay. Sometimes I wonder if the politicians are doing this to hoodwink people and make them believe that they are actually working for the democracy. I find it quite amusing.

While in high school I used to visit Ahmadabad during Christmas break and play cricket with my brothers and their friends. They had a friend called Rajendra (Raju) Ghai who visited his parents in Ahmadabad during Christmas holidays from the northern part of India a town called Lucknow. Raju's father used to work as an administrative officer in the same Institution that my father worked and my father was senior Mr. Ghai's boss. I used to play cricket with my brothers and him as I was quite a tom boy. I was 12 then and he was 16. When Christmas vacation got over he would go back to Lucknow a town in the State of UP, and I would go back to Madras in the South a distance of about 2500 miles. That was that. Little did I realize that this would be the guy I will end up marrying?

College Years

After my high school graduation it was decided that I would indeed go and be with my parents and go to college there. Instead of going to college as a day scholar staying at my parents' home it was decided that I should attend Maharaja Sayajirao University, M.S. University for short, which was in a town called Baroda 60 miles away from Ahmadabad. Both these towns are in a state called Gujarat on the western side of India. So, I stayed in the student housing also known as hostels in India. The girl's hostel that I stayed in was named Sarojini Devi Hall. There were no coed hostels and it is still the same either for girls or boys. There were strict rules regarding code of conduct and curfew times. We had to be back in our hostel by 8pm latest. The main doors would be locked at that time. If you were late you had to find ways to sneak in or be reprimanded and disciplined.

I did my first two years of college in M.S. University with a science cluster. I did make some good friends. The food in the student housing was awful. I used to look forward to going home on the weekends and bringing back home cooked shelf ready food as we had no refrigerators in our rooms. We had a ceiling fan to cool off no air-conditioners either. No one minded all this as our bodies were used to these conditions. These two years were uneventful and I cannot remember much happening, other than my participation in a dancing event at the hall. I did kind of ok academically but not well enough to be given a chance to get into medical school. So my parents and I decided I would go back to Ahmadabad and complete my B.Sc. (Bachelor of Science) with a major in chemistry and minor in botany at St. Xavier's college, a four year catholic college. The first two year credits were transferred from M.S. University of Baroda and I was going to complete my last two years of a four year undergraduate degree at St. Xavier's.

Notice how decisions are always made with parents in the picture. This is a basic cultural difference between the US and India. May be such is the case in most of the Asian countries but I cannot authoritatively say that although my instinct tells me that. My friends from Asia who live in the US concur with this belief. This theme of decision making continued for quiet sometime in my life and it did not bother me a bit. Such authoritarian parental attitudes still exists among many Indian immigrant families in the US leading to parent child conflict. Many of the US Indian immigrant families are still 20, 30 or even 50 years behind time in their thinking about a child's independence. I can see the trauma in a house hold created by such cultural differences. I will share some of my experiences on bringing up children in this fiercely independent society.

My undergraduate education was entirely funded by my parents. Living expenses were too. There was no question of a part time job or any such concept. I used to walk to my college from home a 15 minute walk. My role was to concentrate on my studies and not get distracted. However, that was not the case. In the last two years of my college life is when I had a crush on Rajendra Ghai, who ended up being my husband later in life! By then he was going to college in Ahmadabad and lived in the same colony as I did and his father worked for the same institution as my father did and my father was Raju's (his nick name) father's boss! My twin brothers and he used to play cricket together and Raju was a year senior to me in college. We had this secret romance or so we thought. Little did we realize that it was not such a secret?

I need to clarify the attitudinal belief among Indian families of pride in earning a wage while studying. For me this is a valuable US (Western) concept. However, In India majority of the middle class and wealthy families would not want their child to work while studying. Many Indian families believe that working would distract a child from studying and the earning potential will deter them from further studies. I never understood this attitude, although, there may be some truth to it. In a progressively thinking family if a child is allowed to work then this act will be ridiculed by friends, neighbors and family. Thereby, causing great deal of anxiety for the family and provoking a sense of shame. I have been told by my Japanese and Chinese friends that such an attitude is also common in their countries. In India, most routine chores are performed for you by servants not just in the wealthy families but also in the middle class families. In my opinion this is a major reason that many Indian students have great difficulty adjusting to the life style in the US. Of course, this adjustment to say the least is multi-pronged starting with food, to missing family to being totally reliant on one self. Sometimes this leads to a lot of psychological trauma. I will talk about this later.

I did ok in my undergraduate classes. Botany was by far easier for me to grasp than chemistry but I stuck with both. In 1963 when I took the botany practical examination, as botany was my minor subject, I had to identify plants by their physical characteristics and I aced that part of the exam. Two years later in 1965, I had my final exam for the major subject chemistry. During the practical examination (laboratory was called practical in India) I had to identify a miscible liquid- liquid mixture and make a derivative of one. Identification was easy and I knew one of the liquids was toluene. I decided to make a nitro derivative without giving it a thought. Can you guess what the nitro-derivative is? It is TNT (Trinitrotoluene). Well, when the first nitro group attached it was ok when the second attached there was a big bang. The proctor walked up to me and asked me what I was doing and I explained that I was making a nitro-derivative of toluene. You should have seen his face. It turned horrific and he yelled "Please shut the burner I will give you full marks". I did and walked out. Till I was out I did not realize what I had done. This is what exam fever does. The way grading used to be done in the Indian educational system it was all or nothing. In other words, grading in the Indian system in those days did not take any notice of your performance for the entire year. Only my final examination mattered and my knowledge would be judged on that alone. So the pressure was always intense. In that intensity I did not have the sense to think what I was doing. I could have blown a hole in the laboratory and worse yet be dead and taken a few souls with me. It is funny now but then I felt so scared.

During this time I started playing the piano an instrument that intrigued me. I did have some knowledge of South Indian classical music known as, Karnatic music, and wanted to learn Western classical as well. I grew up listening to classical music both western and Indian at home. My mother a wonderful Karnatic music violinist and a vocal singer, and father a connoisseur of western and Indian classical music created a strong urge in me to be involved with music in some form or another from an early age. Along with listening to music I wanted to play an instrument. Learning to play the violin from my mother though an option was intimidating. I tried but we ended up in fights as she was impatient with me. So began my journey on wanting to learn to play the piano the instrument that awed my imagination. Thus, in my life music, dancing and arts forms went hand in hand with formal education in sciences.

Family Moves to Calcutta

All in all I did alright in my undergraduate education and graduated with a bachelor of sciences (B.Sc.) degree in 1965 and I was 19. One day after my exams we had a family meeting. Dad announced that we were moving to Calcutta (known as Kolkata), in the state of Bengal on the east coast of India. My mother and father had planned the move and decided it would be a good move for the family and his career. He had accepted a position to further the jute research, as jute was a very important cash crop for the country. Today, jutes importance in India has reduced, just like in the global environment as synthetic fibers have gained a greater importance compared to natural fibers. Then my parents looked directly at me and said, "If you and Raju are serious your relationship will last this separation". I was in shock that they knew and I had not given a thought about the seriousness of the relationship. I thought wow! My parents have extra sensory perception. I have no idea how they knew perhaps by my behavior that might have been instrumental in revealing my feelings or there were others who informed them about my secret rendezvous.

I was still in two minds about what to do in terms of my education. Part of me wanted to go to medical school and the other part wanted to be in the laboratory tinkering with research. I did not think that obtaining a medical degree will allow me opportunities to conduct research. I had admission in Stanley Medical College of Madras but the thought of leaving my parents was painful so I decided to ignore that chance. My non-emotional rational was that medical profession was mechanical and I needed to satisfy my curiosity by doing research and asking specific questions to answer them. The most important thing was the attitude of my parents about my education that no matter what I should get higher education with a professional degree. For Indian society my family's views on educating women was an anomaly and far-fetched.

In the sixties in India, it was common for women and very often men as well not to leave home for an undergraduate or for that matter graduate education. Even today, many times women stay in their birth home, get educated work and so on till they have a family of their own. This tradition is very unlike those in the US where expression of independence is of outmost importance. Indian tradition demands conformation for the good of the family over ones' own. Many of my American friends find it weird and unusual. These cross cultural differences cause a clash and cultural divide between the younger generation born and brought up in the US with their immigrant parents of Indian origin. I will share my experiences and thoughts on this subject later.

Graduate Education

By the time my family and I moved to Calcutta, now called Kolkata, I had made my mind up that I wanted to complete my Masters degree (M.Sc.) in biochemistry. I was curious and eager to learn about the processes of life, the fate of metabolites and the role of enzymes. Enzymology was an intriguing field at that time much like genetics today and was being talked about all the time. I wanted to know everything about enzymes in plants, microbes and animals. I still can sense the excitement that I felt then and wanted to be accepted in the biochemistry program at Calcutta University. I applied for the Masters program in biochemistry at Calcutta University and waited. About two months after the application, which seemed like an eternity, I was called for an interview. I remember one of the panel members at the interview asked me why should they admit me as my father could afford to send me anywhere in the country. I was taken aback and retorted, "would he like his daughter to be sent anywhere in the country and not stay at home?" In Indian culture this was an acceptable answer and the panel laughed and gave me admission. This experience thought me to think quickly and not get irritated at a question that seems absolutely irrelevant. I was so excited went home jumping with joy.

Metropolis of Calcutta was a novel experience. The suspension bridge Howrah over the river Hoogly would always give me goose bumps. This bridge was much like Tapanzee and when I cross the Tapanzee I never fail to think of my days in Calcutta. There always was a strong smell of dried fish over the bridge. This smell while crossing the bridge made me, a vegetarian, nauseous. I was brought up as a vegetarian and have maintained that eating habit till today. I do consume cheese, milk and eggs.

I used to be driven around places in a car with a driver. Unlike the majority of the population that took the overcrowded public transportation of Calcutta like buses, trams and trains. At that time it did not bother me as I was not exposed to the reality of life. Thinking back I wonder how this privilege did not bother me or make me think harder. There were fleeting thoughts of this privilege but it never caused any distress. On the contrary, today I feel a pang and some shame for not being more sensitive to the poverty around me.

Our driver's name was Aziz and he was a Muslim. Unlike now then people of all religion co-existed without too much tension within India. I still can remember him smiling and waiting patiently for me. He was employed by the Indian Jute Research Association (IJRA) for my family. According to Indian standards he got good wages and benefits. It was a luxury that feels like a dream. In addition, we had a cook, gardener and a cleaning person. All domestic help was male. I also continued my piano lessons at the Calcutta School of Music and participated in Trinity School of Music Theory exams for two years. Not once did I think this to be a privileged situation; lack of exposure I guess.

My master's (M.Sc.) biochemistry program consisted of 23 students. I used to dress very traditional like most Indians. Western clothing was there but not very common. For college, I wore a well starched and ironed sari. I do remember an incidence after a month of classes when a female classmate commented that I had not repeated a sari in 30 days and that my saris were always well laundered. I was taken aback as I never even thought of that privileged aspect of my life. Another incidence that traumatized me and still bothers me today was a day when one of our professors did not show up to class. It was a 2 hour class. The entire 2 hours my classmates who were all from Calcutta talked in their native mother tongue Bengali, I sat there the entire time feeling lost and not understanding a word. Towards the end one classmate turned towards me and said in perfect English that if I want to be in Calcutta I need to learn Bengali. This deliberate act of exclusion did teach me a lesson. I did pick up some Bengali to get by but also realized how it felt to be excluded. Emphasizing how marginalization and exclusion from a group traumatized me; ever since I have tried very hard to be inclusive. To this day when I think of that day it brings out very complex emotions that I cannot describe.

This incidence shows the diversity of India, a country with 30 plus languages and many more dialects, all varieties of religion and cultural beliefs co-existing sometimes in harmony but always in Brownian movement. I cannot imagine how difficult it must be to govern such a complex country but the democracy goes on. Mathematical and biological sciences continue to grow and scientists educated in India hold their own. This does say something about the education system does it not? I will let you be the judge

My Marriage and Ph.D.

I completed my master's degree (M.Sc.) in biochemistry in 1968. My final exams were postponed twice due to some social unrest in Calcutta but did eventually take place in November of 1967. I got my results in 1968. This year marked another life changing event I got married to my wonderful husband Rajendra Ghai.

After marriage, I moved and set up home in a town called Baroda, known as Vadodara now, in the western part of India in the state of Gujarat where I started my two year college education and my husband, Raju as he was called in India, was doing his Ph.D. in clinical biochemistry in the medical school. I was keen to get my Ph.D. as well and did not want to waste too much time. I applied to the microbiology department at the Maharaja Sayajiroa University know as M.S. University. I do not remember the process but was accepted for the doctoral program. One thing I do remember clearly that there was no talk about being a female or male. I was accepted on my educational merit. Although I talk about gender now, in those days it never even crossed my mind that there would be a gender difference till I came to the United States.

My research was on mangoes (Mangifera indica); conducting research on post-harvest ripening of mangoes in a microbiology department. I did not give this unique situation of conducting research of fruit ripening in a microbiology department any thought. In retrospect it seems odd! Mango was an important Indian fruit and the goal was to understand the post-harvest ripening to preserve the fruit so it will be available during the entire year, ultimately leading to better methods of preservation while providing an economic benefit for the farmers. In Gujarat, the state that Baroda was and still is a part of, the mango variety alfanso was the most important one and our department concentrated its efforts on that. The microbiology department was funded by a PL 480 grant from the agricultural services of USDA to conduct this work. A collaborative arrangement was also worked out with an Institute in Israel where a wax had been developed to store fruits. This wax was being used to prolong the shelf-life of alfanso mangoes. Waxed unripe mangoes were stored in the cold between 0°-5°C and prolongation of ripening was monitored. We also evaluated the taste qualitatively of wax coated mangoes ripened at room temperature after different time intervals of storage. I have to say the taste was horrible and did not taste anything like freshly ripened mangoes!

The vast majority of the work in our department was centered on understanding the post-harvest ripening process of alfanso mangoes. Post-harvest physiology and biochemistry was compared during the ripening process by studying the metabolism during ripening. My thesis was on carbohydrate metabolism of ripening alfanso mangoes. In retrospect, I wonder why my thesis title was so broad and not specific to one enzyme or a specific pathway.

This was a very innovative time. To be on par with what was going on in the universities in the US we had to improvise many techniques. I remember developing a disc electrophoresis unit with a colleague V.V.R. Reddy and analyzing the proteins during ripening. I used to stand outside the cold room and collect fractions from a column every 10 minutes and then analyze these fractions for enzyme activity. There were no automatic fraction collectors as in the US. Even though frozen half to death by this very tedious process the thought of giving up never crossed my mind. When the results were obtained like a protein band in unripe fruits that disappeared during ripening I was filled with excitement. This process taught me self-discipline and patience. I used to put in 12 to 18 hour days during the purification procedure. One specific time at 2 am in the morning when I finally finished collecting the last fraction I walked out of the building to go home and was astonished at the scary sight of fallen trees and realized a powerful storm had occurred and uprooted many trees. My husband escorted me home as we climbed over the fallen trees.

Another colleague A.K. Mattoo who had done work on ripening mangoes as well had found a protein inhibitor in the unripe mango that prevented ethylene formation. Ethylene is a ripen-ing hormone in climacteric fruits. Continuing that work I found that similar inhibitors were present in unripe mangoes that pre-vented carbohydrate metabolism specifically conversion of starch to sugars. In addition, I found that there was a feedback control to regulate sugar breakdown in ripe mangoes, by beta carotene and fatty acids that are found in greater concentration in the ripe fruit, thereby allowing sugar accumulation and makes the ripe fruit sweet. Some of these findings were published in a Biochem Biophysics Research Communication in 1970. Being a first author and a first publication was a major accomplishment, especially, from the developing world with limited equipment. I got excite-ment about research developments and findings however small they were. It created enormous satisfaction during my entire doc-toral program. I used to imagine how much more I could accom-plish if I was working in the US with all the modern equipment. The thought of difficulties that arose from laboratory politics, including pecking order as a junior lab scientists and gender poli-tics never crossed my mind. I was oblivious to all that and was just interested in research and completing the task that is my Ph.D.

In 1971, I had my first child and completed my dissertation as well. I graduated with the doctoral degree in 1972. At the same time I obtained a postdoctoral fellowship from the Council of Scientific and Industrial Research (CSIR) organization of India for 3 years to continue the same research. I worked till 1974 June and then followed my spouse to the US in November of 1974 on a dependent visa which used to be called J2. Our first home in the US was in Buffalo, New York. Imagine us from a tropical country never seen snow arriving in Buffalo in the beginning of winter.

Visa process was a unique experience. My three year old son Pranav, and I went to the US Embassy accompanied by my father to get the visa. Some documents from the US were not correct we had to call the State University of Buffalo (SUNY) and get the documents to obtain the visa. Which took about a month or so. I was booked by the Belgium airlines (Sabena) to travel to London spend a week with my father's friend Dr. Kenneth Woodford and family then leave for the US. There was an air of excitement mingled with anxiety and sadness. I did not have a job which was a new phase in my adult life. I also was given only eight dollars as foreign exchange. In those days India was not open and had restrictive economic and market policies. Today, things are much better. Dad had arranged with Dr. Woodford, who was supposed to give me $200 when I landed in London? He and his family were going to prepare me for my life in Buffalo, New York, starting by educating me on how to dress for the severe winter weather.

I had never worn a heavy coat, boots, gloves, hat or scarf. Here I was ready to globetrot in a sari, open sandal on my feet with a 3 year old. It felt like I was traveling to the moon. The day of departure was very sad leaving all my family, friends, and everything I was familiar with behind. The excitement of a new beginning gave some comfort. We (my son Pranav and I) boarded the plane in November to Heathrow via Brussels. Pranav was excited looking at the plane, cars and other wonders I was fearful as to how I will manage this journey. On the whole the first leg of the journey was uneventful with familiar food. As the journey progressed emotional uncertainty, fear and million thoughts took over. I was wondering what will happen if Dr. Woodford did not show up in London, where will I go with eight dollars so on and so forth. There was no end to crazy thoughts. When we landed in London there was Dr. Woodford waiting with two warm jackets one for me and one for Pranav. Was I relieved to see him standing at the arrival section with a warm and beautiful smile.

Dr. Woodford was the director of the Grassland Institute and his family lived in Henley on Thames. It was a beautiful breath taking setting. The next day he took me to the institute where some pioneering work was being done on grass, weeds, entomology, animal husbandry and others. Due to my jet lag, naivety and lack of wisdom I did not ask many scientific questions. It is a regret I will have to live with for the rest of my life. I do remember the scenic beauty around Themes. A week flew by with shopping and lessons on living with snow from Mary Woodford, who was from Toronto, Canada.

After a week, my son and I boarded Sabena the Belgian airline again to New York. The mixed emotions of leaving home resurfaced. Although, I was 28 years old my life in India was fairly sheltered with exposure to work and family. This was my first trip abroad. With no Google or internet most of my exposure to other cultures came from books or by talking to family, friends and acquaintances. Even though my exposure was better than most Indians I felt quiet unprepared for what lay ahead in the US. Child rearing with no family support was a big fear factor for this insecurity.

Trip across the Atlantic

Sabena being a Belgian airline had to stop in Brussels en route to New York. There was a wait in Brussels and we were asked to deplane. I was carrying Mary Woodford's heavy fur jacket that she had so generously given to me for use in one hand while holding my son on the other. I was afraid to go too far away from the gate so stayed around it for the layover time. When the boarding time arrived I walked in with my child and my coat on to the plane. The steward looked at me and said courteously and seriously "Madam we have sweepers cleaning the plane, no need to clean it with your coat". It took me a few minutes to understand his sense of humor and realize he was referring to my coat trailing on the floor. I pick up my jacket smiled and walked on to my seat. When I think of this incidence I wonder if I looked like Linus in Denise the Menace dragging his blanket all over the place. Funny as it is today it was traumatic then.

My first trip across the Atlantic emphasizes the fact on how food is an important part of one's identity and culture. It certainly was for me. I was and still am a vegetarian. I do eat eggs and milk product. In the 1970s, vegetarianism was not popular even in the US a country which is open to all cultures. During meal time the flight attendant gave my son Pranav, his "kiddie" meal and kept saying that my special meal was coming. After a while I got a whole head of steamed broccoli with no salt, pepper or spices. I had never seen a broccoli and thought it was a spoiled cauliflower! I could not eat. For dinner I got the same head of broccoli. This insipid head of broccoli is how I began my first trip across the Atlantic. Other than that my flight was uneventful till I landed in New York. So began my eventful journey to the US from the east to the west, with cultural difference in food, clothing and language.

We landed in Kennedy as it was getting dark on an early November evening. The aerial view was breath taking. We were spending the night in Manhattan with a friend and she was supposed to pick us up and then drop us the next morning at LaGuardia for the flight to Buffalo, New York as my husband was not able to leave his research and laboratory. We got of the flight cleared customs and waited for our luggage. I had never seen a luggage belt and had never carried heavy stuff on my own. The entire scene was like a movie setting. When the luggage arrived I just stood rooted to the spot not knowing what to do. A tall African American cop was watching me and he walked over and asked me what my problem was. It took me a few minutes to understand what he was saying as his accent and language was quiet unfamiliar. Finally, I found my tongue and asked him how was I to get my luggage of the belt. He must have thought well here is another fresh of the boat case. He pulled it off and said get a red cap. Who knew what or who a red cap was? In any event I managed to take the luggage out of the customs area into the arrival lounge.

Guess what I did not see my friend! There was some announcement over the PA system calling some name which sounded vaguely familiar. In my state of panic I did not pay any attention to the announcement. After a while I decided to go to the information desk and ask them how to get to LaGuardia. I thought I will spend the night at the airport. At the information desk they looked at me and asked me if my name was Mrs. Ghai. That is when I realized I was being paged. The guy said he had been calling me for about 30 minutes and said that my friend was delayed and she wanted me to take the bus to Grand Central Station and she will be there to pick me up.

This left me in a complete panic about all the logistics. A kind porter who was watching the entire scene came up to me and said that he will take my stuff to the bus stop and then ask the driver to inform me when the bus arrives at Grand Central station. The porter took my luggage and set it at a place outside and said the bus will come here. I did not know how to count the money so naïve or stupid that I was. I pulled the money and said take what is due. He was very honest and took one dollar. The bus came and I kept asking the driver does this go to Grand Central. He grunted just picked my luggage and put it on the bus while not saying a word. So began my journey in the US. I had no choice but to get on the bus. In the mean while Pranav my son was so excited looking at the lights and the traffic. He was talking a mile a minute in Hindi the language he was fluent in then. I wanted him to shut up but just let him babble as there was no point in making him panicky as well. Once the bus was outside Kennedy terminal the driver came over to collect the money. Again I pulled the money and he took my fare and told me he would let Pranav ride for free. I kept asking him does this go to Grand Central and he did not answer that question. I do not know why. A couple sitting two seats behind was watching all this. They tapped me on the shoulder and said they get off at a stop before Grand Central and they will let me know so I can get off at the next stop. I was so grateful for this kindness. I did reach Grand Central and there was no friend! Another young couple who got off the bus stood by me and said we will wait for your friend to show up. This is not a safe place for a young woman and child. Then I saw my friend walking across. The reason for this story is to share how strangers are kind and trustworthy even in a metropolitan area like New York. I was also questioning my judgment on wanting to come to the US. At night, the view of Manhattan from my friend's living space, a pent house in mid-town, was just beautiful. She did take me to LaGuardia the next morning and we boarded the flight to Buffalo. The bottom line is it was an experience that I will never forget and finally I landed in Buffalo with a thud the next day.

Buffalo, New York

I regress in personal experiences beyond science to share some of the challenges a foreign student had to cope with in the 70s. Not only was I a foreign student I was a woman, a wife and mother with limited cultural experience to the west. It was not just culture but also the harsh weather, the environment as well as the system in the University and outside. These experiences can make a person strong or break them. I have heard stories where individuals decided to return back to their home countries. Especially, in those days there was not much support in the universities for foreign students and family's acculturation. Coming from India even the pots and pans in the kitchen looked different! Every evening at home it was the same story my husband wanting to go back and I holding him up saying we need to complete at least 3 years before we go back. Here we are after almost 4 decades.

I was expecting my second child and decided to take a break from science. I wanted my son to get adjusted to the new society while I deal with my pregnancy. My daughter, Nirupa, was born in 1975 at the Children's Hospital in Buffalo. My gynecologist's name was and is Dr. Nirmala Mudaliar, who was of Indian origin. Early on in my life in the US comfort level of finding an Indian physician was really important to me as I felt that these physicians will understand my psychology. Today, medical schools are giving educational courses on "cultural competency" whereas then I created my own without much fanfare. She understood that I may opt for no epidural, and may follow some natural home remedies, which was very important to me. Finding Dr. Mudaliar was a blessing in disguise, I found her through a neurosurgeon in Buffalo, who was a friend of my family in India. I had a vast network of doctors educated in India who knew my family and wanted to help me which was serendipitous. I cannot say enough about the kindness of all those people and why strong networking among the Indian community was so important almost a life saver for my family's survival and sanity.

SUNY Buffalo had maintained that their medical coverage for postdoctoral fellows was very good. Based on that information we chugged along with no fear about money, living hand to mouth, on one postdoctoral fellowship which was $8,500/year. A few weeks after my delivery the hospital called me and said I owed the hospital $2000. We did not have that money now what are we supposed to do. I told the hospital I will do some voluntary work and the hospital can use my hours to pay for it. The hospital said that was not necessary and that I could pay it over a period of time which we did. Dr. Mudaliar, waived all her charges. This is the type of economic difficulty that foreign students had to deal with. Not understanding the system with fear of being on the wrong side was a strong emotional stress and thinking that a wrong move will send you back to the country of origin in essence deportation. These experiences helped me build a strong character along with empathy for new students from other countries.

During my pregnancy I was going crazy without any intellectual stimulation. It was like solitary confinement in the small apartment. I started looking for postdoctoral positions. There was a position that came to my attention at Roswell Park Cancer Institute. I called the investigator and talked with him. The first question he asked was if I was married and the second question was if I had a child. I said yes to both. His reaction was he needs somebody who will work 12 to 14 hours a day preferably a bachelor not a woman with a child. I hung the phone up and cried. I had never been exposed to such blatant discrimination. I decided not to look for a position till after the baby was born. Little did I realize all that was going on in woman's liberation movement in the US at that time and the male dominated workplace, in retrospect, it was a good thing, because if I understood the sacrifices required to pursue my ambition of doing a postdoctoral fellowship. My goal was to be in the laboratories and develop my skills. With that in mind I just went about looking for a postdoctoral fellowship.

My First Job

Help comes in unexpected ways. My husband had a colleague Dr. Sylvia Christakos. She used to take me to the doctors every week during my final stages of pregnancy as Raj could not leave the laboratory! She mentioned that a new faculty in the Department of Pharmacology, Dr. J. Craig Venter, yes today's famous human genome person, had a postdoctoral position available. I applied and got it. At the same time our green card application requirement for permanent residency was being processed. We feared that if I asked for a change in my visa status our immigration process may be jeopardized. We did not talk to the lawyer who was handling our case but decided that I will work without pay 3 days a week in Dr. Venter's laboratory. He was a gentleman and did not want that but decided he will go along with the decision and give me an opportunity to be in the laboratory. I did not know who to approach within the university system to ask about the procedure and there was no information or guidance like a well-organized foreign student's office in those days.

Dr. Venter's lab was working on cardiovascular beta adrenergic receptors. I had to learn a whole new language in cardiovascular pharmacology, which was exciting and intimidating; from plants to mammalian physiology. An interesting contradiction between plant and mammalian cells is the protein concentration. For example, very high water content in mangoes was a problem during protein purification. I had to think about how to remove the water while conserving the protein during the concentration process without denaturation. The opposite was the case with cultured mammalian cells or heart tissue and we needed to dilute the protein without denaturing it.

I was the first postdoctoral fellow in Craig's laboratory. He kept asking me if I wanted to be a research associate or a postdoctoral fellow. I did not know the difference but was told that postdoctoral positions could be short term while research associate positions could be longer term. Of course I wanted the latter. In his laboratory I learned tissue culture. How to grow, maintain and store them in addition to conducting research on beta adrenergic receptor binding assay and cyclic AMP the second messenger assay. I used to radioactively iodinate the ligand for the entire laboratory. This radiolabelled ligand was then used for binding assays by all those working in the laboratory, which consisted of a technician and two graduate students. This procedure was done behind a lead shield in a hood. Every so often the university radioactive safety officer would want to screen me to be sure that I was not glowing! After a year or so I was told that I was a live wire and my thyroids were quite high in radioactivity. I am not sure why that was the case although I followed the necessary precautions and safety procedures; perhaps, the lead shield was not as well made as today or the hood was not well exhausted, I will never know the answers to these questions but I sure can speculate.

The time in SUNY Buffalo was a great deal of fun. I was learning new things and Craig was a great teacher as well as an ambitious scientist who wanted to be successful. I remember when we sent the first abstract to the 1977 pharmacology meeting ASPET (American Society for Pharmacology and Experimental Therapeutics). We wanted to celebrate so Craig bought a bottle of Champaign that we chilled in our refrigerated centrifuge. During the entire time, the centrifuge was labelled out of order, such was the fun we had. We opened the bottle in the evening. I am not sure if it was against university rules to drink in the laboratory however that did not stop us from celebrating.

The 1977 ASPET meetings took place at the Ohio State University, Columbus, Ohio. Craig's entire laboratory attended the meeting. This was my first scientific poster presentation in the United States and the feeling of pride and accomplishment was great. Craig was a great motivator and he believed in having fun. All of us from the laboratory decided to play tennis and I was Craig's partner in a doubles game. It took Craig a very long time to serve so I looked back at what was going on and he served. The ball came right at me and hit my left eye. For a moment I thought I had lost my vision in that eye. I was rushed to the hospital emergency room and my family was notified. Joyce our baby sitter took care of the children while Raj rushed to Ohio. They were evaluating me and said they might have to operate on me. The team will decide what was necessary in the morning. Luckily I did not need an operation. I had a retinal tear and peripheral loss of vision for which no remedy was available. Once back in Buffalo, I was directed to an ophthalmologist, Dr. Dilip Patel. Little did I realize that his wife was my classmate in India, what a small world! Dr. Patel, who was considered an expert in cryosurgery sealed my tear, and his handy work is admired by ophthalmologists even today. This incident did not deter me from continuing my quest of research in the US.

I was in Buffalo during the blizzard of 75 and 76. I remember working with the tissue culture cells on the morning of the blizzard. Craig walked in and said I should go home. I was very adamant that the cells will be ruined if I did not split them. I stayed till 12 noon when Raj came and said the University is closed and we should leave. We left as our children were home with Joyce our baby sitter. Boy was it difficult to drive, almost zero visibility we travelled very slowly, and the five miles to home took more than two hours. We were lucky as those people who left later than us had to either go back to the university and spend the night there. This was an experience to remember.

Our baby sitter Joyce O'Conor lived in the same apartment complex and she was such a lucky find. Her parents had a small farm outside Buffalo and the kids loved going there. Her parents were the grandparents that my children missed having. Between my friend Sylvia and Joyce we made up for the missing family. Sylvia thought me to stand up for myself in science and against dominating individuals. Now, she is a faculty at the Rutgers Newark Medical School and we have maintained our friendship. Another postdoc who worked with Raj, Dr. William (Bill) French, knew I liked ballet and classical music and took me to cultural events by the Buffalo Symphony Orchestra. I cannot state the importance of these two people in my life. Without them I would have been a basket case. Notice my husband's nick name changed from Raju to Raj in Buffalo. It has remained as such since.

Move to Mobile, Alabama

Two and a half years into my Research Associate position Raj wanted to move to the Department of Biochemistry in the Medical School at the University of Southern Alabama (USA) in Mobile in 1978. I was quiet upset and said that I will only move if

I found a position as well. I learned from my past experience on how difficult it was to find a position. So Raj was forced to say that he will move only if his wife, i.e. me, found a position. It caused a great deal of tension and frustration among the family but I stood my ground. Within two months of the ultimatum, I found a Research Associate position in the Department of Pharmacology with Dr. S.J. Mustafa, whose laboratory was working on cardiac and coronary artery adenosine receptors. All that I learned in Craig Venter's laboratory laid the foundation for this part of my life.

After my eye accident Craig kept telling me to sue him. I was a bit disturbed by that. I kept saying it was an accident and did not realize the litigation oriented society that we were living in. There was a lot that we had to learn not just science but about the democracy that we were living in and the system at large. In any event Craig made me apply for workmen compensation. I still remember the ophthalmologist who examined me in Mobile Alabama who had to send a report to the workmen compensation and had to put a price on my eye. He kept saying, "How can I put a price on your eye". I did not quite understand the implications of that statement then. In any event, I got $2000 for my peripheral vision loss. This peripheral loss of vision made using a reverse phase microscope very challenging.

Raj and I were somewhat familiar with the history of the US but not to the extent that we should have been. There was a lot that we had to learn from science to society and living in the USA. Our colleagues and friends in Buffalo thought that we were out of our mind when we told them we were moving south, that too deep south. Any further south we would have to swim across the ocean. They kept asking us whether knew what we were doing. In other words they were saying we did not know what we were doing! We did not realize the implications of these comments. In our innocence we just did what we thought was the best move without thinking too much. If we would have been savvy enough to know the scientific landscape of the nation, the nature of funding and all the other aspects of what it takes to be successful we might not have moved and who knows what my future would have been. Looking back the move to Mobile was a good one. It made me an independent thinker and researcher. I learned to stand on my own feet and not be afraid to speak up.

My stay in Mobile from 1978 to 1983 was really a phenomenal experience. Apart from receptor binding which was a routine procedure in Jamal Mustafa's laboratory, I set up the coronary artery strip assays to show the relevance of cell membrane receptor binding to physiological action. In the pharmacology department Dr. Stanley Greenberg was an expert in vascular strips and I learned these techniques and methods from him and set it up in our laboratory. These assays would run for 12 or 13 hours and I used to work way past midnight. When the correlation between receptor binding and physiological response was found the excitement was euphoric.

After two years in Mustafa's laboratory, I felt the need, for an independent position with my own laboratory. At the same time Mustafa decided to move. I took this as an opportunity and approached the Department Chair Dr. Thomas (Tom) M. Glenn. He negotiated a deal with the Department of Biochemistry and I obtained an instructors position shared between the two departments. This was a crucial step for my future. I worked in collaboration with Dr. Joe McCord who was one of the discoverer's of super oxide dismutase (SOD). I learned a lot from Dr. McCord who had a laid back personality while setting my own laboratory up at the department of pharmacology. A year into my independence I received a two year grant for new investigators from the Pharmaceutical Manufacturer's Association (PMA). This was the first PMA grant in the Department of Pharmacology at USA. My title was elevated to Assistant Professor. It was a non-tenure track soft money position nevertheless a big step. In 1981 I got a grant as a co-investigator with Dr. Karen Burton from the American Heart Association. I thought my career in academia will be fine.

My colleagues in the pharmacology department Drs. Eugene Palmer and Gesina Longenecker were true mentors. They showed me the ropes and helped me with grants and thinking things through. Remembering a time when Gesina actually rewrote a grant that I had written makes me realize how fortunate I was. She was a true friend, mentor and savior all wrapped in one to me. I cannot express my emotions in words for the support she gave me. I wonder what I would have done in life without support from all these wonderful people.

Serendipity played a great role in my scientific career development. None of these happenings were planned. The one thing I wanted was to conduct research. Another essential element was my family. It was clear in my mind that I should be an equal provider along with Raj. As long as these two goals were met I would go with the flow of life. I was not particular weather the research was with microbes, plants or animals. In retrospect it seems very naïve but in the end it worked out well for me. Perhaps, if I had stuck with one system and built a reputation in a specific area it would have been better for my scientific accomplishments. However, the path that I took a multi-disciplinary one shaped my future. Each move thought me something apart from scientific skills that is about human nature, negotiation and the politics of science including gender issues. Among the many interesting people that I met, meeting Tom was a career and life changing instance which was not apparent to me then.

Dr. Thomas M. Glenn was the chairman of the Department of Pharmacology at the medical school in USA. Rumor had it that he was the youngest chair in the nation. All I saw was a charismatic and energy filled individual. His vision was to build the department to a national stature in par with other big name state schools. There was a great deal of excitement and at the same time an unsettling anxiety among the faculty that were afraid of change. Not realizing all this I was just interested in building my laboratory and career. Tom was also an impatient individual who wanted change to happen soon. When change did not occur as soon as he perceived in 1982 he moved to head the research division in Ciba-Geigy pharmaceuticals in New Jersey.

The change of leadership caused a lot of anxiety for me. I did not get any certainty from the new chairman about my position. This uncertainty caused many insecure feelings and I wanted a more secure situation. However, I decided to ride the wave and just stay put till the dust settled. My heart was in academia, I had made friends in Mobile, and I liked the small town environment and the south. I had just begun to build my own research identity and did not want to really move. I did not get my wish.

Move to New Jersey

Again Raj was recruited by Tom in February of 1983 to run and build a high through put screening facility. There was no mention of hiring me. I decided not to move unless I found a position. Although, my position was insecure on soft money I decided I had to take a chance and put my foot down. Such a decision was risky but I thought I had nothing to lose. If things did not work out in Mobile I will move, if it did work out I was in a stronger position to negotiate. Splitting up the family was a difficult situation but I had to contemplate on my move. I mentioned to Tom that I will not move unless a position was found. He kept assuring me that there are many opportunities in New Jersey. Finally, I was called for an interview in the cardiovascular research department of Ciba-Geigy by Dr. George Weise.

I came for the interview and was hired in May and I joined Ciba-Geigy in June. I had no clue about the role of women in Industry; to top it all a woman of Indian origin. When I came for the interview at the seminar that I gave I noticed there were very few women. I was hired at the Ph.D. entry level position. Although an entry level position there were no women at that level running a research laboratory in the cardiovascular research department of Ciba-Geigy in New Jersey. In spite of having the same level of experience as Raj I was hired a level lower. I just noted these issues and did not say anything as I needed the job more than anything else. I remember asking Tom what he said to Dr. Weise. His response was that he gave my resume and George liked it and asked me for an interview. Although, I want to accept this explanation at face value I am not sure if that was the case. Remember this was 1983 and Ciba-Geigy was a Swiss company and Switzerland was not known to be friendly towards working women. I noticed many subtle and not so subtle gender related issues but just ignored them since it would have been a useless fight or so I assumed.

I was setting up a laboratory to conduct work on adenosine receptor antagonists. There was an ongoing chemistry program for adenosine antagonists already in existence and the chemists ruled pharmaceutical industry and Ciba-Geigy was no exception. They were the king of the roost from establishing programs to getting a higher pay than biologists in those days. Most often chemists would synthesize analogs that were insoluble and biologists had to find ways to solubilize them before testing those analogs in the biological assays. I used to tease my chemistry colleagues that they add methyl, ethyl, butyl and then futile and send the analogs to be evaluated for biological activity. Chemists wanted to make analogs that would easily precipitate which meant that they were almost insoluble in water or buffer whereas biologists wanted them soluble. Such kind of basic disciplinary differences created challenges sometimes insurmountable.

In 1984, we had an international Ciba-Geigy cardiovascular workshop in England. George wanted me to make one of the presentations on receptor binding as a screen. Since it was a workshop I decided to talk about pros and cons of such a system. I remember how all eyes went up as there were no scientists especially from the US branch of the company who had challenged the system. It caused quite a commotion among the research hierarchy. My role was to speak the truth and say what a binding assay screen was worth, especially, if there was no correlation with some physiological or biochemical response. I had fun doing it and wanted the screen to provide positive hits that would make it all the way to the clinic and eventually to the market.

The excitement to see if the synthesized compound had any biological activity was really great and the chemist would walk into the biology laboratory demanding results. In those days membrane receptor binding assays were used as a first method of screening while I was developing a physiological response which would take greater time. It was a very difficult task to explain that such biological assays took longer time. So, I devised a system to keep some of my aggressive chemistry colleagues at bay. I realized that the more aggressive colleagues were afraid of blood and gut, so, I used to leave these biological samples at the entrance of my laboratory which stopped some of them from entering the laboratory. This gave my lab mates some breathing time.

There was some evidence that adenosine was involved in allergy and asthma. And theophylline a weak adenosine antagonist is and was used as a therapeutic agent for asthmatics. I started thinking of the strong insoluble adenosine antagonist that we had developed. If only we could test it as an aerosol spray it would be a novel medicine. It was a monumental task convincing all the higher ups to agree. Finally, it was accepted that the concept was worth testing. In 1986 we assembled a multi-disciplinary team which included individuals from research, clinical development, formulations, marketing /sales. Such a team approach was unheard of in those days. Guess what I was the team leader. Some eyebrows went up but I was oblivious to all that and was intent on the task at hand. We as a team developed an aerosol spray within 4 months. I had identified a pulmonary specialist in Hamilton Canada who was very excited to test the product. Safety and other FDA mandated criteria were all completed in a record time and the product was ready for testing. International collaboration at such a level was a rare occurrence in a multinational pharmaceutical company during the mid-80s. Excitement was very high. The product was tested but we did not continue with it as Tom decided to move to Genentech! One man's departure made it difficult to continue and changed the course of direction.

I was ambitious and dissatisfied with my progress and did not want to be a second class citizen in the research division. Hence, I decided to take advantage of an eighteen month rotation program in the company through marketing and sales. I was told I needed to carry the detail bag to doctor's offices for six months and then rotate in the marketing department for a year. I started the rotation in 1989. None of the political implications crossed my mind. I just wanted to learn the business. Thinking back on it I wonder what made me do all this - a research scientist in sales and marketing.

I was given sales training for 3 to 4 weeks, with information on the products, doctor's behavior and sales decorum were thought. During this sessions we, all the members in the training session had to role play the fast talking sales pitch. I wondered how sales representatives with half-baked knowledge on science and medical concepts of products did such calls. I was quite impressed by my colleagues fast talking and agile minds, which was very different from researchers in the laboratories. At first, I found sales rather intimidating waking into doctor's office to pitch a spiel on the cardiovascular products of the company. I believe my cultural and family background had something to do with my negative attitude towards selling as I was brought up with the belief that dignified women do not get involved in sales. I think such preconceived notions create a barrier to progress. Eventually, I was able to go beyond that belief and I did get use to giving a sales pitch without embarrassment. As time went by, I realized I had to market myself and sell my scientific ideas. These marketing and sales concepts were not openly taught when I was a graduate student and I had to learn from my own experiences. Moreover, I realized that I had to market and sell my ideas to management as in writing a research grant for a review committee. In the long run, the skill sets that I had acquired were invaluable.

It was decided by my sales manager that my business card should not have any credentials! I found this quite unsettling but just accepted it. The thought never crossed my mind that my manager may also be intimidated by my education. I had a territory from Red Bank to Toms River, New Jersey. I used to drive 250 miles each day from home to do my job. For 3 weeks my sales manager accompanied me. After that I was let loose on my own and it was quite a lot of fun. Some doctor's wanted to know my background because they sensed I was not the usual sales rep. During that time the sales force was being modernized with computers. Once a week on Friday nights I had to upload my weekly activities to the main computer in the company. The entire sales experience was a novel learning experience. During this time I also learned about gender issues in the sales force. These issues spanned across the board inside the company but also on the outside in various doctors' offices. I remember a doctor who would not see female sales reps. I never did find out what the reasons were. I also learned that establishing a relationship with office personnel allowed me to see the doctor more often. Many such nuances running across the spectrum of human psyche taught me practical and valuable lessons which came in handy in many situations later on in life.

Marketing on the other hand was very different from sales. Strategizing on how to present the product, identifying the best advertising agency to represent the product and to deliver a message of evidence based development were all part of the theme. I was assigned to work with a new palliative oncology product, Aredia™, which was considered an orphan drug, for the company with low sales projections. To me the dollar value seemed large. I was told that for a multinational large pharmaceutical company the projected dollar value for this product was considered low but the company was ready to use this product as a stepping stool to learn about the intricacies of the oncology market place. The budget to market this drug seemed very big especially to a scientist. I wondered why pharmaceutical companies did not put that kind of muscle behind research and considered money spent on research as sunk cost. I viewed it as short sighted and that was the scientist in me speaking without giving thought to the nature of our short term culture.

I was involved as part of the marketing team in selecting an advertising agency. The team selected three agencies and they were asked to produce marketing brochures. These brochures would be evaluated by oncology opinion leaders for the effectiveness of delivering the message. My task was to select and motivate opinion leaders to participate in a round table meeting designed as a workshop. These workshops/meetings would be conducted in different zones of the country to capture attitudes and differences in medical practices. There were two workshops per zone northeast, south, mid-west and west. I was able to convince opinion leaders from many reputable medical schools to attend the workshop/meeting. My persuasion skills along with the incentive package that the team had designed allowed for this success. This experience thought me a lot about human motivation even those that sit on an ivory tower do succumb to monetary incentives!

I completed the marketing and sales rotation at the end of 1990 and moved into Operations Planning and Research Strategy (OPR&S). I thought this was a great move as I could put all the skills that I had obtained to full use in helping develop strategies for research. Soon I realized obtaining a formal business degree would benefit my personal growth. I got admission into the Rutgers University executive MBA program and Ciba-Geigy was willing to support me to obtain this degree. This was a two year program designed for full time working individuals. So here I was a full time employee, a wife and mother becoming a full time student. I am not sure how I did it. The only reason I can think of is a family that supported me 100%. I cannot emphasize enough on how lucky and fortunate I was for being in a company that was education friendly towards its employees. I do not think otherwise we as a family would have been able to support both a child and my ambitions of obtaining a MBA degree at the same time.

One important motivating factor was to set an example to my children. At that time, my oldest child was in college and the younger one in high school. I wanted them to know that education can be achieved at any time in one's life especially in USA. To the best of my knowledge only in America age does not pose a limitation to attend institutions of learning. In India for example the system puts age related barriers to obtain admission to an educational institution. Europe has similar age related hurdles. I truly admire the American system of education where there is no discrimination against age while the American education system may have other issues that need fixing.

Challenging Times

In 1990s economic woes caused Ciba-Geigy to tighten its belt. There were wide rumors that the company is going to start laying off people. I do not have experience with another company but I must say this company tried to do it as humanely as possible. In the end when it came to cutting corners during difficult times, downsizing personnel decisions were based on functioning of by the company and the bottom line. The research organization had to take its share of cuts and my department PPR&S was considered a redundant staff function and not part of the line management by the "lay-off team" so the unanticipated thing happened. My department OPR&S was shut down and I was laid-off. My time in Ciba-Geigy taught me to be confident, try to bring the intellectual curiosity into applied research, be a team builder, stand back and let people take credit, and enjoy the process. There were many awards that the company gave our team collectively and me personally. Although difficult, I tried not to take it personally. This phrase "not to take it personally" has always made me wonder why when challenging unanticipated things happen to you, people say so not take it personally. However, at the time, not taking the situation personally, made the situation tolerable and less bitter!

It was a week before I graduated from my MBA degree and I was laid off and without a job. I had a month to complete my 10 years to be vested into the pension system. What will I do? As luck would have it I was vested into the pension system. I also had a ten month severance pay to look for a job. I decided to take full advantage of the outplacement benefits along with personal coaches that the company was kind enough to provide the laid-off employees.

I was overcome by challenging emotions. At times, I viewed this situation as a setback and sometimes as an opportunity. I started soul searching about my future career goals. I was toiling with philosophical and practical issues on the role of education. Did I improve my situation by diversifying my breath of vision or did I create a road block by adding a business angle to my research. Now was the time to figure out how well I could market myself and see if the training and formal business degree would help me. At that point, I was not sure if I wanted to stay on in Industry or go back to academia.

Philosophically, my upbringing was prompting me to believe that knowledge for the sake of knowledge was an important life achievement while material wealth was not and also knowledge plus wisdom gained from experience can never be removed from an individual. However, the situation that I was in contradicted that philosophy. At that moment it seemed like I was a victim of the circumstances. I wanted to put all these thoughts aside and concentrate on my next career move. Today, contemplating on those thoughts I must say as a society we have begun to view education as a tool only for economic benefit while there is no room for wisdom. Hopefully, in the future, our society will strike a balance between practical living and knowledge for the sake of knowledge and the value for aged wisdom.

My coach and outplacement agency were both extremely efficient. I tackled the situation like a marketing project. Waking up in the morning and going to the agency when it opened and staying there till it closed. It took me a few months to polish my resume and start making phone calls some were cold calls and others to friends, mentors and associates. I was mainly concentration on Industry as I thought academia would look down upon my industrial experience. By chance I happened to talk with a Rutgers University Professor George Walters who taught technology management at the executive MBA program. He suggested that I visit some of the Advances Technology Centers at Rutgers University and that my broad based skills would be useful to them. The Center for Advanced Food Technology (CAFT) was one amongst those that was established in 1989 by the New Jersey Commission on Science and Technology (NJCS&T). The goal of CAFT was to spur economic development through science. This model was based on the National Science Foundation Centers.

I sent my resume to CAFT and followed it by a phone call. Later on Tom Glenn mentioned that Dr. Rod Sharp was the Dean of Research at Cook College, Rutgers University. Based on Tom's suggestion I called Dr. Sharp and made an appointment. I remember that fall day in 1993 very clearly when I walked into Rod's office and had to wait for a long time as there was some crisis or another. I waited patiently for the meeting. When I eventually got to meet him there was an apology and I could tell he was embarrassed. After our conversation he picked up the phone and called Dr. Myran Solberg the director of CAFT and introduced me to him. I left Dr. Sharp's office and went back to my job searching mode. Two days later I got a call from Dr. Jozef Kokini, who was the Associate Director of CAFT at that time, inviting me for an interview. I am not sure how all this came about but I think George and Rod surely had an influencing role for this interview. I was excited and nervous and prepared for the interview. Coaching that I received from the placement agency was invaluable.

My interview at CAFT was very interesting. Jozef Kokini had planned the day and I was to meet Myran Solberg at the end. I have to regress on that meeting. Both Jozef and Myron had noticed that I was a research associate with Craig Venter. As luck would have it Craig Venter had visited Food Science Department and CAFT a month before. This was an added plus to my experience. It looked promising till I met Dr. Myran Solberg. I have to share this really hilarious experience. Dr. Myron Solberg was a small gruff man with a brutally blunt personality. Growing up in a rough neighborhood had made him a straight shooter blunt to the core. What you see is what you get kind of person. He walked into the room shook hands and talked for a few seconds and then asked me point blank his words exactly, "Do you have the thing that Indians have"? I felt my blood pressure rising and anger taking over. Controlling my temper I tried to think fast. My brain was telling me to calm down get the job offer and then reject it if it did not suit you. So I decided to probe further by asking questions and not walking out. I looked at him straight in the eye and said Dr. Solberg I do not understand what you mean. Did you mean the color of my skin, genetic make-up, educational training or something else? Please clarify? He said none of those and added that he needed an aggressive individual not a meek person. I presume, his experience with Indians, mostly students, had created an impression of Indians being rather quiet and meek. I was relieved and saw the humor in the situation and answered back saying that if CAFT needed an aggressive individual they have to look elsewhere but if they wanted an assertive person they had found one. I was offered the position as an Assistant Director.

The coincidence of the job offer and the last day of my severance pay was as if it was planned. I joined CAFT January 1993 and my last severance check came in December, 1992. It was a bit spooky but a welcome start. The winter of 1992-93 was really sever. My commute to New Brunswick from where I lived was only 20 miles but the icy roads made the commute treacherous and long. Normally, the commute took 45 minutes but the icy conditions made it two hours and nerve wrecking. The beginning months were hard.

I was nervous, ambiguous and excited about my academic surroundings after ten years in industry. Furthermore, I was joining an advanced technology center as a staff member not a traditional faculty position. The thought of how departments interact with this center and what the faculty thought of first level applied research before a company would show interest were big questions that crossed my mind. Even now these are highly contested views in academic research. I just plunged myself into the job that needed to be done and wanted to come up to speed leaving all the politics behind. I was the new kid in the block and had to prove myself to the CAFT staff and the faculty which was a daunting challenge.

When CAFT was established some feathers were ruffled and many departments did not want to collaborate with the center. The mission of the center was to take the innovations bundle them up and make them attractive to industry in particular food industry. CAFT was set up with 3 major arms. One, the center had a highly specialized GC-MS and MS-MS laboratory that performed fee for service for many companies not just food companies. This was a revenue generating side for the most part and it held its own. Two, Food Manufacturing Technology (FMT) facility was intended to stimulate innovation in food manufacturing. This also was self-sufficient as it had contracts from Natick for the army on meals ready to eat. Third, was the Basic Research Program (BRP) for which food companies paid a membership and basic research was conducted which the companies felt was important for them and could not be conducted within the company. BRP was also self-sufficient as long as the member companies saw a value and continued to support the research projects. I was hired to be a part of this BRP and facilitate its growth. On the outset, the structure of CAFT seemed very well thought out and sound. I still believe it was a very sound and exciting idea that somehow was miss-managed due to egos and academic politics. Looking back on it I must say that it was the most exciting, as well as, challenging time during my career. I defined my role as to maintain and grow the BRP beyond what was in existence when I joined. This was exciting and my creative juices started to flow.

When I joined there were about 22 companies that were members of the BRP. Most of the research projects were centered on material sciences, physical and chemical structure, and food safety. I used to wonder why there was not a strong emphasis on biological sciences especially nutritional aspects. It was too early for me to rock the boat so I just continued learning about BRP. Intuitively, I felt that a single lab approach was not the way to go and BRP programs would really benefit from a multi-disciplinary approach. With so many creative minds working together in a flexible academic environment, I thought it would be able to propel the member company's internal inertia and help develop future generation products. I also realized a lot of resources will be necessary to build such an environment to mobilize the academic researchers to think in this fashion i.e. multi-disciplinary team work.

I am sure when New Jersey Commission on Science and Technology (NJCST) established centers like CAFT the intension was to stimulate cross disciplinary research. However, in reality these centers were neither functioning across disciplines nor across institutions in New Jersey. So, in 1994 NJCST decided to change its tactics. Instead of providing base funding to the centers like CAFT, NJCST decided to change course and provide programmatic funding. This change also created an opportunity to build new multi-disciplinary programs for CAFT with an economic development component. It also meant CAFT would lose NJCST base funding completely in 3 years translating into job losses. At that time I was working on a proposal for NJCST on gene nutrient interaction and the development of functional foods. Rod Sharp was instrumental stimulating this idea bringing the concept of epigenetics to the forefront at Cook College and NJ Agricultural Experiment Station. This strategy required molecular biologists, geneticists, food chemists, food engineers and material scientists to work together. I was convinced that it was a great direction for the future but within CAFT there was a mixed reaction as some felt it was too biologically driven and a pie in the sky idea. Nevertheless, I was given the academic freedom and a small amount of funds to develop the idea. I built a cross-disciplinary team across 3 institutions - Rutgers University, Strang Cancer Center of New York, University of Medicine and Dentistry of New Jersey. We had some preliminary data on orange peel extract and gene expression. Although the research finding was extremely exciting we had not given enough thought to the economic development aspect. I did put a proposal together for NJCST which was not funded the first time around. I will describe the development of the entire program later on.

Now the job threat became obvious and I felt the new kid on the block, that is me, will get the axe. I was informed that I will have a part time job for three days a week. I started thinking on what to do next. I was told that the Biotechnology Center was looking for a part time business development person and I applied for that job. I was interviewed by Dr. Peter Day the Director of the center and was offered the position. Now, I had a full time position with three days at CAFT and two days at the Biotechnology Center. I remember Peter saying that I was too expensive for them. It was a difficult transition but I managed to function between the two centers. I made a conscious effort to physically be at the 2 centers so each center felt that I was doing my part and pulling my weight. After a while both centers were comfortable with my performance and dedication. I viewed this as an opportunity to bring the two centers to collaborate with one another. At the same time thoughts like -- what is happening to curiosity driven research, are there sources of funding? How do we build that? These thoughts were crossing my mind! I thought by building a program of epigenetics and nutrient gene interaction at Cook College we will fill both the needs one, curiosity driven and two, application oriented research.

The next round of NJCST funding came along. Rod Sharp wanted to pursue the idea of epigenetics and as the Dean of Research his office took my initial proposal and expanded it to include many of the researchers in the New Jersey Agricultural Experimental Station, expanded the economic development component and we submitted the grant again for funding. This time we were funded but the funding was the part that contained my initial proposal. Go figure how these decisions are made. So here we were at CAFT excited and confused. Now we had to follow through on all that we proposed-- conduct the research, develop enough intellectual property and establish a company. The excitement was euphoric.

This was 1995 and the funding was for 5 years. I was the Principal Investigator (PI) and had to bring all these high ego centric and opinionated people together to accomplish the task at hand. Although the job at hand seemed enormous the anticipation of building a new program that had the potential of putting CAFT on the cutting edge made me very excited. The goal was to build a successful program, a strong team, and a self-sufficient CAFT. Such opportunities come by very few times and I was lucky for having been at the right place at the right time. I envisioned the natural product/food chemists as the nucleus of the program, as they were in short supply in academia. Geneticists and molecular biologists will work with them. We wanted to stay with natural products from the agricultural industry. Each member of the team had enormous experience and knowledge on different aspects of agricultural products. Between Chi-Tang Ho, Robert T. Rosen (Bob) and myself we had knowledge on natural product chemistry, herbs and spices, to traditional medicine that included Chinese, Indian and native Indian. Plant extract and products to be tested were based on these experiences. Our first knowledge based selection was orange peel extract! Which we affectionately called OPE. We chose orange peel extract because it was a by-product of the beverage industry, orange peel extract was used in animal feed and marmalade so it was in the consumption chain. If OPE was turned into a consumable product we would have found use for a waste product. Human geneticists Charles Boyd and Katalin Csiszar of the team decided to evaluate the effect of the orange peel extract on gene expression. They chose a gene they were familiar with and their lab was working on, lysyl oxidase that had implications for a Cardio vascular disease. They were astonished at the gene expression results of the orange peel extract. K.Y. Chen another team member used COX 2 gene expression. Such results authenticated Chi-Tang, Bob and my credibility for selecting plant extracts among the team members, and that we as a team were on the right track. After that no one questioned our reasoning for choosing a specific herb, plant or food product for testing although we always shared that information freely. We were ahead of the times and had made a big step on trying to bring the agricultural science and medical field together.

We as a team were successful in being granted a very broad patent on the use of cellular pharmacological screens for determination of the role of plant extracts in the regulation of gene expression. Within two years, we collaborated with the private sector in the establishment of a company, Wellgen, Inc. and successfully transferred many of the patents to the company. Rod Sharp helped us identify Mr. Richard (Dick) Laster, the former executive vice president of General Foods Corporation. He was appointed to the Chairman position for the company. All in all, it was a very exciting time. Dick and I traveled to venture conferences, food and nutraceutical companies making presentations to finance the company. Our first financing was derived from Angel Investors. The two arms of the program involving scientific research and economic development appeared to be working well. Such a model was new to Rutgers. I remember spending enormous amounts of time educating the university personal and explaining that a program of this kind does not curb intellectual freedom and in fact encourages publication. The more we published research findings and filed patents, the better for the startup. The Company appointed David Evans to the CEO position, former Vice President of Business Development. Dick Laster choose me to bring him up to speed for assuming the reins of the business. After five years, we submitted a research extension in CAFT Scientist Bob Rosen served as the principal investigator which was funded. Research was going well and Wellgen had four employees and beginning to fund collaborative research programs at CAFT.

In the mean time we had an idea of developing an omega 3 product and conducting a clinical trial for inflammatory bowel disease. We did get that grant funded as well and started working with sensory testing and investigators at Robert Wood Johnson Hospital to test the product and its efficacy. The vision was to market these products, through WellGen Inc., and create enough intellectual property for the company. One year into the omega project things started changing at the state level. These changes were affecting NJCST. As we were beginning to start our omega clinical trials the grant was terminated in mid-stream. It was a big blow. We still had our other grant funded for 2 more years. I kept things moving and Rod decided to take retirement. One thing lead to another things started falling apart.

At that time there was an opportunity made available through the Gates Foundation which had announced a request for pre-proposal for a future generation agricultural product. I put a team together with faculty from many centers and department with the aim of producing vitamin B_{12} in plants. A humongous research task but futuristic and we were invited to put a full pro-posal. As a team we worked very hard but were not successful in our efforts. The excitement was created and the team was there for the future.

In 2003, Dr. Adasoji Adelaja an agricultural economist was appointed Executive Dean for Cook College and the New Jersey Agricultural Experiment Station. He asked me to move over to build programs and I decided to do so. I kept my three days a week at CAFT and moved on to NJAES for 2 days. I was not very happy with that decision and I moved back to CAFT. During this time I secured a USDA higher education grant to teach students about food, nutrition and health and economic development. In 2006, I was told there was no funding for me and I decided to take early retirement. I am not sure if that was the right decision or not but that is what I decided to do. Once I made my mind up there was no turning back. Now, I needed to figure out my life as to what I wanted to do next. This was a premature unplanned retirement and I was not ready for it. Here I was trying to reinvent my career in my sixties. I tried to find a position with not much success. I thought perhaps being a consultant would be the way to go. All this was not exciting me and while soul searching I realized that one of the activities at Rutgers that gave me immense satisfaction was when I shared my experience and knowledge with others.

During my time at Rutgers I participated in many activities including many committee involvements. I got the perception that some of the committee involvement was to fill a quota for minority and women. I satisfied both the needs, however, most of the activities truly required my expertise and experiences. Those included conference planning, key note speaker for Johns Hopkins Talented Youth Program, scientific speaker at the Institute of Food Technology, motivating students at the New Jersey Public Schools, presentations to the dietetic associations, one on one career guidance to many students and others. Not only did I get immense satisfaction each time I participated in these activities I thoroughly enjoyed them. Basically I liked being a giving person rather than taking although I must confess, at times, taking was fun too. So my strategy was to find ways to get involved with voluntary work. I wanted to build my experience and get to know the nonprofit community. So, I decided to use the next few months to figure out my strategy. The goal was to use the nonprofit experience as a stepping stool to get on the boards of not-for-profit foundations which will help me give back to the community.

Although giving back to the society appealed to me I did not know how to get involved in those activities especially suited to my experience. I was struggling to figure out how to do that. At that time I was made aware of two organization Community Access Unlimited (CAU) and drug abuse prevention organization

Prevention Links. I participated in marketing and fund raising activities for both these organizations. These were valuable experiences however these experiences did not satisfy my emotional needs. I must confess that I am not able to put my finger on why these activities are not fulfilling. Perhaps, the dissatisfaction was due to the fact that my experience of food nutrition and health was not being utilized. I was groping in the dark on how to bring that satisfaction. Serendipitously, around the time I happened to meet New Jersey Assemblyman Upendra J. Chivukula at a meeting and he casually mentioned that I could stop by at his office. This opened the door for a volunteering possibility in his office and to learn firsthand on the ins and outs of an elected official's day particularly an engineer, as Assemblyman Chivukula is an engineer by training and had worked in AT&T prior to seeking elected office.

It was an eye opener for me on the business of the office and how people from all walks of life demand attention from their elected official. By interacting with him I understood an engineer's perspective to politics. We talked about the new immigrants and educating the new wave South Asian population in New Jersey about preventive health. Many of the new arrivals were not familiar with the health care system and did not take care of themselves with regular check-ups, proper diet and exercise and how my experience with food health and nutrition would really be beneficial to such a population. I met a few like-minded medical professional and we formed a program entitled South Asian Total Health Initiative (SATHI) which was based out of RWJ medical school under the leadership of Drs. Sunanda Gaur and Naveen Mehrotra. I was one of the founders and on the advisory board of SATHI. Incidentally, SATHI in Hindi means companion and we feel the name is symbolic and appropriate for the cause. In 2009, I was able to obtain funding for SATHI launch. In 2010, I was involved in a team that conceptualized and organized the first conference on health disparity and health literacy in South Asians.

Around this time I heard about Natural Gourmet Institute in Manhattan, a culinary school that supports healthy eating. I found the concept very interesting and contacted their founder Annmarie Colbin. She suggested I talk with their public class director Judith Friedman. This was an unexpected direction but I wanted to be involved in some structure. Judith wanted me to demonstrate my cooking skills to her and set up a date for me. She liked what she saw and asked me to put a proposal for a class. I did and was asked to teach a class. I was excited and intimidated about the possibility of teaching healthy cooking to people from many walks of life. This was a completely new venture. The public classes ran from 6:00 pm to 10:00 pm. The school gave help with preparation prior to the class and I had to be there at 4:00 to get ready for the class. The students all of whom were working people arrived after work at 6:00 most of them were not scientifically trained. This meant I leave New Jersey at 1:00 and talk about science to a lay public audience. I did two cooking classes which was physically grueling. Along with how to cook healthy I thought scientific concepts on heat labile substances, good fats and bad, and on the use of herbs and spices. There were about 25 to 30 students and I had to cook for 40 as the staff would also eat what I cooked. It was like the cooking show in front of a monitor, a unique experience to say the least. At this time Judith wanted me to expand and include an Indian market tour class. In this course I would take the students on a trip to the market and teach them about Indian spices and foods; again a very novel and intriguing idea. I put a syllabus together on Indian kitchen utensils, spices and foods while weaving in the health aspects. I prepared the handouts. The handouts contained the names of herbs, legumes and spices in English plus two different Indian languages Hindi and

Tamil, as well as their botanical names. I took the students to Kalustyans in Manhattan that carried many Indian spices and foods and then to an Indian grocery store across the street. Thought them about health, agricultural produce, differences between preventive and curative foods, cooking utensils and how to weed out not so authentic hyped up information that appeared in the press. Mostly, the class contained common sense with a scientific twist. After a 2 hour and 30 minute class I would take the students for a sumptuous Indian meal at a local restaurant and share information on the dishes. These classes were a lot of fun with 20 to 25 students. It was a challenge to teach students from 18 years of age to 70 years of age in the same class. There were MDs, lawyers, teachers, high school seniors and others who signed up for the class. I had a lot of fun but the fun had to stop as my arthritis started acting up.

During these classes I continued my volunteer time two days a week in Assemblyman Chivukula's office. One thing led to another and I was nominated for a position as an advisory board member to the New Jersey Office of Minority and Multi-cultural health. It was an honor to serve on the board for 3 years as a representative of the South Asian community. In the meantime Naveen Mehrotra started Sri Krishna Nidhi (SKN) foundation that would address and deal with many health issues in the Indian community. I was asked to be on the advisory board of SKN Foundation. An important project at the SKN Foundation is called "Move It to Lose It". Through this program we teach children about regular exercise by dancing and provide instructions on nutrition. At the same time Center for Management Development (CMD) at Rutgers University offered me a part time position to increase student enrollment as well as awareness for the center. I decided to tap in on the healthcare industry and create a clientele through that source. I managed to create a joint program with the Hospital Association of New

Jersey. Rational being that hospital administration needed management skills. This position lasted for 18 month. I enjoyed the mix of activities some paying and others voluntary. On the whole I realized my heart was clearly in building the voluntary side like SATHI and SKN Foundation.

During my aging process, I did not want to neglect my health and wanted to practice what I was preaching that is to exercise regularly. I had been practicing yoga by myself at home for 10 years but wanted more human interaction. With this goal in mind I started going to an exercise class for 2 days a week for seniors at a senior center called SAGE in Summit, NJ. At this exercise class there are participants aging from their mid-nineties to late sixties. These classes are an inspiration and our coach Claire Marsh Bay is a real motivator. Whenever the opportunity permits I teach aspects of food, nutrition and health to my fellow exercise mates. This unassuming exercise class has provided me with an opportunity to be on SAGE's Board.

Looking back at my life I must confess science training made me strong, analytical, fact finding, tenacious, and the attributes that have pushed me ahead. For those who read this I want to say that the road was not straight not just life but my scientific career as well. When I started my education and training in science all I wanted to do was research and be in the laboratory tinkering with life science tools. Every little success was exciting like a protein band on the electrophoresis, new protein, a positive biochemical reaction, learning about a new plant, herb or food ingredient. But the twists in life were not factored into this equation called living. With these twists I became more flexible, nimble and adaptable. I tried to take every punch that was dealt and twisted it to suit me and my science. I tried not to dwell on any setbacks along the way. I did not think about either being a minority or a woman that does not mean I was not aware of those facts. Most of the times when I fell I got up dusted myself, cleaned my knees with peroxide and started marching again. I realize now that my happiest moments were when I was building and sharing and giving back to society.

I think I have learned the meaning of not to take it personally. This is an American cliché and I never understood what it meant for quite a while. Now, I say it all the time and think I do understand the implications of this. If I took things personally I would be concentrating on myself and not on the task at hand. However, if I say to myself it was the situation not me the blow does not sting as hard and in life there are special people that help you fall back on your feet. Sometimes such help comes from unexpected quarters.

Bringing children up in a culture that is so different than what I was used to was a challenge. I had to put my thinking cap on be on the guard all the time. The most important thing was not to give too many liberties and have a spoilt brat. I always was in a dilemma on how to draw the line between being a liberal parent with enough discipline so that they can develop their passion and yet be practical and self-sufficient. We wanted to instill some of our Indian culture but not make it like a punishment. This was a difficult task as the environment in the 70s was not like in the 90s and there was not much exposure to India or Indian culture like it is now. It was always a dilemma for me. No matter what, if for any reason my children had gone astray society and family would have blamed me the mother and pointed a finger at me and would have said it was because I was so career minded. Somehow I had to find a way to balance the act of being a mother, wife, homemaker and a career woman. I was always worried about that finger pointing and wanted to be a good mother a self-imposed pressure. In the end things have turned out alright. Neither of my children chose biological sciences one is a mathematician and the other an epidemiologist. We as parents never forced them to go into a specific field but we did say they had to get a college degree. Something we did right in spite of being in a foreign land. Sometimes I do have doubts if we, my husband and I, could have done things differently. All in all as a family we are alright and we did try to blend the lifestyles of east and west. I would like to believe we succeeded.

This has been an exciting journey with many ups and downs. Along the way I have learned that the only constant in life is change so I welcome change. Throughout my life I did not dwell upon my ethnicity and/or being a woman and did not allow these facts to get in the way of my progress. My motto has been not to look at myself as something different while - making (formal or informal) presentations, talking to someone or in any given professional situation. I tried my best to be open to ideas while believing that others may have better ideas than mine and that there are many solutions to an issue. Last but not least this fun journey would not have been possible without the help from many people including my loving family, friends, supporters and mentors. I want to conclude by saying I love science, evidence and facts, while intuition is great I prefer to look for facts that support those intuitions.

I would like to dedicate this to my children, husband, grandchildren, my friends and mentors. Who knows what else is ahead but I am ready to face it. My next phase in life is dedicated to seniors as I am one of them today. My dream is to age gracefully and make my final exit without any trouble to society or loved ones.

GG1- Department of Operation Planning and Research Strategy (OPR&S) Ciba-Geigy Pharmaceutical 1991 –Standing 2nd from right is me.

GG2 - My son Pranav and I 1973 in my parent's summer house Kodaikanal, Tamil Nadu, India.

GG3 - Ghai Family in Portland, Oregon 1987
from left Rajendra, Nirupa, me and Pranav

GG4 - With colleagues in OPR&S 1991. I am 3rd from
right along with Lois Won and Barbara Graziano

GG5 - My parents Dr. T.S. Subramanian and Mrs. Meenakshi Subramanian 1945 in front of their first house in Kanpur, India

GG6 - Microbiology Department during my Ph.D.
I am standing on the 2nd row first from left

GG7 - With my executive MBA classmates and
faculty in Cambridge England summer of 1992
sitting in the front row 3rd from the right.

CHAPTER 7.

Excursions in the Science Arena

Joseph Hamburger

My relentless mentor and dear friend Julius Kreier kept coaxing me to write a chapter in this book. I tried to resist because digging in piles of history is not a desirable activity for an old guy like me, who feels evolution of finality is the real name of the game, rather than the progressive decline of deceptive eternity. Eventually we agreed on a free-style chapter rather than a rule-determined style which may look like a CV or a scientific paper extended to the private domain. Over the years I have been trained to dance in the circus of political correctness adorned with well styled, rule-submissive, scientific papers expressing mostly self-exciting scientific-truths of rather limited general interest and shelf life. One of the pleasures of old age is the relief from imposed political correctness. Not that the resulting product is a more relevant one, or a more justified one, but at least there is a degree of associated relaxation.

I have seen but rarely read chapters or books written by scientists or by their ghost writers seemingly aiming at adding life stories to their already published scientific work after it was reviewed and accepted by peers. Admittedly, I have not completely understood the drive to produce these stories, nor their possible interest to others. Those who produced scientific work which clearly provided significant steps forward in human understanding are usually satisfied, I believe, by the mere activity which brought fulfillment to their curiosity about nature and its puzzlements. Usually other writers took upon themselves to write about major developments in science and added some stories about the scientist whose creativity and findings caught massive attention by the public. So, whether writing briefly about my own experiences leading to scientific work will raise some interest remains an unanswered question.

I remember that tiny flowers on the slopes of Mount Carmel looked quite big and impressive when I was kneeling and watching them closely as a little boy in Haifa. These were true wonders, especially when accompanied by a procession of busy ants. I could watch them for a very long time. Was this the time when I was first imprinted by curiosity about nature? Probably yes. But it was also a refuge from troubles and harsh maternal treatment at home. My parents left the convenient European life-style of the middle class and upper class, and landed in Palestine just before WWII, robbed of their cultural heritage. They barely escaped the shock of this physical and cultural displacement into each other's arms and hardly endured the life-style of the working class and the never-ending struggle for acceptance. The education they enabled me to receive was costly and for them a burden, but a true contribution which I should never forget. For me this was a gift of life long value.

As I see it, curiosity is indeed the most powerful drive for scientific research. This is probably an inborn trait which can be improved upon, by the right education, if the basic tendency exists. I try to think back on its manifestations and remember my disappointed mother who persistently complained about my rather low average grades at school. "How come you do not strive for better grades?" she was demanding. I made the mistake of telling her that my daydreaming was far more interesting and creative. Curiosity and its satisfaction sometime did not go well together with indoctrination called "education". Fortunately, some of my teachers recognized this dichotomy.

During my studies at both primary and high schools, I was captivated by biology as classes on this subject included outings to nature and collection of plants and creatures, as well as drawing (a favorite activity) of plants and creatures. Fortunately, the subjects of biology and natural history were given much attention in both elementary and high schools. In sprouting Israel (formally established when I was 8 years old) Excursions to nature were an important part of our education to foster "love of country", its nature and ancient history. Still is. These outings likely contributed to my future desire to work in the field in Africa later on.

A group of highly capable teachers operated in our high schools at that time, many were Ph.D. s and Professors, refugees/ immigrants from Europe before and during the Second World War. They were far more capable than the average high-school teacher, both academically and educationally. Rarely could any of them find a Job in Israel's single university, The Hebrew University of Jerusalem, already gorged with famous European Academic figures attracted by Zionism or driven out by Anti-Semitism. Many high-school students at that time profited profoundly from their teaching and educational capabilities.

The first university years were also reinforced by European-educated Professors whose scientific research previously contributed a great deal to science in the US, Germany and Europe in general. The academic system was European-based with German academic tradition dominance. Microbiology and parasitology were the first departments operating in the newly established Hebrew U Medical School. They played a principal role in eradication of malaria, schistosomiasis, and other infectious diseases of humans and livestock in Palestine/ Eretz-Israel -The Land of

Israel. They also played an important role in assisting the Military Medicine Forces of the British Army in the Middle East during WW II.

Prof. Saul Adler FRS, who discovered the life cycle of *Leishmania tropica*, was the head of Parasitology Department when I joined the laboratory of Avivah Zuckermen as a student, and there I completed my Masters and Ph.D. degrees on vaccination against rodent models of malaria. Fortunately, and a demonstration of the role of chance in scientific development, a lady technician working with me in Aviva's laboratory was a personal friend of L Orenstein, who together with JB Davis invented, in the late 1950's, polyacrylamide gel electrophoresis at Eastman Kodak Company. The Tuperware-made prototype of "disc electrophoresis ", the first version of acrylamide gel-electrophoresis, was donated as a gift to this technician and awaited utilization. Disc-electrophoresis was performed in glass tubes into which the gel was cast, and in later versions, developed by others; the electrophoresis was performed using slabs of gel, for running proteins and DNA in an electric field for separating components by size and charge. This was the technological approach which started molecular identification and later used for DNA sequencing.

As part of our studies on identification of plasmodial components by disc electrophoresis we embarked on identification of components of *Plasmodium falciparum*. For this purpose I joined Aviva on a trip to Gambia, West Africa for three months of work at the British Medical Research Center (MRC) at Fajara, near the Capital Buthurst (now called Banjul), where pioneer studies on immunity to plasmodia took place under the directorship of(later Sir) Ian McGregor. We collected infected erythrocytes from blood of patients and from placentae collected from infected women, where infected erythrocytes are highly concentrated, and took the harvested plasmodial extracts back to Jerusalem for disc electrophoresis analysis.

During my stay at the MRC compound in Fajera, Gambia, I met with a prominent schistosomiasis expert Dr. SG Wright from the London School of Tropical Medicine and Hygiene. He kindly took us on an expedition for collecting snail hosts of *Schistosoma haematobium*. If I have to trace my earliest curiosity about schistosomes I have to point out the interesting time I spent with Dr. Wright.

The studies on disc electrophoresis of plasmodial extract were a good preparation for my Ph.D. where fractionation of plasmodial extract by their component was done by disc electrophoresis in an enlarged version of the instrument. Other than that my MSc thesis concentrated on vaccination of rats with a crude extract of *P. berghei*, a malaria parasite of rodents. The positive results obtained led to my Ph.D. thesis on fractionations of plasmodial extracts in search of fractions active in inducing immune responses.

For my Ph.D. work a technological development was required for fractionation of plasmodial extracts by acrylamide gel electrophoresis. The university had a technology-mechanics workshop, an important unit at our, far from affluent, Medical School. This unit constructed instruments too expensive for purchase or not available on the market. Workers in this unit helped me build an enlarged Disc Electrophoresis unit, employing large buffer tanks and Plexiglas tubes. In each tube a large gel cylinder could be cast. This enabled running large amounts of plasmodial extracts in the electrical field for a predetermined distance, and subsequently cutting equal slices of the gel cylinder for further analyses. Buttons bored-out from each of the gel slices were subjected to cytochemical staining and to serological examinations by gel diffusion (positioning antigens in the button vs. immune sera and viewing the precipitation bands formed). Moreover, since acrylamide gel is a potent promoter of the immune response (adjuvant) individual slices could be macerated and injected to rats as a vaccine. Fractions inducing protective or enhancing immune responses were thus identified and found to correspond to fractions containing species-specific antigens. This was before culturing of plasmodia was possible and thus before separation of the immunizing blood stages (Merozoites) was possible. Nevertheless, these results from the early 1970s still stand relevant for comparison with recent results regarding the degree of immunity (or enhancement) they have exhibited.

During my studies I found that it is usually a challenge to uncover new data and then convince teachers, colleagues and peer-reviewers of the value of a new truth in order to facilitate its incorporation into the general pool of scientific knowledge. The dominance of dogma is powerful because new truths are always a challenge to the well- being of thought habits and of established authority structures. A most recent example to this notion is the new truth in crystallography discovered by the Israeli Noble Prize winner Daniel Schectman who discovered a pentagon-symmetry in certain metal crystals referred to as quasi-periodic crystals. This revolutionary finding was strongly opposed by many. He was even shamefully dismissed by his Post-Doc mentor, and later ridiculed by Linus Pauling, who remarked at a scientific conference in front of a multitude of top crystallographers that "there are no quasi-crystals as much as quasi scientists". But there are much earlier examples that are well known… and others less known or altogether unknown due to their relatively limited impact on science. I have to mention in this regard my Ph.D. mentor Avivah Zuckeman, a Ph.D. from Chicago University and a close friend of Golda Meir. Avivah was critically ill and I came from the US (where I spent my first Post-Doc training period with Julius Kreier) to visit her. She said quite forcefully "Yossi, you have to keep your biting teeth hold like a bulldog to the truths you believe in". This notion was not a new revelation but a reinforcement of what was becoming clearer and clearer over years of activities in the academic arena, and of interaction with teachers, colleagues and fellow students.

Yet, bulldog teeth are not an incentive for scientific work, as is curiosity, but may overtake the agenda due to the potential dichotomy between scientific work and academic career. An example for a heated debate in my early scientific development came up after I presented in a scientific meeting in Israel my first results on molecular (DNA-based) identification of schistosomes within snails, which evolved after my return from Post-Doc in the US. An established and authoritative senior scientist ridiculed this approach claiming that there are sufficient morphological and immunological differences to stick to for identification. Another senior colleague came to my rescue by saying that "we can thus continue to use pencils although computers are coming in". I could not have expressed it myself in such a politically incorrect, yet eloquent, manner. I felt that criticism at home, sometimes a result of demonstration of authority or, worse, an attempt to block the development of a young scientist who is preparing for the struggle for tenure, can potentially seriously harm one's career.

Regarding acceptance of new findings, as one who was aware of his own tendency to take dogma to unknown fields, I had to be prepared much later in my career to experience colleagues' opposition to publications which presented findings which took the subject matters beyond the prevailing dogma and the interest to maintain it. In one of these cases it was found that the 200KD circulating antigen targeted for diagnosis of infection with *Wuchereria bancrofti*, a lymphatic filaria, exhibits features of human connective tissue proteoglycans, results which raise questions about the source of this molecule. In another case, a repeated sequence of *W. bancrofti* targeted for diagnosis by DNA amplification was found to show features of repeated sequences in SAR/MAR, a DNA region binding to scaffold/matrix nuclear proteins, thus, raising the possibility of identifying SAR/MAR sequences of a wide variety of various eukaryotic pathogens as a source of species specific DNA sequences for diagnostic purposes.

Much of the time and struggle involved in years of studies and later-on in years of scientific work, was spent rather alone, despite the enormous and justified efforts of scientific authorities and establishments to foster interaction between scientists as in seminars conferences and the like. To me as for others, the most helpful factor in connecting between scientists, and between them and the general knowledge were scientific papers. Thus, comprehensive but critical reading emerged as a must but not enough for keeping ahead. The rest may perhaps be described as originality vs. hunting-gathering of ideas, insights and technological hints. Some of it resembles "intelligence work" at scientific meetings, whether in corridors, coffee tables, bars, clubs etc. I found out that the right proportion between external input and creativity/originality should always be on the line of examination. The differentiation between super-quick in-takers who often become quick possessors of ideas created by others is a constant struggle and confusion in the world of priority determination (regarding patents, significant scientific papers and the like). Over the years I had to learn that when coming up with significant findings which may impact large-scale operational decisions in public health, as well as economical outcomes, one should better have a strong backing by the right colleagues and institutions. The clash between creativity and entrepreneurship turned out to come with the territory, as was the realization that it is better to have them both when large-scale recognition and/or economic advantages are on the line.

Critical assessment, sometimes open debates in various forums is usually an essential component in scientific interactions as it promotes multifaceted scientific evaluation. Students who attend such debates could gain a broad outlook on the subject presented. As a teacher, a broad mind gained from debates can help identify and apply somewhat different approaches for following a student's thesis when required. Students' work, after all, is the main component in scientific research. Not only are they the main working force, but should hopefully also present a challenge to dogma. In any event, it helped a lot to keep an open ear and mind for occasional and sometimes unexpected and co-incidental inputs that turn out, in the long run, to be very important for one's scientific work.

During my Ph.D. work, my mentor Aviva Zuckerman, contributed a great deal to my discipline, an important factor in scientific work that requires a due balance with unorthodox approaches leading to innovation. This is a thin line which carries the danger of leading to confusion or to a downfall (and I am talking of my own experience as I was never disciplined enough). Aviva also encouraged the use of the new technical approach of fractionation by disc electrophoresis. When completed I took my Ph.D. thesis to London, where Aviva spent a year of a sabbatical leave, to have her review it. She admitted she was not completely aware of the innovative results included in this Ph.D. Her frankness enabled taking this notion for my prospective relations with my own students later on. The interaction with students and colleagues had an important role in my excursion to science land. The association between verbalization of thoughts, mostly casually, and sprouting of new scientific ideas becomes apparent as retrospection creeps-in. The interaction with students, whether in form of questions, seminars, dissertations and so forth, is particularly interesting, if one pays sufficient attention and avoids dismissing unripe and less disciplined or less eloquently expressed thoughts.

Here I note something about the importance of a close interaction between a scientist and his colleagues and students: A short time ago a revolution was imposed on departments belonging to what used to be our Institute of Microbiology. The younger staff members moved to a new building and the retirees remained in the old building separated from younger colleagues and their students. This was a blow to a continued interaction. A lot of trash was thrown away during rearrangements. A colleague scanning the piles of trash in search for "reusable" items found a little clay statue of an owl with the word "Professor" inscribed on its base. I had received this statue about 11 years ago from a lady student who completed her M.Sc. with me on molecular copro-diagnosis of *Echinococcus granulosus* in surveys on transmission by dogs. Her work involved identification of species-specific repeated DNA sequences of *E. granulosus*, a small tape-worm living in dogs intestines, with eggs transmitted by fecal contamination to herbivores and humans, and causing the development of large cysts in liver, lungs and other organs including brain. Based on these sequences she developed a PCR assay, as we have previously done for Schistosomes (bilharzia parasites-see below). This student first graduated in mathematics, and decided to slow down the excessively intensive turning of the wheels of her mind by moving to biomedical research. She started her Master's project a couple of times but it did not work out. I did not hesitate to accept her as a student in my laboratory. She completed her Master Degree with distinction and a big ceremony was held at the Medical School for her and for several other excelling students who received prizes from the Dean. A lot of attention was required for instructing her, for which a close interaction was required. But I have not paid sufficient attention to the nice owl she gave me when she left, and I put it aside for mere decoration. But this time my colleague who found it turned the statue upside down which I never did myself. This enabled me for the first time to read her message to me, the heart-warming personal note of appreciation she has written on its base 11 years ago. I missed a heart-beat and then I thought that this episode reinforces the rule of paying close attention to students in all aspects of their development, as scientific development is closely linked with other aspects of personal development.

Now back to the time line. I started my Post Doc training in August 1973, shortly before the Yom Kippur war erupted, with associated worries and grief for fallen friends and acquaintances. The broad mindedness of Julius Kreier was both strengthening and educational. Julius, to my impression, was fore and foremost a philosopher. Our scientific discussions did not concentrate only on microliters of a component in a reaction, nor on numbers and function of cells of this or that sort, or on tables, diagrams and graphs of results etc. The question of the broader meaning of results was paramount for planning further steps ahead, for assessing relevance, and for mining new insights. Although "god is in the small details", as the famous cliché goes, the wider overlook is a must, for recognizing the full potential of a new knowledge. Once a new knowledge is attained the matter of its dissemination requires due attention. Julius taught me that decisions for delaying a publication should be examined carefully with attention on whether this delay might be a demonstration of insecurity or unrealistic perfectionism. After all, much of what we do is a "here and now version" of a new knowledge, with us or others to contribute further to its development in due time later on. Also, competition is always a thought to be taken into account when timing of a publication is concerned.

I spent a couple of years at Ohio State University, working on Malaria which resulted in several papers on penetration of blood cells by parasites, and on the immune response to plasmodia. Instructing students was part of my job at Julius Kreier's laboratory and this prepared me for my later involvement in instruction at the bench when I returned to Jerusalem. It was then made clear to me that I will not be able to return to malaria research at the protozoology unit in Jerusalem, but an opening at the Helminthology Unit was proposed as an alternative. Considering that peace talks started with Egypt I decided to accept the challenge (frankly, no other option was available) and sought a place to study schistosomiasis (bilharzia), a disease very widespread in Egypt. Julius Kreier played a pivotal role in facilitating my movement to Case Western Reserve Medical School at the Department of Geographical Medicine (DGM), and thus was a key factor in the development of my scientific work. Kenneth Warren headed this department. He was a cornerstone in schistosomiasis research, the one who coined the term Neglected Tropical Diseases (NTDs) for the diseases of the tropics not given sufficient attention due to the limited resources of the masses afflicted, and thus raising less interest by drug and biotechnology industries. The term NTDs is still part and parcel of the definition of these neglected tropical diseases (excluding the major tropical diseases: malaria, tuberculosis, HIV). My research concentrated on schistosome egg antigens and resulted in the identification and characterization of an egg antigen which induces granuloma formation, the main immune-pathological feature in schistosomiasis.

Interaction with colleagues I have met at Case, like Ron Pelly, David Grove and Jim Kazura were key factors in my scientific development as we shared lines of thought and I have acquired the intensive work habits at Case. A profound impact on my future work was the result of my interaction with Adel Mahmoud who later, after Kenneth Warren became the head of the Science Section in the Rockefeller Foundation, became the Head of the Division of Geographical Medicine at Case, and subsequently the Head of the Vaccination Unit at Merck Co. The depth of thought and high capabilities of this excellent Egyptian Scientist, went together with a refreshing sense of humor. The feelings of potential collaboration lost by wars in the Middle East became dominant in my thoughts sprouting from this interaction and led me later on to collaboration with regional scientists. I still entertain the hope that a wide-range regional collaboration will be realized in my lifetime. Adel and I continued to work and publish together on granuloma-inducing antigens of schistosomes after I have returned to Israel, and his advice continued to accompany my long time work with Charles H King a member of the Department of geographical Medicine at Case, later named The Center for Global health and Disease.

Although I was offered the opportunity to remain at CWRU, my wife and I have decided to return to Israel. The division of Helminthology was headed by Prof. Guta Wertheim, a morphologist-systematics expert. We did not see eye to eye on the direction of research as my decision was to concentrate on immunology-immunopathology and later on molecular (DNA-based) diagnosis and monitoring of schistosomiasis and other parasitic diseases. Fortunately, I could follow this line of investigation when the three Units of the Department of Parasitology (Protozoology, Helminthology and Entomology) were united, and the inclusion of the united department in the newly established Kuvin Center for the Study of Infectious and Tropical Diseases.

From my experience at the helminthology unit I remember leafing through a memoir-book of a Helminthologist, whose name I forget. Well, in this case a memoir –book did help the development of "out of Dogma" scientific thoughts. My attention was caught by a "funny" sentence in that book saying that the British Monarch can talk in the "royal plural" (e.g. "we are not amused" as for Queen Victoria) only when referred to together with the population of worms living within him/her. This led, years later, to the realization of how close can humans and their worms be considering, and that certain worms (like tapeworms and schistosomes) can live in the human body for many years, indicating the existence of a wide antigenic compatibility between host and parasite. Later on, further thoughts based on information in the literature led to the understanding that unisexual infection by schistosomes can go with minimal pathology, as eggs, the main cause of pathology, are not laid. Thus, schistosomes can be related to as universal implants in the infected vertebrate host, and if so perhaps genes expressing beneficial products (hormones, antigens etc.) can be implanted transgenicaly into them so that they can function as vectors producing important therapeutic molecules as an in-vivo factory, without significant pathology. The picture of transgenic schistosome expressing green-fluorescent protein has hung for years now in my home (as well as in the Dean's wing). Schistosomes producing human acetylcholine-esterase or growth-hormone were also produced but soon the story of the use of stem cells became more acceptable and considering the Yick-factor expected to be involved in implantation of worms for therapeutic use the idea was dropped, although its potential for treating some orphan diseases, and for immunizing against malaria was considered in depth and in consultation with Key scientists working on anti-malaria vaccines. The potential of schistosomes as vectors of gene-product factories in farm animals is still being examined. This is to demonstrate the danger of potential failure when taking creativity too far. But taking this risk is unstoppable when curiosity prevails.

Plunging into research on a subject somewhat removed from my main line of research, was also encouraged by a conversation I had, years after joining the Kuvin Center, with Graham Mitchell from Walter and Eliza Hall Research Institute, who also served for several years as the director of the Sidney Zoo, indicating a somewhat unorthodox line of thinking. Graham, a prominent immune-parasitologist, visited the Kuvin Center and I asked him directly how he divides the time of his scientific work. His answer was 40:40:20. "Forty percent for bread and butter", namely work on the lines drawn when writing the relevant grant proposal, another "40% for next bread and butter", preparing the next generation of grant proposals on the subject, and finally "20% for flying high", the non-conforming track. Remember the transgenic schistosomes mentioned above? they sprouted from this notion too.

Adel Mahmoud was a key element in my connection with Africa, starting from the late 1970's, after my return to Israel, and continuing until after my retirement. My African experience was a key factor in my scientific development and an eye opener to human development, cultural incomprehension by Western standards of thoughts, strengths and needs of African people. An eye opening story I read in The Nation, the leading newspaper in Nairobi, described a Jeep tour taken by offspring of Stanley who travelled to Africa in search of Dr. Livingston ("I presume"...). They took the rout he took and visited the villages he visited on his search for Livingston. The "elders" in each of these villages told stories about Stanley. Stories passed orally but which maintained their accuracy as made evident when compared to Stanley's letters and diary. The significance of "story telling" with the accuracy involved for maintaining knowledge, before writing was invented, was unveiled to me by this story, and made me think that we have lost quite a bit of primordial abilities when moving ahead to writing, monotheism and other habits encouraging dogma and central control.

Over the years I also became aware of attempts to eradicate schistosomiasis by making technological changes which disrupted local cultural habits. Regardless of these imposed developments, kids did not go to snail-clean swimming pools built for them because this involved losing freedom from adult inspection, and women refused to use water of boreholes dug in the center of the village for bathing and laundry because doing it at the riverside enable interaction with other women an exchange of information on food, health, babies, marital life and who knows what else. This notion was recently reinforced in me when I watched a program about Africa where the Safari instructor and driver told about the difficulties of malaria control in remote villages as the knowledge passed on by the elders is disrupted by tools and habits introduced by Westerners. For example, the boiling of bones and offal and using this liquid to extract the bark of the Cinchona tree (containing quinine) for preparing an antimalarial remedy was almost abandoned. By contrast the adoption of the old Chinese antimalarial remedy based on extracts of *Artemisia anua*, now serves for producing Artimesinines as modern antimalarial drugs. Anyway, we probably have to accept that holding on to old traditions will rarely compete with modernism and technology, yet when modernism fails, returning to old knowledge might help. For me the realization of the contribution of old/traditional knowledge/experience and the evolution of human thinking is very helpful in critical evaluation of thought processes.

My connection with Kenya, with the help of Adel, was actually a turning point in my scientific activity when I conceptually and practically moved from being a parasitologist to become mainly concerned in tools for promoting public health related to NTDs. Adel introduced me to the medical system involved in public Health and in related field studies in Kenya. The Division of Vector-Borne Diseases (DVBD), now called the Division of Neglected Tropical Diseases and Vectors, played a central role in this activity. It included offices and a central laboratory in Nairobi and field laboratories distributed in rural hospitals in Kenya. This arrangement enabled taking laboratory work and sample collection to the field and close to patients and transmitting hosts (for example snails and mosquitoes). Dr. Arap Siongok who headed DVBD, and later Dr. John Ouma were my Kenyan partners to whom I have a long standing debt for enabling my work in Kenya and for their essential input. Later on I made contact with Davy Koech who later headed the Kenya Medical Research Institute. I first met Dr. Koech at Case Western Reserve University where he took some training and later became the Director of KEMRI. This enabled expanding collaboration to enable on the one hand basic research on diagnostic tools more akin KEMRI's lines of work, and, on the other hand, applying new tools to field-collected biological materials from humans and intermediate hosts under the hospices of DVBD.

During my first trip to Kenya (repeated by about 25 others), Adel was running a large-scale research in the Coast Province to explore the relation between the number of eggs found in the urine of patients with urinary Bilharzia (caused by *Schistosoma haematobium*) and the degree of pathology as revealed by injecting i.v. contrast material and then taking X-ray examinations (IVP) to reveal obstructive damages. Eggs in the patients' urine were counted for his study then collected by us and taken to Jerusalem for extraction, fractionation of the extract, and identification of the egg antigen responsible for inducing granuloma *by S. haematobium*. My first Ph.D. student, Sara Lustigman, was a key assistant in my studies on schistosome egg antigens. Sara and other students were of course also exposed to these activities by trips they took with me to Kenya. Later on Sara herself was infected by the "African experience virus" and incorporated this experience into her studies in the NY Blood Center.

My Kenyan experience prompted me also to share it with my wife and 3 children and show them some realities of life when water comes from open water bodies and where connection with the environment and the diseases it carries affects the history of human development. A memorable visit with my wife in 1982 was interrupted by a local *coup d'etat*. Our hotel was located just 500 meters from the President's (Daniel Arap Moi) palace, thus close to the target of automatic and mortar fire. Several days of curfew imposed on the night crew at the hotel to take over and run it to their best ability. Fortunately, Igor Mann (previously head of veterinary services in Kenya), a colleague and friend lived just over the hotel's fence. We crossed the fence, and as the curfew was lifted took to Mombasa by train to avoid further trouble in Nairobi. Igor's unique African artifact collection was later donated by his widow Erica to the Israel Museum. Igor is dearly remembered by us as he took our daughter Sharon, during her visit to Kenya, under "his wings" after a severe crisis she experienced. And so, parts of my scientific work radiated to our private life in more than one way.

Later on, my research expanded to lymphatic filariasis and hydatidosis in parallel with regional collaborative work initiated by the American Foreign Office and NIH, after the Peace Treaties were signed between Israel and Egypt and Later Jordan. This was really fortunate as it matched my strong will to be involved in regional collaboration, as a result of my interaction with Adel Mahmoud. The aspect of regional collaboration became central in my scientific work as mentioned below. African students (several from Kenya and one from Zimbabwe) spent training periods in my laboratory and returned home with, I hope, strengthened capabilities. Two of them actually evolved from technicians to Ph.D.s.

Before proceeding to describe the development of my scientific work, I have to make some reference to challenging disturbances. Episodes of health problems potentially having a temporary impact on one's work is part and parcel of life. But undergoing an aortic valve replacement, at the very time when the discussions on my tenure took place (1980), was a more serious problem, and a real dilemma. The evaluation of the postoperative life expectancy in this case was not very promising at that time, and my individual capacity to handle this situation was completely unknown (to myself as well). The verdict was pretty obvious. Thus ended my academic career at the age of 40, and the struggle for continuing my scientific work became a dominant factor. The strenuous conditions involved in this situation led over the years to the development of a chronic cardiac problem requiring repeated invasive procedures and routine medication. This became a given, and the decision had to be taken to incorporate the required health-routines and avoid making this condition a major component of my existence. Scientific curiosity and the strong will to teach were saviors in this regard, and my wife

Judith became a full associate-combatant, were a true support. Opponents took advantage to gloat and to improve positions, and proponents including my friend Dan Spira, a Senior Prof. in our Department, worked really hard to find a plausible solution which enabled continuation of research, competing for grants, instructing research students, teaching and organizing multi-teacher courses. I am grateful for their efforts. The chase thus proceeded with intensified research and teaching activities, yet without the Academic stamp, the associated self-confidence, and without the academic standing required for enabling active influence on academic decisions taken in the system. Yet, the bulk of my scientific work and of my scientific experiences was accomplished under these conditions. This included my African experience, my collaboration with scientists from Arab neighboring countries, and my shift to molecular parasitology, all of these went on in parallel.

Molecular Parasitology, a term introduced about 20 years ago, crosses the borders of the traditional subdivision of parasitology by groups of parasites, namely, Protozoology, Helminthology and Entomology. Development of tools for detecting infected snails caught my interest considering that the spread of schistosomiasis and its persistence in poor communities emerges largely from the intensive multiplication capacity of the parasites within snails resulting in persistence of transmission in these communities even if subjected to repeated treatments. The incorporation of molecular tools became essential for my work on identification of infected host snails of schistosomes which started as a collaborative work with Kenyan scientists from DVBD. Initially monoclonal antibodies were prepared for detecting infected snails. However, although monoclonal antibodies could easily identify infected snails much earlier than they shed infective, fully developed larvae this was not early-enough for covering the entire rate of snails with unripe infection (prepatent infection). Molecular tools were introduced for this purpose through several stages of tool development. It was initially found that total schistosomal DNA marked by a radioactive tag could identify snails infected with schistosomes from the 1st week after infection. This suggested that the total schistosomal DNA contained a large proportion of parasite-specific sequences. A paper by Steven A Williams, from Smith College, later showed the way to proceed. It focused on highly repeated genomic sequences in the genome of *Brugia malayi*, a filarial parasite causing lymphatic filariasis in some areas in the Far East. This was a new approach for DNA-based detection pathogens as previously this was accomplished by targeting ribosomal and mitochondrial genes, rather than repeated genomic sequence, for detection. The targeting of highly repeated sequences and the use of the polymerase chain reaction (PCR) for DNA amplification enabled very sensitive detection of parasite DNA in biological materials. We subsequently identified and cloned highly repeated sequences in the genomes of *S. mansoni* and *S. haematobium,* the human schistosomes of Africa, the Middle East and South America. These sequences occupy 12-15% of these schistosomal genomes, and when targeted for detection by using PCR very sensitive detection become possible from the first day after snails become infected.

Later on, identification was focused on species-specific sequences for differentiating human schistosomes from animal schistosome which can infect the same host snails and thus distort the quantitative results. Specific detection required further characterization of suitable PCR primers. In practice however, these molecular tools were not usable in field laboratories in Africa having minimal equipment available, and lack the training of the laboratory workers in molecular biology. Further developments in sample collection and storage/shipment were introduced in order to enable large-scale application of detection tools based on DNA amplification. Making molecular detection available widely in field laboratories in Africa is now considered a technological breakthrough which will impact mass-diagnosis and point of care diagnosis in regions where various infectious diseases are prevalent. Recognizing these requirements we went on to develop Loop Mediated Isothermal Amplification (LAMP) which does not require high technology instruments or intensive training. Using these tools for integrated diagnosis of several diseases at once is now a goal that molecular diagnosis and monitoring can help accomplish. Here comes the need to be supported by major research groups and institutions which on a final analysis will determine whether research of individuals in very small and peripheral small research groups (like in our case) can gain due and rapid application for mass diagnosis/monitoring. Thus the time factor hangs over the scene. When the baton goes to the hands of younger scientist the outcome of the race will no more be in the hands of technology developers but will become available in public domain (as I thought they should when the same is helping developing countries). As in this case it is more like a song becoming somewhat popular.

These developments took place in my laboratory from the time of my initial connection with Africa and through many years of collaboration with African scientists. A very important factor in accomplishing these results was my collaboration with Charles H King from the Center of Global Health and Disease at Case western University (headed by Jim Kazura, my old friend and colleague from the time of my Post Doc Training). I am very grateful to Charlie and Jim for their support over the years. Joining Charlie's project in Kenya enabled my access to field laboratories and to collaboration with snail collection teams. It all culminated in using LAMP for identifying infected snails in a field laboratory, and examining the possibilities of product development quit recently. Quite recently we have established collaboration with Clive Shiff from John Hopkins School of public health, for developing a LAMP-based diagnosis of infection in humans by examining urine. It is also planned to extend this approach for integrated diagnosis. This is now still under development.

Official collaboration with scientists from Arab neighboring countries was mainly (but not only) the outcome of the American involvement in Middle-East foreign policy. Shortly after the peace treaty with Egypt was signed a large collaborative project was signed between Ain-Shams University and The Kuvin Center. Protozoal and viral diseases were initially included and my participation with studies on bancroftian filariasis (caused by *Wuchereria bancrofti*) was included later on, as this disease was quite prevalent in Giza near Cairo. This collaboration involved meetings with Arab scientists of whom Reda Ramzi, from Ain-Shams University, became a close friend. We collaborated and published together over the years on the development of molecular tools for monitoring/diagnosis of schistosomiasis, filariasis (cause by *Wuchereria bancrofti*), and hydatidosis (caused by *Echinococcus granulosus*). Initially we both joined the summer course on molecular biology/genetic engineering at Smith College under the tutorage of Steve Williams and his team and sponsored by New-England Biolabs Company. Although we have started with cloning and characterization of highly repeated sequences of *S. mansoni* before I took the course at Smith College. This course is an outstanding learning experience which helped a great deal in my introduction into using molecular tools for developing diagnostic/monitoring assays for various parasites. Steve will always remain in my memory as the ultimate teacher and source of ideas. The exchange of ideas between us continued over the years. I suspect Steve was more than generous in (quietly) helping us develop the molecular tools for diagnosis/monitoring in which he was already an expert. Later on I spent a sabbatical period at Steve's laboratory which led to the discovery of highly repeated sequences of *W. bancrofti*. These sequences were later targeted for diagnosis by using PCR for DNA amplification in human hosts and in mosquito vectors. The concept of Xenomonitoring (monitoring infection in vectors as a tool for assessing transmission of a human infectious disease) was developed based on these studies, and used for evaluate transmission residual post-control transmission of filariae by mosquitoes and schistosomes by snails.

The studies on filariasis were carried out in collaboration with Reda Ramzy from Ain-Shams University, Cairo. Later on these were extended to collaborative studies on other parasites. Our collaboration extended beyond the time when scientific relations between our universities were officially active. This collaboration then extended to studies on developing molecular tools for detecting schistosomes supported by the European Community in collaboration with Dr. Andreas Romig from Heidelberg University, Germany and Dr. Joseph Jourdane from Perpignan University, France. The collaboration with Reda extended even further to studies on developing molecular detection tools for *E. granulosus* in collaboration with Phillip Craig from Salford University, Manchester, England. Reda is a true friend. Over the years we had the opportunity of visiting each other at home and make acquaintance with each other's families. I am afraid Reda's connection with me did not do him very good regarding his relation with colleague at home. For me this turned out to be an experience of friendship and an interaction not many Israelis had the opportunity to experience. My hope for future normal, friendly and thought-fruitful relations with Arab people still lingers despite the turmoil and pain currently involved.

Further on a personal experience regarding relations with Arab scientists: This happened before I have joined the, NIH funded, collaborative project with Ain-Shams University. During a meeting at the American Academy of Science in Washington DC, I have encountered problems with the slide projector and a Jordanian scientist, Sami Abed-El Hafez (a prominent expert on

E. granulosus/hydatidosis) came to my rescue. This was before a peace treaty was signed between Israel and Jordan, and this gesture was much appreciated. Sometime later a Palestinian student of Sami el-Hafez, Ibrahim Abbasi, came to look for me at the Kuvin Center. Ibrahim, who completed his Master thesis with Sami at Yarmouk University in Jordan, was looking for a place for doing his Ph.D. thesis. I proposed to him to join my laboratory and since then Ibrahim participated in all of our studies on schistosomes, filariae, and *Echinococcus*. Ibrahim completed his Ph.D. thesis on detection tools for *W. bancrofti*. These studies included characterization of a relevant repeated sequence, development of a PCR assay, looking for *W. bancrofti* in patients sputum and in mosquitoes, identification of repeated sequences in *E. granulosus* and developing a very sensitive PCR assay and then carrying out a study on copro-PCR in dogs, transmitting the disease, and later quite recently of the development of a LAMP assay for this purpose. After completing his Ph.D. Ibrahim (now Dr. Abbasi) became a close partner for carrying out our collaborative work with the US, Kenya, Germany and France. Dr. Abbasi should now take over when I am fully retired

A last line of scientific curiosity centered on *Toxoplasma gondii* a protozoan intracellular parasite of man and many species of mammals and birds where it tends to develop in the brain, but also in all other tissues. In women pregnant for the first time, and thus unprotected, fetuses can be infected in the uterus and suffer from severe brain damage associated with abortion or birth of babies with severe brain damage. *Toxoplasma* undergoes sexual multiplication in the definitive host, the cat, and cysts present in cats' feces can transmit the disease to farm animals thus affecting the meat industry. Since infected meat is a major source of transmission to humans, examination of meat for presence of Toxoplasma is a mandatory test, traditionally done by bioassay (infection of mice and examining mouse pathology several weeks thereafter) and taking much effort, animal facilities and time. My Ph.D. student Harold Salant participated in the testing of cats and farm animals for infection by *Toxoplasma*, and in replacing the bioassay by molecular detection using PCR. Later on Harold participated in the development of a LAMP assay for detection of *Echinococcus*. Molecular diagnosis is now becoming the gold standard for identifying infected animals. Although this test competes with the traditional bioassay, carried out for years by Dr. JP Dubey, one of the finders of the life cycle of *Toxoplasma gondii*, Dr. Dubey participated in some of our studies although he was aware of the coming competition. I think this was very magnanimous of him.

On a final note, it turned out that the bulk of my scientific work was done in collaboration with Arab scientists. This should, hopefully, become the prevailing standard all over. I hope it happens sooner than later, and still in my lifetime. I am still waiting.

I am practically retired now with some last strings yet to tie. Since my work was motivated by curiosity, I believe there was less than usual ego and glorification involved, hence a degree of acceptance towards real old age. On a retrospective account I can point out mistakes I have done mainly in my PR, and lack of sufficient discipline and patience, but I can truly say it was interesting. I could fortunately dismiss (but not forget) human malevolence and spare the memories of good experiences.

Motivation led by curiosity is bound to diminish as the volume of knowledge on the subject of one's expertise increases. So other things in life took more of my attention. Abrasion can also affect motivation over time. One such example is the realization that the internet has become a major source of teaching, and the diminished role of student-teacher direct contact (except for the case of research students). Also, it became evident that students are more interested in getting their degree and scholarships and the role of learning avidity per-se progressively diminished. Focusing curiosity on other matters is bound to take over in future, old age, activities. It takes looking into.

Retrospection also leads to the recognition of the price family members, children in particular, sometime pay as a result of the total dedication of a parent to scientific activities, which have to be taken home because they go with you all the way.

CHAPTER 8.

A Brief Review of My Life & Career

Julius P. Kreier

Introduction

I was hired as an assistant professor at the Ohio State University in the summer of 1963. I retired after 26 years as a full professor in 1989 having been a teacher first of veterinary students in the veterinary school then of undergraduate and graduate students in the college of arts and science. I conducted research, guided postdoctoral fellows and last but not least advised students in their struggles to find their way in the scientific and academic world.

I did not enter the scientific and academic world directly after I graduated from veterinary school. I worked as an animal disease control officer for the federal government for several years before enrolling in graduate school and entering the academic world.

The story that follows is largely a personal story of how my career developed. I am writing it with the hope that it will help those to find their way who may be interested in considering a life in science, in academia or in public or private research institutions.

The Early Years

My Family

Like everyone I received much from my family. The gifts were both biologic and cultural. The assortment of genes and other biological characteristics of each parent during the production of egg and sperm and the combination of these characteristics at conception endowed me with the basic attributes of my being which were then much shaped by the environment in which they were developed. This started in my mother's womb and never stopped as long as I have lived.

It should be noted that the attributes I received in this process were a random selection of the attributes present in the gene pools of my two parents. As the gene pools of my two parents were a randomly selected subset of those available in the general population the collection I received was unique and I thus started my life relying on attributes which I had not chosen. In fact one may say we all start as a result of a lottery and lotteries distribute their rewards quite unequally. It is of course possible that my parents and their parents back through time had selected their mates for characteristics they considered desirable and thus decisions they made may to some degree have determined what I got. It is just that I had no role at all in determining what I got to start out on.

In writing about conception I can't resist putting in a word about when during gestation life begins. Both the egg and sperm are alive. If either were not, conception would not occur. Life began sometime in the remote past when the first living cell was formed. It never stopped. The difficult question is when does human consciousness arrive? I believe it is usually thought to occur at or shortly after birth. In passing I should also mention that I found during my youth that my father, a Lutheran, believed in the Calvinistic theory of predestination which I consider to be in contradiction to the concept of free will. I have never been able to reconcile this belief with my belief that the environment and my choices i.e. my free will, affect how my genetic endowment is realized. I however decided quite early that belief in free will is in itself required if you want to deal effectively with the problems you encounter as you move through life. But now I feel that it's time to describe my family so that the reader can understand something of what I believe they gave me and how I dealt with what I received.

My families on both sides were immigrants to the United States from Germany. My father was born in 1885 in Altstadt near Nierenberg in what is now Bavaria. His mother was of the French Protestant stock which fled to Germany when France expelled its Protestant population. His father's family members were wealthy people who lived in Bonn. Beyond that I know little about them except that my father maintained contact with a woman he called Aunt Sherman, a member of his mother's family. Aunt Sherman entered the Sherman house as a nurse to care for Frau Sherman during her terminal illness. She later married Herr Sherman after the wife died.

I did see some of the letters my father wrote to Aunt Sherman. But as they were hand written in a most elegant antique German script I was unable to read them. I do however remember a few photographs of my father taken in his youth. The one that stands out in my memory was of an elegant young man in a tuxedo preparing to attend the Bell Arts Ball at Putsi Hofstangles house in Munich. It stands out in my mind because the man I knew had abandoned all concern for his appearance and was regularly assumed by visitors to the shop who did not know him to be the janitor. I also remember several pictures of the Sherman house on the Rhine. One showed the servants lined up in front of the house and another, taken during the First World War, was of wounded German soldiers sitting on chairs on the extensive lawn between the house and the Rhine River.

I tried later to again see the letters and photographs he had from his pre-immigration life. I found out that after his death my mother had thrown all of my father's papers in the fireplace and burned them.

My mother's family came to America well before my father. The branch of her family with which I am most familiar was the Necker branch. Necker was my mother's maiden name.

My mother said the Necker who came to America in the 1840's had been a student at the University of Heidelberg where his six older brothers were professors in the arts and sciences. The young Necker, like students to this day, became involved in political activity, his in support of the socialist movement. He came to America when the revolt was put down by Bismarck and he fled Germany to save his life.

I know very little about how he made his living in America, but he obviously survived, married, and had children. One of his descendants was my mother's father, my grandfather, Julius Necker. He studied in the Academy of Fine Arts in Philadelphia. After he graduated he joined a company that produced fine arts reproductions by the lithographic method. He continued painting thereafter as a hobby.

Grandfather Necker married a woman of a German Catholic family named VonWitkampf and they had two daughters, the, younger my mother. Grandfather Necker died just before I was born so I had no direct contact with him but he was nevertheless a considerable influence on me, in part through my mother, in part from my father's somewhat negative view of him and, in part through the extensive library he left behind which ended up in our house. He was perhaps as my father said and ineffective socialist intellectual, an atheist who became the friend of the family's Catholic priest and shared many political and religious discussions over bottles of wine with him.

Strangely my political and religious views resemble my grandfather's more than my father's despite the close relations that existed between us. The Necker family was bilingual. They maintained German as the household language while using English to interact with the general non-German society. This pattern stopped after my father married my mother and we moved into the house in which I was born. My father believed that if you came to America you should become American and speak English. My father came to America in 1909 at about 24 years of age. He came after he had completed his education. He attended an art school in Munich or possibly Nierenberg, I am not sure which. After art school he did an internship in stone carving. I am not sure of many aspects of his life in Germany as he was not much for talking and rarely spoke about his past. All I know was he came as a second-class passenger, and left the ship in New York. This information my son found out from an Internet site about Ellis Island and immigrants. From New York he went directly to Indiana where he had lined up a job as a carver. He did one day tell me he did not like the job he had lined up, disliked Indiana and quite soon got on a train to Philadelphia. How he picked Philadelphia, how he met my mother, how he chose to start the business he spent the rest of his life in, I don't know. All I do know is that sometime after his arrival in Philadelphia he started the business in which he made ornamental features for buildings such as movie houses, banks, railroad stations, churches, and etc. He got the jobs by bidding, the bids were submitted to the building contractors or the architects or whoever wanted someone capable of producing the art work required.

He was quite successful until about 1935 when the depression stopped most building. The Second World War then came and there was little building during it and when building started after the war the buildings were bare of ornamentation. I have mixed feelings about the Bauhaus movement to this day. During the depression and the war and for some time thereafter he went into the shop every day and worked alone on statues for churches and other things anyone wanted.

He never called himself an artist but rather a craftsman in the old sense used by the anonymous craftsman who built Europe's churches and palaces. He never retired and continued working almost to the day he died.

The rather abrupt drop in his income resulting from the Great Depression affected him strongly. It also shaped my view of the world. What I remember from the time is my father, in the best European bourgeois tradition, saying he was spending more than he was earning and that we would end up in the poor house if we kept spending so much money. I was just a child at the time but it created a worldview that affected my behavior for the rest of my life.

I know now that he was a careful person with his money. When he was earning a lot he paid off all his debts and saved prudently. He also sent money to his father's family when they were seriously affected by the inflation and financial collapse that occurred in Germany after the First World War. This money which he sent to Aunt Sherman, who was apparently managing the family finances, permitted them to save the family home in Bonn.

When Germany created a new currency after the complete collapse of the old system the house was sold to the Catholic Church and is now a nunnery. The family which included Aunt Sherman and two girls and a boy, who I believe were my father's siblings, lived on the proceeds of the sale. My father never discussed any of this with me and what became of them I have no idea. He never visited Germany either. I do remember that after the Second World War he received a phone call in the shop. I remember him saying no to an offer from someone in the U.S. Army of a job with the occupation.

All I know about my family's finances I learned later when as the last surviving sibling I had to manage what was left and distribute it to my younger sisters and my children as directed by my brothers and elder sisters wills. The reason their wills controlled what was left of my father's estate was because they stayed in Philadelphia. George after getting a PhD in psychology and working for about six months as an assistant professor at Temple University quit and entered my father's business. He then took over the property in the city and the business after my father died. Gertrude, my older sister, moved into the house in the country and cared for my mother and my father until they died. Gertrude who had become a teacher in the Philadelphia school system chose to live in the house in the country. She set up a trust which directed that the younger siblings in order of age administer her estate after she died until the youngest that's me would distribute the assets. When my elder brother died he left his assets to the trust my elder sister set up. The main assets in the trust were the real estate taken over by my brother and sister. It was a surprise to me that there were still assets left after we all grew up, but that didn't affect the worldview I had developed as a child from observing my father who I do believe really feared poverty.

Early Years, My Childhood

I was born on November 30, 1926 in a bedroom in our house attached to my father's shop in Philadelphia. I was the fourth and last child born to the family. The eldest a girl, Gertrude, was born and lived for her first years with my mother's family. The second, a boy named George, came about six years later. After my brother, George, my sister, Elizabeth, and I followed at short intervals of about 1 and 1/2 years each. These last births proved a strain on my mother and after I was born my mother developed postpartum depression, paranoia, and she rejected her last child. As a result my father essentially took responsibility for me. The situation did not change much with the passage of time.

I have very little memory of my life before about five years of age. I recall I spend most of my time with my father following him around when he attended the furnaces that heated the huge house we lived in and playing with or at least watching the rabbits and squirrels my father kept in the yard alongside the house and connecting to the shop at the rear of the house. I also spent time in the shop, sometimes in the evening falling asleep on a pile of burlap while watching my father modeling some figure or other he was making. This memory gives me an opening for describing one of my first thoughts of a possible career. It was a negative thought. I decided I could never make a living doing that kind of work.

There was one event in those years of which I have no memory but which had lasting consequences for my life. I was a very active child. My mother said I was a little monkey who climbed on anything. It turns out I not only climbed I also fell. Some time when I was between one and two years of age, I don't know exactly because no one ever told me, I climbed on a high sideboard and fell off when I reached the top. I knocked out some teeth and developed an infection in my gums. Some bacteria, probably staphylococci, entered my bloodstream and lodged in my left hip joint. An abscess developed and after some time burst, drained, and healed. I understand it was touch and go for some time with periods of high fever.

In the 20's, there were no antibiotics or chemotherapeutic agents known and doctors could only give palliative treatment. I believe my parents later refusal to let me see any doctors about my hip stem from that experience.

In my childhood the hip joint gave me little trouble. I remember not being able to sit Buddha position like the other children but I had no pain in the hip joint and could run around like the other kids my age.

In retrospect I believe I was somewhat behind in my development. I had some problems with my speech and learning, probably as a result of the high fever during the infection. I failed to learn to read at the time the school thought I should and as a result I was held back in second grade. This was a shock to my parents. I had been enrolled in the same select grade school as my three older siblings all of whom were stellar students. As a result of this I was labeled as stupid. In the subsequent years I did learn to read and actually became quite proficient. The only thing that did not change was my family's view of my intellectual capacity. As a result I was not enrolled in the academically selective high schools which my siblings attended and I did not even think about colleges or expect to attend one.

While during my childhood I was not bothered much by my damaged left hip, when I entered adolescence that began to change. Around 12 or 13 a degenerative arthritis developed in my left hip. The amount of motion became limited and it became painful. With the loss of motion by the time I was 20 years old the left hip joint became locked in a flexed and abducted position and was quite painful. As a result of the lack of motion in the hip joint when I stood or tried to walk my spine was twisted and also became painful. The use of a 2 or 3 inch lift on my left shoe did not help much.

You might think that with the hip that caused a limp and was often painful I would've become inactive, but in fact I was quite active. My parents chose to pretend that nothing was wrong. When I complained of pain and asked for medical help I would be told that if I really wanted to I could stand straight and that I should ignore any slight discomfort. I concluded that the situation was what it was and I should make the best of it. I developed stoicism. The stoic approach to life's problems served me well throughout my life. It can be considered a gift from my parents. Its development may have been reinforced by observing my father's behavior. I observed quite early that he was strongly stoic and tolerated injury and sickness with never a complaint.

I mentioned earlier that after I was born my mother entered a postpartum depression, developed paranoia, and rejected her child. She also rejected her husband and most aspects of running the house. In 1936 my father made a decision which in retrospect I am sure was made to provide a distance from my mother. What he did was buy an abandoned farm of about 15 acres which included a very old stone house. He began to rebuild the house and turned the grounds into a garden which occupied all of his spare time for the rest of his life. He still went to the shop daily and worked his eight hours there. He also kept one of the apartments in the house in the city for my mother and supported her. We children also ended up split between the two parents and the two houses. My younger sister and I would meet after school in the house in the city but at the end of the day we went with our father to the house in the country. We also spent weekends and holidays with my father. I don't know how the decisions were made but my brother stayed with my mother in the house in the city. At this time my older sister was off at college and on her own.

It turned out that my spending most of my youth, from 12 to 20 years in the country, did affect my career choices because there I started raising animals, rabbits, chickens, ducks, geese, goats, sheep, and even once a pig.

I attended the Philadelphia school system as my father retained his official residence in Philadelphia. As our house in the country was just north of the city I ended up in a high school in North Philadelphia. My father dropped me off on the way to the shop and the house in the central city.

I never worked hard in high school, I just did what I had to pass and I did all right but it was all a bit of a bore. Despite my lack of effort in high school some of my teachers thought I had some potential and did encourage me to seek post-high school education.

During my second year in high school the Philadelphia school system started a half-time agricultural program. The leader of the program visited the high school I attended where he described the program and passed out literature about the program and application forms. At the time as I mentioned earlier I was quite active raising rabbits, ducks, and other fowl and attending a pet half grown goat who presented me with a baby goat some months after I got her. At any rate as I had no idea of what I would do with my life I thought perhaps some form of agriculture might be a possibility and I signed up for the program. What I got out of the program was I learned a bit about agriculture, I got outside working on the school farm some of the time, and I got summer jobs, the best of which was working during the summer on a large dairy farm filling silos and collecting hay. I also met a fellow student who became my best friend. The negative effect of the program only appeared later when I applied for admission to Temple University, a school near our house in the city, and I discovered I needed a couple of extra high school academic courses to get in. As a result my admission was delayed a semester.

As indicated in the preceding section when I finished high school I applied for admission to Temple University in Philadelphia. Temple was and still is a community college. Most students lived at home and attended classes by commuting to and from home. Tuition was the only cost for those who lived at home like I did. I chose Temple because costs were low and I could live at home. I really had no clear idea of what I would study. The only thing I did know was that I did not want to work in an office. I chose to major in some field of biology and I was pretty sure my father would pay my tuition although I never discussed any of this with him, or anyone else, until after I was admitted.

Temple University had a student body that was made up of the children of the people who owned the local grocery stores and pharmacies and etcetera. At the time, 1946, a large proportion of the students were Jewish although the school was originally founded by a Protestant group. Today it is a city school.

The general academic level was quite high. Most of the students, unlike me, seemed to know where they were headed, medicine was a favorite. I noticed this in part because I had for a period developed an interest in medicine, particularly psychiatry, as a result of observing my mother's behavior especially her periodic descents into insanity. I couldn't help observing the prodromal signs which seemed to me to be physiological and were manifested most obviously by changes in her facial expression, her face would appear swollen and wooden. At any rate I never acted on this interest as I accepted my parent's view of my ability level.

The academic environment at Temple did stimulate me to work fairly hard and I did without plan or declaration follow the premed program. The biology courses, the organic and other chemistry courses, and the physics courses I chose were full of premed students who would do anything to get a high grade so there was a stimulating competition that I never felt in any classes before.

While I was enrolled in Temple University my sister Elizabeth, who was next up for me in age and with whom I was quite close, joined the University of Pennsylvania orchestra. She was in the cello section. While playing in this orchestra she became quite close to a young man, also in the cello section, who was a student in the University of Pennsylvania veterinary school. When he visited us he would talk about the opportunities the veterinary field offered ranging from private practice treating animals, to federal programs in animal disease control often involving public health and zoonotic diseases. One reason I was attracted to veterinary medicine was because it is a profession that would not confine me to an office. Until I met my sister's friend I had never thought of Veterinary medicine. His talk however started me thinking about the field. I realized it would combine my interests in animals and medicine. In addition the University of Pennsylvania veterinary school was near where I lived and getting my information from a student provided me with information about costs. At the time the costs were quite reasonable as the college, despite being a part of a private school, was heavily subsidized by the state government.

During my third year at Temple University I visited the veterinary college and talked with various faculty and was given application forms and some mild encouragement to apply, which I did. Later in the year I was invited to a formal interview. I remember sitting with the admissions committee and after some time being told to wait outside. More time passed until I was asked to come back in and told I could start the following September, that was 1949. I learned about a year or so later from a faculty member who'd been on the admissions committee that my long wait in the hallway outside the admissions committee's room was the result of some members questioning whether a person with a bad limp should be a veterinarian. My informant said he convinced them that there were many areas in veterinary medicine that I would be able to contribute to using my brain even with a severe limp.

In looking over my life I have often been amazed at how large a role chance has in the paths I chose to follow. If my sister hadn't met and I should add later married that veterinary student it would never have occurred to me to apply for veterinary school. If this chance meeting hadn't occurred I have no idea what other path I may have followed.

While we are on the subject of the effect chance encounters have on our life choices I think it is time to mention a couple of other chance encounters and events that played a major role in my life.

I enrolled as I said earlier, in an agricultural program in high school. This experience made me more inclined to be influenced by my chance encounter with the veterinary profession through the veterinary student my sister met. The agricultural school experience also appeared on my application to veterinary school, a use to which I never expected it to be put. The fact that one man on the admissions committee believed that a crippled leg shouldn't be a basis for keeping me out of the veterinary school also was a chance occurrence.

There was another result of my joining the high school agricultural program that had large effect on my life, although not professionally. There I met the fellow student who became a lifelong friend and who some years later played a role in my meeting the girl who became my wife. My wife affected my life in many ways not the least by providing a stable pleasant life we shared until her recent death.

As I mentioned earlier Temple University stimulated my interest in biology and also helped me learn how to study. I went through at the same time as a flood of veterans returning home from the Second World War were entering academic life with the help of the G.I. Bill. They had a beneficial effect on the college environment. They were there to study and didn't tolerate childish nonsense by the fraternity boys and other students just out of high school. I felt comfortable with the environment they created. The experiences at Temple helped me to deal with the high academic demands I encountered at the University of Pennsylvania Veterinary School.

I often thought I could've been one of those veterans. With my somewhat late development I was 18 while still in the next to last year in high school. I was called up for the draft in 1944, but rejected because of my damaged hip. My mother said how lucky I was, as if I had been drafted, I would have been just in time for the landing in Normandy. I remember telling her I would have been glad to chance getting my head blown off in exchange for a painless hip joint.

At any rate at the end of my third year in Temple University I entered the veterinary college at the University of Pennsylvania. During that first year I worked harder than I had ever worked in my life. We were enrolled in class eight hours a day, Monday through Friday and four hours on Saturday. We had assignments to complete in addition to attending class.

I remember anatomy, histology, genetics, chemistry, and many other courses that never seemed to stop during the first three years. The second half of the third year and the fourth year were devoted largely to clinical training. I remember that the program was very strong in the basic medical sciences while the clinical aspects were rather weak. It didn't bother me because I found I really preferred the basic medical sciences to clinical work.

Between my third and fourth years I had a particular learning experience. That summer I got a position with a very successful small animal practitioner in a wealthy suburb of Philadelphia. I took it largely for the money as I was not much inclined to small animal medicine. I was a general helper. In addition to cleaning cages and feeding dogs I did whatever I was asked to do and I was able to observe all aspects of how the place was run. The veterinarian who ran the practice was very smart and extremely skilled in handling people. I will describe one case that typifies how he operated and which helped form my view of small animal medicine. The first encounter with the clients was handled by a skilled receptionist. This receptionist would meet the client and collect information about them and the animal they brought for treatment. She noted any behavior patterns and the animals name and its likes and dislikes. She of course also recorded the owner's view of the animal's condition, particularly information about why the animal was brought into the veterinary hospital. She then returned to her desk and made out a file card which she showed to the doctor and briefed him about the clients and the animal. The card and the briefing system became of great value on subsequent visits, because even if he didn't remember the clients and the patient he would be prepared by the receptionist before he met them and then would be able to astonish the clients with how well he seemed to know even fine details about the patient, such as its favorite treats and etc. In this particular case the owners were elderly people, both overweight and their dog was also overweight and was brought in as the wife was concerned because the dog just refused the kidneys cooked in cream and other delicacies she had prepared. The veterinarian was very sympathetic and told the clients they should let the dog stay with him for a week during which time he would study the situation carefully and institute treatment. He would assure the clients the dog would have a good appetite when they picked him up. After they had left, the dog was brought into the back room and I was told to put him in a cage and put a sign on the cage saying water only. The only person who checked him daily was me. When the dog was picked up there was a dramatic demonstration in which the dog was brought out of the cage and offered a bowl of food which he devoured. After a long description of how carefully he had observed the dog during its stay in the hospital, he recommended a special diet for overweight dogs that he had in stock. At the time I considered the veterinarian's actions unethical. In later years I realized his diagnosis was correct, the treatment appropriate, and if he had simply told those people they overfed their dog and they should put the dog on a diet, they would have found another veterinarian to tell them what they wanted to hear. Nevertheless, this experience made me realize that I would not have been able – and would not have wanted – to do what he did, not matter how much money it would have yielded.

There are many other experiences I had during those four eventful years in veterinary school. I can't possibly describe them all, but I will describe a few that remain in my memory because they had lasting effects on my future life.

During my first year in veterinary school I turned 21 and my ability to walk deteriorated greatly. Perhaps my father recognized how bad my walking had become and using my age as an excuse my father said to me that as an adult I should be permitted to make my own medical decisions and that he would pay the associated costs. I then contacted the student health service which made an appointment for me with their orthopedic surgeon. After he examined me I told him I would like him to do a cup arthroplasty, a procedure in which the joint is opened, the joint surfaces smoothed and a metal cup inserted in the joint. He replied that he didn't do that procedure as it had, at least in his experience, a low success rate. He recommended a simpler procedure called a subtrochanteric osteotomy in which the femur is cut below the hip joint and the leg is held in a more or less standing position by a plaster cast starting just below the armpits and going to just above the genital and anal region in the middle and normal side but to the ankle on the operated side. The procedure required being in a cast for about three months. These were the options as total hip replacement did not exist in 1950. I wasn't too happy with his recommendation and should have at least gotten some additional evaluations but I was desperate and decided that I would just go ahead and accept what he recommended. Surgery was scheduled for shortly after the end of the spring semester.

My elder sister, Gertrude, had become a teacher in the Philadelphia school system at this time and had decided to move into the house in the country, bringing my mother and brother there also, and thereby trying to patch up the split in the family. Gertrude took responsibility for me during the summer I was in the cast and confined to bed. She cared for me carefully and I thank her for what she did. My close friend Harold who always spent a lot of time at our house continued to visit. It was only much later that I found out that while I was confined to bed that he and my sister became friendly. Later when I was working in

Mexico they got married. I was sawed out of the cast just before class started in the autumn quarter. I started attending classes about two weeks late. I got a room in a veterinary fraternity near the veterinary school and somehow or other got through the year. After I had recovered from the surgery I found I could walk and stand much better than before. The long rest gave the hip joint time to heal, but as the motion of the joint was still very limited I had more trouble sitting and I discovered I couldn't drive a car with a standard transmission as I couldn't operate the clutch. It may seem strange but that played a role in my becoming a professor as I will explain later.

As I mentioned earlier it was during my second year in veterinary school that I met the woman who later became my wife. As she became such a large part of my life I feel it is appropriate for me to expand a bit on the meeting. I had been dating a girl named Naomi who was some remote relative of Harold, my close friend. Naomi decided she didn't really find me that attractive and decided to pass me on to another girl named Ruth, a friend of hers. To get things started Harold arranged that the four of us would go to dinner together. Harold provided the car. It was an old model A Ford from 1930. After dinner we dropped Naomi off at her house and decided to take a ride around Fairmount Park as it was a fine warm but slightly rainy night. It was actually weather caused by the tail end of a hurricane up from Florida. The three of us were sharing the front seat with Ruth in the middle. I guess I got a bit carried away and as a result Ruth accidentally hit the accelerator with her foot. The motor raced and the fan came off and the motor stalled. Harold told us to get out and walk around in the park while he fixed the car. It seems a bit strange that despite my behavior we hit it off that night and for the next three years when she came back to Philadelphia from Pennsylvania State University we would get together. The same arrangement persisted after I graduated from veterinary school and went to

Mexico to work on a program to eradicate foot and mouth disease there. She came to visit me in Mexico in the fall of my first year there. My choice of the job in Mexico oddly enough, despite the physical separation, reinforced the relationship. To help explain how this helped bind us together I will jump ahead a bit and describe her first visit there with me.

This first visit occurred when I decided to purchase a car. To avoid the high taxes on cars purchased in Mexico I made the purchase in the United States in the town of Laredo. I arranged with Ruth for her to meet me in Mexico City where we spent several days as tourists and then flew up to Laredo where we picked up the car.

The drive back to Mexico City was an adventure. After crossing into Mexico we chose to take secondary roads to the Atlantic Coast and then head south to reach the area of Veracruz in which I was stationed. Leaving the US I had trouble finding the International Bridge I asked a group of men standing alongside the road how to get to the bridge. They didn't answer. I then asked in Spanish. They laughed and told me exactly how to get to the bridge. They just weren't going to talk to an English speaker.

I was fortunate to have bought the car during the dry season because when we got on the back roads, many of which were unpaved, it would have been impossible to use them when they were wet because of the mud. In many cases there were no bridges and where the roads crossed the streams we had to ford through the water. In one small town where we had stopped to eat, the local Catholic priest recognizing we were strangers said hello to us and when we expressed interest in archaeology invited us to his house to see his collection of pre-Columbian artifacts. It was a wonder and as good as many museum displays. I think we gave him pleasure also as he was an educated man living in a very small town and was quite isolated. I believe he was happy to have had us as visitors.

When we reached the area in which I was stationed we entered a more developed region. The town of Tampico on the northern fringe of the inspection area had a lovely small hotel where we stayed a few days. It had a restaurant that served fish caught in the gulf right near where the town was located. As we continued southward the next town we reached was Papantla a fairly large town which was the district headquarters of the foot and mouth disease eradication program. The town had a very nice market and was located near a rich archaeological site. There is a pyramid called the Tajin located in the site (Fig.1). A large proportion of the population, in that region, are Totonac Indians, who congregate on Saturday nights in the town square all dressed in their tribal costumes (Fig.2).

At the end of my second year in Mexico we had eliminated the disease and I was discharged and reassigned to a disease control program in Maryland to deal with tuberculosis and brucellosis of cattle there. When I received my discharge I invited Ruth to meet me in Mexico City and we drove back to Philadelphia together. We married shortly after we got back to Philadelphia. I came back to Philadelphia during the Christmas holidays in my second year in Mexico. It was a longer stay than I expected as I developed viral hepatitis on the way home and spent most of the visit in the University of Pennsylvania Hospital. Other than a brief episode of vivax malaria hepatitis was the only serious disease I encountered in the two years I was in Mexico. Ruth visited me in the hospital but a visit in the hospital was not what I had been looking forward to.

In a way my career choices did play a role in my marriage. Ruth it turned out, liked animals and that gave us common ground. Ruth also was interested in archaeology so my working in Mexico in an area rich in pre-Columbian ruins provided us with another rich area of common interest.

Now I will turn to events that occurred near the end of my time in veterinary school, events which directly shaped my future.

A professor who taught biochemistry to the veterinary students approached me and asked me if I would be interested in going into graduate school. I replied no because as I explained I had been in school all my life and felt the need to get out of school and start working. This occurred before I had thought much beyond getting my veterinary degree and working as a veterinarian.

I had been following the progress of the US Mexican program for the eradication of foot and mouth disease, but my hope for getting into that program looked dim as in my senior year it appeared that the program had been successful and was in the process of being shut down.

I had applied for several positions with rural veterinary practitioners but had been rejected because they felt I couldn't handle the rigors of a large animal practice. A good friend of mine in the class who intended to set up a small animal practice asked me if I would join him in that effort. We discussed it, did some investigation, but my heart was not in it as I didn't really want to run a business or work as a small animal practitioner. During this period, when I was studying for the Pennsylvania State veterinary license examination and trying to decide on what aspect of the veterinary field I would pursue a notice came to my attention that there was an outbreak of foot and mouth disease in an area of Veracruz, Mexico which had been considered cleared of the disease and that the US government was again setting up a unit to deal with the new outbreak. I immediately contacted the local USDA office, obtained a job application form, and applied for a position with the group.

There were great differences in the hiring policies of the USDA and how you are enrolled in the U.S. Army. They are relevant to this story because if the USDA had given me a physical examination as did the U.S. Army when I was called up in 1944 I would never have gone to Mexico and my entire career would have been different. The USDA didn't even ask me if I had any physical problems and I didn't volunteer any information either. There was another interesting aspect of the government's hiring policy. We were to be sent to a Spanish-speaking country yet no one asked if we spoke Spanish which in fact I didn't.

A short time after I graduated and got my license to practice veterinary medicine in Pennsylvania I received a letter from the Department of Agriculture saying that I was hired and would join the group as a veterinary disease control officer and would report to the office in Mexico City as soon as the negotiations with the Mexican government setting up the reactivated plan were completed. In the meantime, as I was being paid, I was assigned to work with a group of meat inspectors in a slaughterhouse in Philadelphia. I worked in the slaughterhouse for about two months before I was transferred to Mexico I must admit I did not much like working in the slaughterhouse but I stuck with it because I very much wanted to work on the program in Mexico. I also was aware that meat inspection was important component of the public health field which did interest me.

In the next few paragraphs I will describe a few of the things I observed during the months I worked as a meat inspector. I found for example that meat inspection included the inspection of livestock before slaughter in part to control what is called emergency slaughter of sick animals, an effort by the owner to recover some value before the animal died. I also observed that veterinary inspectors assume responsibility for the general sanitation of the entire meat handling process as well as play a role in the grading of the meat. Even I, just out of veterinary school, was surprised to learn how large a role veterinarians play in public health. The Army Veterinary Corps for example plays a major role in the Army's Health Programs. A number of veterinarians with whom I worked during this time were veterans who described to me what they had done in the military. Their activities ranged from supervising base food purchases to overseeing base sanitation and tracing the sources of outbreaks of food borne illness. I also encountered veterinarians who had worked in similar activities in the general society. The largest public health program run by veterinarians is a well-established program to control and ultimately eliminate bovine tuberculosis and brucellosis from domestic cattle. These are animal diseases readily transmitted to humans which before the program started caused much serious disease in both cattle and humans. The program for the eradication of tuberculosis and brucellosis in cattle tied in with my meat inspection experience as cattle suspected of infection by the tests used were branded and slaughtered in the federally controlled slaughterhouses. I learned more about this connection later when I worked for a year in the tuberculosis and brucellosis eradication program after the foot and mouth disease eradication program in Mexico was successfully completed. Working in these programs was a valuable component of my education and participating in actual public health programs did help me later when I started teaching students who intended to enter the fields of medicine, veterinary medicine, and public health. The next part of this chapter will deal primarily with my experiences in Mexico.

Foot and mouth disease was brought into Mexico with some latently infected Brahman bulls purchased in Argentina by a Mexican rancher who believed that cattle of Indian origin would be more suitable to the tropical climate of Mexico than cattle of European origin. Disease appeared in the susceptible Mexican cattle shortly after these bulls were introduced to Mexico. Infection spread rapidly throughout Mexico. The American and Canadian governments had eliminated the infection from their countries. The disease had never occurred in Mexico and Central America and the undeveloped jungles of southern Panama served as a barrier to the disease entering North America from South America. The disease in Mexico was considered likely to spread to the US through the highly porous Mexican-American border.

As foot and mouth disease is highly contagious and spreads rapidly by direct contact or by contact with farmer's shoes or other contaminated items the method of control used in the USA and Canada and most of Europe is slaughter and burial of infected and contact animals. Vaccination is sometimes used to control the rapid spread of the disease, but as there are several antigenic strains of virus and the duration of immunity from vaccination is only about six months the vaccine is only used to slow the spread of the infection. Slaughter of infected herds and direct contact animals is then used after vaccination is stopped to deal with any outbreaks that then occur.

Slaughter and burial of infected and contact animals is a harsh method for the control of disease. Many people have questioned its use to control foot and mouth disease especially as fatality from the disease is usually only five or at most 10% and as foot and mouth disease does not affect humans. Control of diseases of farm animals particularly ones that are not transmissible to humans is justified on an economic base. The cost of living with the disease is compared to the cost of eradicating it and eradication is used when eradication, a single effort for isolated countries like those of North America, is calculated to be less costly than living with the disease. Living with foot and mouth disease is quite costly, as immunity is short-lived, vaccines are not very effective and must be made for each strain of the virus, and spread is rapid among the not immune animals. Infected animals can't walk well or eat easily, milk animals stop producing milk and range animals lose weight and may die if they lose their hooves and are unable to feed because of a sore mouth and inability to walk around to reach their food. The costs are perpetual if the choice is made to live with the disease. Eradication on the other hand is a single expense and continues only for a finite time.

At this time I will describe the organization of the foot and mouth disease eradication program on which I worked. It is necessary to understand the organization of the program to understand how it affected my career development. The title of the program was the cooperative program for the eradication of foot and mouth disease in Mexico. In passing I might mention that among the US personnel the term cooperative was often considered a bit of a joke, but we did cooperate enough to successfully eliminate foot and mouth disease from Mexico.

The region where I worked for about two years was in the north central part of Veracruz State. It extended roughly from north of the city of Veracruz to the South of Tampico. It was bounded on the east by the Atlantic Ocean and on the west by the foothills of the mountains that rise abruptly to the Mexican central plateau. It covered roughly an area surrounding the unexpected outbreak of foot and mouth disease on a ranch near the town of Gutierrez, Samora. The outbreak was in the middle of a region that was considered large enough to cover any infections that might spread from the infection which had occurred. The area which surrounded the unexpected infection was divided into districts to which veterinarians were assigned. Each of these districts had a town where the veterinarians stayed. These districts were subdivided into sectors to each of which a pair of cowboys were assigned (Fig.3). In each section there were also towns selected as bases where the Cowboys stayed. The Cowboys traveled over their sectors on horseback following a route map showing where ranches with cattle were present. The map included a timetable showing the time and day they were expected to be inspecting each location. The town of Papantla was near the middle of the zone put under inspection. It was chosen as the headquarters where the overall supervisors worked and it had a radio system connecting the headquarters to each of the vehicles used by the pairs of district veterinarians.

Practically all of the positions in the program were staffed by two people, one Mexican and one a citizen of the United States.

There were some exceptions to this role. One exception was a laboratory in Mexico City run by an employee of the US Department of Agriculture who supervised the use of serological test to determine if foot and mouth virus was present in samples collected by the field veterinarians. The antigenic strain of the virus was also determined by this laboratory. A second exception was a bilingual aide assigned to each American field veterinarian. He was called an inspector A. He was usually a person who had experience supervising Cowboys. His main duty was to supervise the Cowboys who did the day to day inspection of the cattle and to give the American veterinarians aid and advice when requested. The third exception was a financial officer who paid the expenses encountered in the operation such as the wages of local help to clean and disinfect areas were disease occurred as well as incidental expenses for supplies obtained locally. He was an American who was bilingual and had experience as a paymaster.

When not otherwise occupied the pairs of district veterinarians moved around their districts using the maps of the Cowboys' routes. They would spot check the cowboys at random to assure that all of the cattle in the district were looked at by the Cowboys at the scheduled times.

As noted earlier each of the vehicles driven by the district veterinarians had a two-way radio to connect them to the head office. If the Cowboys saw any sick animals, that is, one's dripping excess saliva or limping badly, they were instructed to notify the district veterinarians or the main office by any means possible. If the main office was called they would radio the district veterinarians. After the Cowboys notified the appropriate authorities they returned to the ranch to await the arrival of the district veterinarians. If the district veterinarians, after examining the animals (Fig.4), considered that the animals may have had foot and mouth disease they would notify the head office, collect tissue and fluid samples from the lesions (Fig.5) and in collaboration with the veterinarians in the headquarters arrange to get the samples to the Mexico City laboratory. If the samples proved to be positive the process of eradication would be started (Fig .6).

In a peculiar way the relationship between the pairs of veterinarians was in some ways similar to a marriage. We both were told what our duties were but neither of us could give orders to the other. We were instructed to actually plan our day's activities by joint agreement. Most of the Mexican veterinarians were not local to the area in which we worked. In many towns to which we were assigned the Mexicans were almost as foreign as we Americans were. We were forced to be together in these small towns as we only had a single government vehicle assigned to us. In the small towns there was not much to do. We would eat together and usually rented rooms fairly close together also. We would usually meet each other for breakfast in some local restaurant and while eating agree on what we were to do that day.

As I mentioned earlier, that after my first hip surgery I was unable to drive a car with a standard shift. The vehicles we were assigned were Second World War type jeeps or four-wheel drive trucks all with standard transmissions. I was unable to drive these vehicles. What made it possible for me to carry out my duties was a result of the way the agreement between the US and Mexican governments was written. Neither of us the Mexican or the American veterinarian were permitted to do our work alone. We were required to travel together at all times while working on the program. While I never discussed my inability to drive the vehicles we were assigned I am sure my Mexican partners figured it out quite quickly and did all the driving. None of them ever complained to me or in any way mentioned their awareness that I couldn't drive. None of them ever said anything to my supervisors either. They just seemed to love to drive and I think were just very kind and considerate people. I of course did all of the other things that were required and I was the one to hop out of the truck and open and close the gates when we entered the ranches or went from pasture to pasture while supervising our Cowboys and examining the cattle. I also was the one working the winch to pull us out of mud holes in the rainy season and of course I examined the mouths and feet of the cattle when necessary. I also rode horses when we had to go to ranches where the roads were inadequate for the truck to pass. In the mountains some of the rides were spectacular and I really enjoyed them. And speaking of enjoyment I want again to say how much I enjoyed working with the Mexican personnel and how much I appreciated their consideration and help to me.

Thoughts about Academia

In the following section I will describe a few experiences that provoked thoughts in me that played a role in my developing desire to enter the academic world. They are randomly remembered from my time of employment by the Federal Government first briefly as a meat inspector in Philadelphia, then as a livestock disease control officer working to eradicate foot and mouth disease from Mexico and later in Maryland on the tuberculosis and brucellosis control program.

I will start off with a few words about my early political views. They will help you to understand some aspects of my behavior. If you wonder where these views came from remember my mother's father mentioned earlier. As a young man I believed strongly that democratic socialism was the most humane form of government and I also believed that people who worked in the civil service were there in part because they felt that their work in that capacity would help bring about a better society. My experiences in the civil service brought about some modifications in my views on the subject, modifications that could perhaps be best considered to be maturation of my thoughts. What I learned from my experiences during that period was that civil servants were people with the same goals, hopes, fears, and insecurities at work in them as in all other people. I began to realize that very few of them even thought about democratic socialism or any other ideology.

From my experience as a meat inspector I realized that some very necessary jobs had to be done even though to me at least, they were quite unpleasant. In fact I worked as a meat inspector for a short time only and only because I wanted to work on another type of job and I had to wait for that job to open up. The people running the branch of the civil service that employed me wanted me to do some type of work as I was being paid.

During the short period that I worked as a meat inspector I met many people who are making a career in that field. On those rare occasions when they talked about their employment they would say that it was a steady, secure, job with reasonable benefits, reasonable pay, and with a decent pension plan and medical insurance. I never heard anyone talk about socialist ideals or other abstract concepts and that certainly caused me to think critically about many aspects of my own beliefs.

When I did finally get to the job in Mexico my views of the civil service and people in it continued to develop.

I was rather surprised to find out that almost all of the veterinarians in the program in Mexico in which I worked were graduates of the veterinary schools at Texas A and M and Auburn Alabama. I don't know how that happened. I was moreover the only graduate of the veterinary school at the University of Pennsylvania in the group. I became friends with quite a few of these people but I always felt a bit like the odd man out as there was a cliquishness based on shared Southern culture which played a role in many decisions made by them and particularly by the man in the central office. For example it seemed to me that the veterinarians who played cards and drank with him were usually assigned to the nicer districts and ones near to the headquarters town. And here also I noticed as I became acquainted with the people in the group that ideology seemed to not play any role in their decision to join the civil service. What they talked about was their careers, their pay scales, security of employment and etc. They were in general good people and they did their job fairly well, idealistic thoughts just didn't seem to be important to them. This seems strange to me as the concepts given to them by the Southern culture they shared obviously did affect their behavior just as I realized my culture affected my behavior.

I will jump ahead here to extend this discussion of views held by people in the civil service to my later observations on why I decided to join the academic world. In academia I thought that idealistic thoughts played a larger role in career choices than I observed in actual practice. The concept that the business world should be applied to universities was strong particularly among the administrative personnel. Idealistic views of the importance of teaching and increasing knowledge by research played a lesser role in the lives of faculty members than I had anticipated. I should note that I felt this more strongly at Ohio State University than at the University of Illinois.

I do wish to add that none of what I have said here should be taken as a condemnation of the people in the two types of organizations in which I worked. It should rather be interpreted as my gaining a better view of reality. I worked to try to act by my ideals, it was just that learning how people and institutions function made me more able to function better myself. To survive you must learn to adjust and not give up.

An event that occurred in my second year in Mexico gave me additional information on the nature of civil servants. Early in my second year, several months after what turned out to be the last infection had occurred, and I may add in which I led the eradication team, I was told by the supervisor of the program that there was to be a reduction in the staff and that I would be sent back to US and offered a position there. I accepted that and started to make arrangements for leaving Mexico. Several days later I received another call from my supervisor telling me I could continue for the second year of the program. I was not told why there was a change but that I should consider myself lucky to be kept on as he knew I wanted to stay. I replied that yes I would like to continue but as I had already made arrangements to return to the US I would therefore only stay if I received a promotion of one grade in the civil service ranking. He then gave me a spiel that I should not make any conditions but consider myself lucky and take the offer. I replied that if I got no promotion I would not stay. I was then told he would have to contact the central office in Washington and he could not say what the outcome would be. I then hung up. Within not more than 5 min. he called me back and said it was all approved, I would get the promotion and could stay in Mexico.

I have no proof of what I will say next but I am sure that there was no way he could have gotten that type of approval from the central office in Washington that fast, so I believe he already had been told to keep me on and wanted to brag that he had saved money for the government. I had never heard from anyone else that there was to be a staff reduction at that time. I knew I was the odd man out in a group of Southerners and I suspect that he had a friend he wanted to bring aboard and to do that someone had to go. Someone in Washington told him they would not be prepared to transfer out a satisfactory employee and spend money to bring in a new recruit and that he should forget the whole thing. So I stayed in Mexico for a second year, but my view of the morals of some civil servants fell quite a bit.

There was another experience during my stay in Mexico that lowered my respect for civil service employees while at the same time helped me to understand the workings of the civil service. I had read a short novel called: *Dead Souls* by the Russian author Gogol sometime earlier. I didn't really understand the novel entirely at the time I read it but the events that I observed while working as a civil servant in Mexico gave me a basis for understanding it better. Bear with me as I explain how that occurred. Sometime during my first year, when the program was pretty much fully operational a new man appeared on the staff. He was a veterinarian, the only one other than me who was a graduate of the veterinary school at the University of Pennsylvania. He was assigned to the program by someone in the Washington office. He was introduced as a relief man, which meant that he was never given a regular assignment in any district, but was temporarily assigned to fill in for anyone on vacation, or ill, or where temporary help was needed. As a result he had the opportunity to observe all aspects of the program. He was treated with some deference by the local supervisor of the program and his staff. Rumors circulated that he was really an investigator who was sent to report to Washington about the field program.

He and I became fairly friendly and we would meet for dinner whenever he was posted to a district near where I was stationed. When the program closed he stayed with the veterinary disease control division of the Department of Agriculture. I met him some years later when he came to give a seminar at the school of veterinary medicine at the Ohio State University. At that time he told me he had been working with US programs in various countries.

I never knew if the rumors that he was an investigator were true. I never had any reason to believe he was anything but the relief man he was introduced as. He could of course have been both.

The Gogol novella centered on the activities of a man who appeared in a town in rural Russia. He went in a rich carriage to the various landed estates there asking about the status of the serfs on the estates. He was particularly interested in the serfs who had recently died and offered to purchase the death certificates of those dead.

The officials of the town, who were all civil servants, appointed by the central government and my colleagues were very much concerned about this man who acted much like an official. They thought he could be an investigator sent to investigate them. In their meetings to discuss him and his actions they couldn't decide if he should be arrested or whether he would be the one who came to arrest them. It turned out he was not an official but a confidence man. His scam depended on a quirk in a program designed to subsidize the owners of the landed estates. Under this program the landowners were given a subsidy each year for each serf on their estates. The flaw in the program was that the census of the serfs only occurred every 10 years so that if a serf died during the 10 years before the census and the landholder did not send in the death certificate they continued to collect the subsidy until the next census revealed the death. The landowners, who were always short of cash, were glad to sign a paper saying they had sold the dead serfs and given the death certificates to the buyer for immediate cash. The holder of the bill of sale would then file the necessary papers in the capital and collect the subsidy until the next census.

What struck me when I observed the behavior of my supervisor and his aides in dealing with the new man in our group was the similarity of their behavior to that of the civil servants in the Gogol novel. The rumors about the man described in the novel by Gogol were also quite similar to those about the new man who joined our group in Mexico.

It is true that some aspects of the workings of the Russian civil service described in the novel were quite different from those of the program I worked in, but the responses of people in the head office to the problem of how to deal with the new man and the nature of the rumors that spread in both cases were very similar.

When I had read the novel I had had no experience working in any civil service position and I really had no basis for judging the story, thus I thought the story pure fiction. What I observed while working as an employee of an American bureaucracy brought to my attention the similarities between what I observed in my job and what was described in the novel. I began to feel that the workings of the present day American bureaucracy and that of the Russian one in the 19th century were not all that different. I came to believe that Gogol's story was not pure fiction.

I may not have had a very favorable view of some aspects of human behavior observed in my job and described in the novel but the dawning belief I had that people in the same fields are quite similar in their responses to similar events and that human behavior has not changed much with the passage of time was strongly reinforced.

And Now to another Story

The Eastern shore region of Maryland was somewhat Southern at the time. In 1956, it had two county agents in each county, a black one to work with the black farmers and the white one to work with the white farmers. When I was first there I went out with these men to meet the farmers, to explain the programs and sign up those farmers who were not already in the program. When it was lunch time on days I worked with the black county agent we had a problem. He couldn't go into the white restaurants, so we went into the black restaurants. It created a bit of a stir until they got used to seeing me there.

This brings up the opportunity to describe two experiences I had with local people, one in Mexico and one in Maryland. They illustrate how relations with the local people affect your ability to carry out your job - In my case as an animal disease control official.

I will start with the experience in Maryland. I went out one day to determine the status of a particular farmer's cattle. I got lost and decided to ask someone for directions to the farm. I saw a black man working in a field by the side of the road. I pulled to the side of the road and called to him and proceeded to ask directions. He pulled off his hat and held it by his chest and said no sir I don't know sir where that man lives. He then looked up and recognized who I was. He put his hat back on, adopted a different accent, and proceeded to give me exact directions to the farm in question.

The experience in Mexico was potentially more serious but it tells the same story about the role of personal relations in determining one's ability to carry out one's job. At the time I was assigned to a small town called Misantla as my base of operation. Misantla was off the main road and had only one road leading into it. I had used my own car to go to a larger town which had a very nice French style restaurant. It was quite late when I returned. I had almost reached the town when I saw a barrier blocking the road. I stopped and was immediately surrounded by a group of armed men. I rolled down my window when the leader of the group tapped on the glass. He looked in, recognized me, I didn't recognize him, and said oh it's you Dr. Kreier don't worry there is no problem for you. He then signaled to one of the men to remove the barrier, wished me good night and waved me through. I don't know who they were looking for, but whoever he was he certainly would not have received such a pleasant greeting.

To conclude this section describing the role of random thoughts and experiences on my career choices I will tell one more story. While I was stationed in Maryland some new rules were formulated which modified the role of the publicly employed veterinarians in the tuberculosis and brucellosis control program. These new rules were designed to bring the private veterinary practitioners into the control program. It took me some time to realize how these rules would affect the program and my role in it.

I will start this section with a bit of background about the program I joined in Maryland. After I transferred to Maryland to join the tuberculosis and brucellosis control program I spent time getting acquainted with the two counties to which I was assigned. I visited the farmers who had cattle to determine their status in the program. Largely in this area they were dairy cattle. When the animals had already been tested I explained the program and its goals to the farmers before I left. If the animals had not been tested within the previous year, I made appointments to test the animals and spent some time explaining the nature and goals of the program and the consequences of positive tests.

Most of the animals I encountered which had not been tested were on the poorer farms with just a few animals and in many cases they were let loose in the pasture most of the time. At the same time I began to get papers from the central office that cattle on various farms had tested positive for tuberculosis or brucellosis and that I was to go to the designated farms to inform the farmers of these results.

As I continued to work on the program and to study the new rules I began to understand that the new rules were largely responsible for the obvious failure to test the small herds and for my receiving the notices of infected animals in herds I had not tested myself. The failure to have the small herds tested was a direct result of how the local practitioners were paid to test cattle in the program. They were paid a relatively small sum for each herd they signed up for the program but they were also paid a fee for each individual animal tested. Animals in large well managed herds could be tested rapidly and easily and returns for these tests were quite good. The returns for testing small herds were poor. The herd stop fee was small and testing of a few often semi-wild animals took time and did not yield much money.

When I began to act on the notices of infected animals in herds I had not signed up or tested I got some real surprises. Very often the farmer would not seem to know that his animals had even been tested. When I showed them the paper with their signature on it, stating that they had agreed to have their animals tested for tuberculosis and brucellosis by their regular veterinarians on a specific date they would say oh yes they remembered that their veterinarian had been there on the date noted and that he told them they had to have their animals tested, but he said not to worry about it and he would take care of the whole thing. At that point I had to spend a fair amount of time explaining to the farmers the purpose of the program, how it worked, and what the consequences of positive tests were. I then had to mark the reactor animals with a cold brand showing a T or B. I then explained to them that the branded animal was to be turned over promptly to a dealer who would take it to a federally inspected slaughterhouse where a final decision would be made as to the deposition of the carcass. There was a federal compensation paid for the condemned animals. I however don't remember the details of how it was worked out, but I do remember that I had to determine that also.

I will note here that as the veterinarian was paid only a small fee for signing up the farmer in the program, but a much better one for testing the cattle based on the number of animals tested. Thus the veterinarian had little incentive to spend time testing small herds or explaining the goals of the program and the consequences of positive tests. Under the rules that brought the private practitioner into the program the practitioner also did not have to deal with the notification of the farmers of the results of the test. They were thus spared the unpleasant task of dealing with the upset farmer who owned the condemned animals. That was left to the civil servant also.

The program to eliminate brucellosis was, at the time, fairly new. Most of the animals that tested positive were actually infected. In addition most of the farmers had experience of losing calves as result of abortion caused by *Brucella abortis* or at least of seeing the disease in nearby herds. They therefore took these results fairly well. The program to control tuberculosis in cattle on the other hand was an old well-established program already at the time I started working on it. As result the actual incidence of infection was quite low. Despite this there was a low but consistent incidence of positive tests. When the results came back from the inspection at the slaughterhouse they usually reported no visible lesions. To explain these results to the farmer was a part of the unpleasant job reserved for the publicly employed veterinarians. It was not easy. To start off the discussion it must be said that there is no diagnostic test that does not have some false positive and false negative results. When the disease incidence is high it is easy to accept the small numbers of false positive tests and they are not even discussed in many cases. When the incidence of infection is very low and almost all of the reactors are false positives it becomes a real problem and examinations are required. As bovine tuberculosis still exists at a very low level if animals are not tested regularly the disease could spread thus testing must continue. This concept is difficult for the farmers to accept. Scientists are actively looking for better tests than the tuberculin test for tuberculosis but so far have not found one. Some of the false positive tests are not really false in one sense. Chicken tuberculosis for example does not cause disease in cattle or healthy humans but it does cause sensitization to the test for bovine tuberculosis. The human strain of tuberculosis does not cause disease in cattle but does sensitize them to the test. These facts provide good reasons to tell humans not to spit in cattle barns and for keeping chickens confined but not everyone wants to hear that. I should add that bovine tuberculosis causes a severe disease in humans.

But I have gotten off the track again, this chapter is supposed to explain how and why I ended up in academia. In a sense all that I have been describing has played a role in my decision to leave the civil service. For example I resented having the public actions belittled and bringing in the private sector even though the private sector only creamed off the profitable components. The time-consuming activity of explaining the nature of the program, why it was necessary and the consequences of testing were also poorly done or not even done at all by the private practitioners mainly because they couldn't make much money doing them. The practitioners also rejected the unpleasant duties, such as notifying the farmer of positive tests leaving that necessary but not profitable and disagreeable action to the publicly employed veterinarian.

When the program in Mexico ended because it was successful I was offered several positions in the United States. I chose a position as a field veterinarian in Maryland in the Tuberculosis Brucellosis Control Program.

I chose this course of action as it was the easiest course to follow and would provide me with an income while I decided what to do. In addition to uncertainties about my career choices I had decided to propose marriage to my longtime friend, Ruth, and I was not sure what her reply would be. If it were yes I felt she should have a role in choosing a joint future for us. She in fact did say yes and shortly thereafter we moved to Maryland. The year we spent in Maryland was when we actually made the final decision to look for a job in academia. As I noted previously there were many factors which caused me to undertake making the change and they were drawn from a hodgepodge of experiences and conditions in my work. Overwhelmingly however two things finally determined the decision to leave the civil service and to join academia. They were my problem with my left hip bringing about a desire to enter a field and my desire to enter a field with lower physical demands and with a greater intellectual component. I will discuss this matter more fully in the next section.

Factors Influencing My Decision to Become an Academic

I realized at the time that my left hip was again beginning to degenerate. It wasn't yet so bad that I had trouble carrying out my duties but I thought I should start looking for less physically demanding job well before I had a serious problem. My problem with not being able to drive the fleet car kept the issue alive. I explained to my supervisor my problem and he said I could drive my own car with an automatic shift and turn in a mileage record for when I used it in my work. But at that time there were no laws defining special consideration for people with physical problems and the financial office in Washington repeatedly sent me letters questioning why I was not using a fleet car. I would show them to my supervisor and he would say he would deal with it and he always did but it was nevertheless disturbing to me.

As I was almost 30, I thought I should not delay in starting my search for an academic job. I started to seriously search for a less physically demanding and more intellectually challenging work sometime after I had been in Maryland about six months. I had by then become fairly familiar with the area for which I was responsible. I had started bleeding cattle to obtain blood samples to be sent to the laboratory and tested for brucellosis. I had also started doing tuberculin tests. This work was quite routine and required little thought. Once I became proficient at bleeding the cattle and skilled at doing the tuberculin test I began to doubt that I would wish to do these tests for the rest of my life.

These kinds of thoughts were suppressed in Mexico as the program there had been an exciting new world to me. Every day there were new experiences. Rural Mexico and the Mexican people were very different from anything I, a middleclass American from the northern city of Philadelphia, had ever experienced. The goals of the program also excited me and every day yielded new exciting experiences. There were some negative factors, I for example did not really like the bureaucratic organization and the somewhat arbitrary decisions by my supervisors even though I recognized that the type of program we were engaged in required the almost military organization used if we were to achieve our goal of eradication of the disease.

The domination of the group by Southerners also made me somewhat uncomfortable. This bothered some of the Mexican personnel also who commented to me about the race views of some of the Southerners in the program. My inability to drive the fleet vehicles was also always a worry to me even though the unspoken understanding of my Mexican partners prevented it from being a real problem it still caused me to worry. In the final analysis the excitement of being in Mexico, working in a foreign country in a difficult situation, and doing demanding work, kept me from boredom and left me with little time for thoughts about the future. After I left Mexico and started working in Maryland much of the excitement was gone and thoughts of a less physically demanding and more intellectually stimulating work came to the fore.

I discussed the matter with Ruth who had said yes and was now my wife. She was sympathetic and said she would like to move to a university town and undertake a study program leading to a career also.

As I already had a degree in veterinary medicine and some experience in the profession I decided to limit my search to an academic position in a veterinary college. With the aid of my wife who had a degree in English we composed a letter and *Curriculum vitae* to be sent to the 18 veterinary schools than in existence in the USA. The college of veterinary medicine at the University of Illinois was the only one to reply to my letter with an offer. The position I was offered was as a research associate and included the opportunity to work for a graduate degree. I was to be paid roughly what I was earning as a civil servant. The position was in the Department of Physiology and the funding came from a research grant from the U.S. Army to a professor in that department. The funding was for several years and the research was to determine if radiation, as a means of preservation of meat and flour, caused toxicity in the irradiated product.

I accepted the position offered and resigned from the civil service. My wife and I packed our things and we started our drive to Champaign-Urbana in central Illinois.

In passing I will mention that Ruth was pregnant at that time, an event which occurred shortly after our marriage despite her earlier insistence that she wanted to wait to have children until she'd gotten her career started. The event like most of my previous experiences was not planned. I was not upset at the time nor was I upset when a second pregnancy followed shortly after the first birth. It wasn't planned either. Looking back I am happy it all happened as it did for I now know you need to be young and strong to raise children. Now in my old age, I am glad I have two young families to which I am connected. Having children was another unplanned event which like most of the other unplanned events affecting my life was good.

When we arrived in Urbana, Illinois, we rented an apartment and spent a few days settling in. Then I went to the school of veterinary medicine to meet the people with whom I would work and study with. The professor who held the grant introduced me to two other veterinarians who were hired as I was to carry out the project. He then briefly described the nature of the project. In the main part of the project the experimental animals were to be beagle dogs. I was told I would work on the part of the project using the dogs. There was also to be a section of the project using rats but I was not involved in that section. The project I was involved in would start with 24 puppies and would continue for three years. The dogs were to be divided into two groups, one fed a diet containing irradiated beef and flour the other an identical diet but with beef and flour that had not been irradiated. I was not involved in the preparation of the diets nor was I told which dogs were in which group. I was told I would be responsible for monitoring the blood of the dogs. This included total red cell and white cell counts, percent volume of cells in the blood, differential counts of the white cells and the determination of the percent of immature red cells in the blood. The latter determinations were made by microscopic examination of thin blood films stained by the Giemsa technique. Sometime later I undertook in valuation of bone marrow obtained by needle puncture of bone in the hip.

I don't know why I was chosen for the hematologic evaluation of the dogs as I had no more training in that field than what I had received as a student in veterinary school. Perhaps my experience handling animals and bleeding them during my year in Maryland and my time in Mexico played role in that decision.

I don't remember clearly, but I think I'd bled all of the dogs monthly starting on the day the 24 little wriggling puppies arrived and just after we put ear tags in to identify them. I chose to bleed the dogs from the jugular vein which involved my first teaching experience, teaching a veterinary student how to hold the dogs while I bleed them.

It was a busy time as I had to learn all of the techniques I would use and establish the record-keeping system to keep the data I would collect so that I could prepare regular reports of my observations. I also enrolled in the Masters Degree program of the college and helped my wife to get us settled in. She needed some help as the unplanned pregnancy was advancing rapidly.

We started to look for a place to live. The apartment we had rented was very small and we didn't feel it would be adequate when the child arrived.

One of the younger faculty members in the school told me they had a small house near the college that they wanted to sell as they wanted to move to a larger house. It was a small one-story house in a group of houses built just after the Second World War to house returning soldiers. It was built on a concrete slab. Central heating was by a gas furnace in the small room in central part of the house. Heat was blown into the living room and to heat the bedrooms, you had, it turned out, to keep the doors open. The house it turned out, as we discovered that first winter, was poorly insulated. When winter came it came furiously with a strong wind from the Northwest which explained why all the trees leaned eastward. Urbana-Champaign had hot muggy summers and very cold windy winters. It was great for corn and soybeans as the central region of Illinois had some of the best soil in the world. For people like myself, on the other hand, it had a marvelous University.

The classes I took in my master's program were mostly in the colleges of arts and sciences. One of the best courses I ever took was one in virology offered by Salvador Luria, the only Nobel Prize winner I ever had the pleasure of meeting. When I much later was a faculty member at the Ohio State University it often surprised me when the young virologists we interviewed for jobs would talk about the exciting new concepts they just discovered. I remembered that Dr. Luria had covered most of these concepts in his beginning virology course, many years earlier.

I often had thought I would want to enter the field of infectious diseases as that is what I worked on previously. However as the only clear offer I received from an academic institution was not in that field I didn't feel strong enough to continue looking although Dr. Luria's course did stir some thoughts about what I would do when I finished my Masters Degree in Physiology.

When both my Masters program and the study of toxicity of irradiated foods was very near completion I started looking for a place where I could study infectious diseases and virology. A faculty member in the Department of Pathology and Hygiene told me that the University of Wisconsin had a strong program on infectious diseases. He had some information about that program as his brother was a professor of virology at University of Wisconsin. I wrote to the people at the University of Wisconsin and was told I could join them when I had completed my Masters program. As an aside I will note that the irradiated foods were not toxic in any way for the dogs and that I had become fairly comfortable working as a hematologist.

I did learn something more than just hematology while working on the job. I had developed a relationship with two senior faculty members, one a pathologist and one a parasitologist, while seeking help in identification of the leukocytes by their morphology. Knowledge of their morphology is necessary for doing the differential counts on the blood films and later on thin films of the bone marrow. The parasitologist was very self-confident and would make rapid decisions about the cell type. The pathologist on the other hand would study the cell in question carefully and then explain at some length why he thought the cell fit the criteria to warrant its placement in a particular category. He would then discuss reasons why he could be wrong in his classification. At times he would recommend listing the cell is not identifiable. At first all this discussion of what was what confused me and I was impressed by the rapid and certain approach of the other man. As my experience increased I finally concluded that the uncertainty of the pathologist fit what I saw as opposed to the certainty of the other man. This was particularly the case with the cells in the bone marrow which were rapidly changing their morphology as they developed from stem cells to highly differentiated cell types.

As time passed, and my confidence increased, I became somewhat unhappy with the professor with whom I worked and who was also my academic advisor. He was rather dull and had little imagination. He was quite suited for a long-term feeding study, collecting data carefully and regularly. He didn't seem to want more than that.

Sometime after I had received an offer from Wisconsin and before I had formally accepted it the chairman of the pathology and hygiene department walked into my office and said in a cheery voice I hear you are not too happy with your advisor and are going to leave us and go to Wisconsin to study virology. We have a virologist in our department and if you want to you can join our department. I replied let me discuss it with my wife and I'll tell you tomorrow what we decide. We had settled into life in Champaign-Urbana, our little girl was in daycare and my wife was enrolled in the Masters of social work program and had a job at the University teaching their football players to read and write. In short we decided to change departments and stay in Illinois.

We have now arrived at the next-to-last stage of my search for an intellectual career. I may add the search included getting a PhD because with rare exceptions you don't get anywhere in academia without one.

The decision to stay in Illinois after I got my Masters degree turned out to be a good one but it didn't get off to a good start. After I moved to the new department I was introduced to the man who was to be my advisor. He had gotten a grant to study and hopefully produce a vaccine for prevention of shipping fever of cattle. This disease is assumed to be a viral disease that causes a serious sickness in cattle shipped from ranches and farms to feed lots were they are to be fattened before slaughter. It also develops in other cattle following long and stressful travel for any reason. I never got any real grasp of the project during the less than a year that I worked with him. I however started taking courses for my PhD program and also going to farms with him to administer vaccines. He however kept the vaccine production in his own hands and told me very little about what was actually being done. One day he called me into his office and said he had taken a position at the University of Colorado and was leaving Illinois almost immediately. I soon found myself without an advisor and helping a junior faculty member trying ineffectively to keep the research project going. I was kept on as a research associate but spent most of my time taking courses required for the PhD program and wondering what would happen next. In several of the courses I took I became friendly with a somewhat older man, also a veterinarian, who was a professor in the veterinary college at the University of Florida. He said he was taking a leave of absence from his job to get a PhD. We became laboratory partners in a biochemistry course and in general got along quite well. At the end of the year he went back to Florida and I assumed I would never see him again. Sometime during the following summer he came into my office and announced he had completed his PhD program and had been hired as a full professor in the college of veterinary medicine at Illinois with full authorization to advise graduate students. He then said to me he was bringing several large grants to Illinois and would need help in carrying them out and that as we had gotten along pretty well as fellow graduate students he would like me to join him and would serve as my advisor. He then told me that one of his projects was a study of *Vibrio fetus* and the other of *Anaplasma marginali*, a parasite of the blood of cattle. It is tick-borne and semitropical in occurrence. Even though it all seemed a bit strange I decided then and there to go for it. I had spent a lot of time learning hematology so I said I would like to work on the research project concerning the blood parasite *Anaplasma marginali*.

As time passed I learned a bit about how he was brought to Illinois. It is an interesting story that tells a lot about how universities actually work. It isn't always just like they say in the literature you get when you apply for a degree program or a job at a University. I do believe it is a story that belongs in a book for young people considering an academic career. The key to understanding it is realizing that the people running the college will bend the rules as much as possible to recruit people they want to implement their efforts to develop their programs and bring in research grants and projects to enhance their units reputation.

A short time before he showed up here to get a PhD we had gotten a new Dean of the veterinary college. He was an ambitious man and I believe was recruited to improve the scientific level at the veterinary school. This man knew of Dr. Rustic's achievements and his record of getting grants to finance his projects. The University of Illinois had a rule that all faculty must have a PhD. Apparently the University of Florida at Gainesville had no such rule.

Dr. Ristic was a Serbian national who had spent the bulk of the time during the war in a German prison camp. He and several other Serbian men escaped the camp as the Western Allies approached the area. They were being marched to a new camp when allied planes spotted the lines of marching men and strafed them thinking they were German troops. During the strafing everyone scattered and jumped into the scrub along the side of the road. When the planes left Ristic and several other Serbian man found they were huddled together in the ditch along with a German soldier who had been their guard. Ristic who spoke fluent German, persuaded the guard to unload his rifle and pretend he was escorting the prisoners back to the camp but actually they walked toward the Allied lines. The next morning they encountered a column of Canadian troops. The German guard surrendered and Ristic, who also spoke English, spoke to the officer in charge of the Canadian troops. The officer decided that here was what they needed, a multilingual translator and recruited him into his unit. He spent the remaining months of the war with the Canadians and was rewarded with a stipend after he was discharged.

After the war Ristic decided to stay in Germany at least long enough to get a professional degree. He applied and got admitted to a dental school but didn't like it in part because the practical work was on preserved heads of people who had been killed during the war. He then transferred to a veterinary college. After he completed his study, he emigrated to the United States where he got a job at the University of Florida veterinary college. We met as I mentioned earlier when he came to Illinois to obtain a PhD.

My years with him were exciting and stimulating and strangely for a graduate student in a demanding graduate program pleasant. I learned a great deal from him not just science but to the day I retired I treated my graduate students largely as he treated me. He was constantly optimistic and cheerful and constantly spilling out new ideas and discussing projects. We were all encouraged to talk with him about our projects and we discussed all of the projects going on in the laboratory with him almost as equals. One interesting aspect of his leadership was his ability to pick up students who for some reason were in trouble, invite them to his laboratory, and get them back on track. I guess I was one of them. I continued a relationship with him until his death.

I personally am not always optimistic and during my years with him I learned to at least partly overcome my periods of depression or to hide them from my students.

During my period as a doctoral student and actually for the rest of my academic career I expanded my work with blood parasites. I started projects on *Babesia* and later on its close relative *Plasmodium*, the parasite causing malaria. As my projects were largely on tropical diseases of animals including man I attracted many students from tropical countries including Africa, South America and India. I was funded by the Army malaria project and on programs concerning Trypanosomes and Leishmamia.

The information gained during my years learning hematology in the physiology department became a major component of my research on infectious diseases. This was another example of chance events turning out well.

I will now explain how I got to Ohio State University and how I ended up in a microbiology department in the college of arts and sciences

After I completed my PhD program with Dr. Ristic I continued to work in his laboratory and was told I could continue with him as a postdoctoral fellow. It was a tempting prospect but I wanted to get a regular academic position and start teaching and developing a laboratory on my own. I saw an advertisement in a Journal saying there was a position in the Ohio State University Department of microbiology for a veterinarian with a PhD who would teach veterinary students in the veterinary college but be actually employed in the department of Microbiology in the college of arts and sciences. It was an unusual arrangement and it was unique to the Ohio State University. I later found out that a similar arrangement existed with the medical and dental schools. I sent my vitae to the department of microbiology with a letter expressing my interest in the job. Some time passed and I received no answer until one day a professor from the Ohio State Department of microbiology came to my office, introduced himself, and said, he was visiting his son, who was a professor of zoology at Illinois and had decided to look me up while visiting his son. He inquired if I was still interested in a job at the Ohio State University. I replied yes and we talked for some time. He then asked if he could use my phone to call his department chairman. He made the call and I heard him ask if the chairman still had my vitae and application. Some time passed, he then hung up and turned to me and asked if I had another copy of my vitae. I gave it to him and he said he would call me later to set up an interview for the job.

A few days later he called and we selected a date and time for the interview. My wife and I both went to the interview. We were introduced to a member of the veterinary school faculty who turned out to know my wife from social connections in Philadelphia. I never mentioned it earlier but my wife's family were members of the Jewish community in Philadelphia and I met her through my best friend Harold who also belonged to that community. The children of that community with whom I associated were largely students who went to schools I attended and I became attached to them as we seemed compatible. We were mostly unaffiliated to the religions of our parents and were trying to become intellectuals as many young people do. In reality these connections didn't directly shape my career choices but my desire to enter an intellectual environment did push me to strive to work in a university.

But I am off the track again. The bulk of people I met during the interview were faculty of the department of microbiology in the college of arts and sciences. They were not veterinarians they were however members of the department that actually was to employ me and set my pay levels and determined my promotions although I was to be stationed in the veterinary college and teach microbiology and immunology to veterinary students.

I learned during the interview that the person who had had the job for which I was interviewing felt strongly that a second person should be hired to join him in the teaching of the veterinary students. He was the one who had placed the ad. When the department didn't cooperate in this he took a job at another institution and resigned. The faculty then realized they would have no one to teach the veterinary students in the coming autumn semester. Dr. Stahley who was the man who visited my office in Illinois took it upon himself to take steps to resolve the problem. He did this because the chairman, who was near retirement, just didn't seem to be able to act on the matter.

I learned later that the college of arts and sciences was in some turmoil at the time and was trying to decide whether the microbiology department should go to the medical school or whether they should create a division of biological sciences to include microbiology and several departments including zoology, botany, biochemistry, and entomology which at the time were in the college of agriculture

It took several years for these issues to be resolved. When I was offered and accepted the position I had no idea of how the reorganization would affect my career. In fact it never occurred to me that I should even worry about the matter.

It was late spring when I accepted the position. I was pressured to come to Ohio as soon as possible so that I should be able to prepare for my teaching and receive help in getting started. I arrived in Columbus in early summer. I was shown the very nice office and laboratory and classrooms in the veterinary college where I was to work. As the veterinary college was on the west side of the campus and the microbiology department in the central campus there was quite a distance between the two units. I was told to get ready to teach in the fall. When I tried to obtain course outlines from the faculty of my department no one seemed to know anything. My chairman was in India for the summer and the whole department seemed fairly inactive. I found out where my predecessor was working and then called him and asked if he could provide me with course outlines for the courses he had taught. He told me he had no course outlines and could not help me at all. He appeared quite angry with the department and didn't want to help me or anyone else hired to replace him. I continued to search for information about what I was to teach and finally encountered a young man, a veterinarian who was a graduate student with my predecessor and who had been teaching assistant for the courses my predecessor had taught. He gave me copies of the course outlines and course descriptions. They provided me with what I needed to prepare myself for the courses I would teach beginning in the fall. He was a mature young man who had almost finished his degree program having only some research needed to complete his PhD thesis. My predecessor continued as his advisor and the student chose to remain at the Ohio

State University. I told him he could continue to work in what had become my laboratory and offered him help with whatever administrative problems he, as a result of having an absent advisor encountered. He remained in my laboratory for about a year until he completed his PhD. He was always pleasant and helpful when I requested information about the courses I taught. Interestingly I never again spoke to his advisor, my predecessor, nor was I ever asked by his advisor to serve on committees to guide or evaluate his research.

These first years were very demanding. I had to prepare my lectures and develop laboratories associated with each course. I wrote grant applications, started my own research programs, and recruited several graduate students to work with me. When I obtained grant money my wife joined me doing the literature research for my projects and helping me to write grant applications, research reports, and papers for publication. She continued doing this until she started work at the Ohio State University for a Masters Degree in Educational Psychology.

When she completed the degree she went to work in the public school system as a school psychologist evaluating students about any problems they had and placing them in special programs available to them for help.

My work in the veterinary college proceeded quite well. I obtained research funds, attracted graduate students and developed the courses I taught in veterinary microbiology and immunology. The only problem arose from my providing service to one academic unit, the veterinary college, while being employed in another, the microbiology department of the college of arts and sciences. If I needed help with anything and asked my chairman he would tell me to speak to the Dean of the veterinary college who would tell me to speak to the microbiology chairman. It wasn't a serious problem as I had grant money to support my graduate students and research and a teaching assistant to help in running the laboratories attached to the courses I taught. Everything went well. I was happy with my job and I just didn't ask for much from either group.

The first hint that there might be a problem came several years later when I heard that the medical school was setting up a Department of medical microbiology. I did nothing about it but did anticipate that it was possible that the veterinary college would soon do the same. At the time I was still an assistant professor and had not yet been granted tenure, an action that usually occurs when one is promoted to associate professor. The evaluation leading to promotion and the granting of tenure occurs at the end of your probation, which period usually lasts about five years. If the faculty committee evaluating you does not approve your promotion you are notified that within about one year that your employment at the University will end. I bring this up because my status was a major factor in the decision I soon had to make.

The Dean of the veterinary college was quite a decent man but his chief assistant, who was the chairman of the veterinary pathology department was not much liked by the bulk of the veterinary faculty in part because he used his relationship with the Dean to build up his department and himself at the expense of the other departments and faculty. The Dean did control his excesses in that respect but as he was an energetic and intelligent man the Dean depended on him extensively and thus did not want to lose his services. When the veterinary college Dean retired the president of the University chose the assistant Dean to replace him. The bulk of the veterinary faculty did not want him to be Dean and wrote a letter opposing his appointment and calling for an outside search for the old Dean's replacement. Most of the veterinary faculty signed the letter and I was asked to sign it also, which I did. The university president appointed the man Dean ignoring the faculty's advice and even gave the man a copy of the letter. After he was appointed the new Dean called for a college wide meeting in which he showed his copy of the letter given to him by the president, but otherwise nothing much happened right away. One thing however did occur soon thereafter. He came to my office and said he wished to create a veterinary microbiology department and asked me to help him but not discuss the matter with anyone. He said I would be head of his team, which is a nonexistent academic title. He also said he would take care of me which I doubted. I replied I couldn't do that as I was employed by the microbiology department and they would consider such actions to be a betrayal. He then abruptly walked out of my office. I was perplexed for a few moments and then decided to call my chairman, tell him what had occurred and ask for his advice. He was after all, the man who would play a major role in my promotion and tenure. After listening to me, my chairman, replied by asking a question. He asked what I wanted to do. I replied that I didn't really trust the Dean of the veterinary college and as I was happy with how things were going I thought it would be best to stay with the department of which I was a member. He then said he had to ask the rest of the faculty of the department and get their opinions. He added that he would let me know the results the next day. When he called back he said the faculty was agreeable and that as there was an empty office and laboratory available I could move in at my convenience. As this all occurred during the summer break I was free to move as I was not teaching courses and just working with my graduate students doing research. I asked the man who was taking care of my research animals, largely rats and chickens at the time, if he could help me move. The next day he brought his pickup truck and we transferred all the equipment belonging to the microbiology department and my research materials to the microbiology building. This was the last move shaping my career I made and I remained in the department of microbiology until I retired in 1989. In one sense this is a possible point to end the story of how my academic career developed. With the passage of time I became a professor, I taught students, ran a seminar program, did research, published papers describing my research, published books on my subjects of interest in parasitic protozoa, and organized several international conferences on blood inhabiting parasites. Despite this being a possible end of this paper I will nevertheless now tell a few stories to give the reader a feeling about some personal aspects of my life in academia.

In general my life in academia was good although the stories I will now tell often describe things not so good because I want to present reality and not create an unreal picture. After all life is not all perfect and to be happy and survive you must take some aspects of life as they come. My experience as a professor was in medical sciences. A university hires you and tells you about your responsibility to teach students, set up a laboratory and use the laboratory to train students, however they provide limited monetary resources. The professor is expected to apply for external grant funds to support research, a percentage of which they take to cover what they call administrative costs. The constant search for grants to fund your research and support your graduate students was for me the most unpleasant part of being a professor. I enjoyed teaching undergraduate students and working with graduate students who in my case did most of the actual laboratory work under my guidance. I met with my graduate students weekly in a formal way requiring each one to present a complete report each week of what they had done on their projects. The group then discussed each report. I of course would also work with the student when I felt it necessary or they required help. Each student would work with me to develop their project. What they would do was of course shaped by the nature of the grants I received. If one didn't do what you had proposed to the granting agency you had real problems. An unexpected type of teaching that I became aware of as I worked with my graduate students arose as I began evaluating their research reports, and later the drafts of their theses. I found I had to become an English teacher. I had to teach English writing skills and how to prepare coherent reports written in logical form. No one entering the academic world, even those hoping to work in science should not underestimate the importance of being able to speak and write in a clear and logical way. I later found I had to use my skills in expository writing not just to help my graduate students but also in preparing papers written by many of the people who contributed chapters to the books I published and papers from conferences I organized. I was constantly surprised to find that many established scientists sent me drafts of papers that needed extensive editing before they could be published.

I will now discuss a subject, the nature of which affected my career strongly and will in some form surely continue to affect people working in science in the future. I undertook studies of medical science at a time when the study of disease processes consisted of attempts to unravel the gross histologic changes, and the nature of the immunological processes developing as the disease progressed. These studies would attempt to explore the nature of the damage and how the damage was brought about. We all know that invading organisms may secrete toxins that destroy tissue, but I worked on the assumption that much of the damage occurring during infections is caused by the immune system itself. My thoughts can be thought of in terms of war and collateral damage. When the immune system is activated it revs up the complement system and antibody production and phagocytosis. These actions destroy parasites but also do damage to body cells particularly body cells to which are attached parasite components. They may also affect tissues existing in proximity to invading parasites. Just as in war the battlefield is devastated, and must be cleaned up and the area rebuilt. Often damage cannot be repaired leaving the host body devastated or even dead.

The nature of the damage caused by the cells involved in the processes of defense and the products produced by the invading organisms, all of which must be dealt with by the body's defenses was a central part of my research. I also studied products of the invading organism as they were important in the development of diagnostic tests and vaccines. In addition study of methods to identify organisms and the use of these characteristics in classification was important to me as I recognized that you have to know what you're working on

All of this type of work was done using gross morphologic characteristics, serological tests, histopathology, and organic chemistry as tools in the research. Animals were infected and the course of the disease produced was studied to define the course of the infection and recovery. The organisms causing the diseases with which I worked were identified using classic criteria. There are problems with these classic criteria but, in general, the systems of study which I used worked and were what was then available. But I have gotten a bit off the track again. It wasn't the nature of what I was trying to do that caused trouble for me but rather how I tried to do it. A whole new technology for examining biological processes was in the process of being developed. I didn't feel that the techniques I was using and the type of work I was doing was no longer of use and I continued working as I had been working. I will explain this further in the next few paragraphs. At the time I was a graduate student at Illinois a whole new set of procedures for the study of biological systems was being developed in the microbiology and biochemistry departments there. These studies became the basis of what is now called molecular biology and molecular genetics. I was close friends with a graduate student working with Solomon Spiegelman. I remember him telling me how RNA copies of DNA left the cell nucleus, entered the cytoplasm, and bound to mitochondrial RNA and then programmed protein production. This work opened a window to understanding how genetic information in the nucleus was able to program what the cell did. This item is only a small sample of the many studies being done there that led to new techniques for study of biological systems which would later sweep through the biological research field.

The research in the veterinary school where I was working on my PhD was concerned with more immediate problems tied to prevention, treatment, and cure of diseases. It used the older more conventional techniques. By the time I was a full professor and nearing retirement age the new molecular biology and genetics had swept through the field of biological research and if you had not adopted those techniques you had trouble getting research funds from granting agencies.

In my own laboratory I was involved in the study of malaria vaccines. I was supported by the Army malaria project and worked using a rodent malaria infecting rats. In one of my projects I worked on immunity to the blood stages of the parasite using a preparation of the blood stage parasites as the vaccine. This crude vaccine contained a complex of all of the components of the blood stage parasites. This vaccine like all the other similar vaccines produced in other laboratories from other malaria species gave a low level of protection to infection. At this time the molecular biologists entered the field. The rationale for funding them was that with their techniques they could produce large volumes of antigen of great purity. These vaccines were produced by yeasts or *E. coli* in which the genes for certain components of the parasite were introduced. This was an important development as the various human malarias essentially grew only in humans and certain apes although culture of the blood phases of *Plasmodium falciperum* had been developed. The molecular biology techniques for producing antigens were very valuable. There were however errors in the basic justification for using the procedure to develop malaria vaccines. The belief that the poor quality of the immunity to malaria vaccines had resulted from the mixture of parasite components in the vaccines was the argument being used by the molecularly oriented people who argued that they could improve the vaccine by providing pure preparations of the appropriate components of the parasite, i.e. those which stimulated immunity and not those that inhibited it. The reality was that the same molecules were responsible for both of the actions. In passing, I should note it has also been shown that in many cases pure preparations of single molecular types do not produce as good immunity as mixtures as it is easier for the parasite to develop resistance to a pure preparation of a singular molecular type then to a mixture of types. In reality of course the immune response must overcome the inhibition of the immune response to be able to protect the host. As the molecules which protect the parasite are the ones that must be overcome to generate immunity you can't separate the immune generating molecules from those inhibiting immunity as they are the same molecule. It was also true that the young molecular biologists were not always knowledgeable about biology in general and in the particular case of malaria vaccines most of them did not accept the fairly well-established concept that if you don't get good immunity from infection you won't get it from a vaccine either.

Molecular biology and molecular genetics has begun to produce some remarkable results but it is simply a new tool and can only be used effectively by people who also know biology in general and where the new molecular biology can produce data that can fit in with the existing body of biological knowledge. Some of the young molecular biology people were a bit arrogant and pushed their activities while failing to appreciate the importance of understanding the nature of the larger biological systems.

I was getting a bit old and didn't really feel like learning a whole lot of new technology. I must here add that I avoided the whole problem by retiring when the university offered faculty members a five-year increase in the number of years used for calculating their retirement payments. I believe the administrators expected to save money by replacing the older faculty with younger ones starting out at lower pay scales and that they would not lose any valuable experienced people by doing it. I must say I remember listening to our newly hired faculty proposing what to them were new ideas that some of the older faculty remembered having tried and decided they were not good. Knowledge of the past stemming from experience does have some value.

I have tried to follow the careers of some of my graduate students. I tried to provide them with a broad knowledge of biology and even passed to them some veterinary knowledge. Most of my PhD students did get good jobs in universities, government laboratories, or pharmaceutical companies. Some of them were able to add new techniques to their repertory and I tell myself that they were able to use these techniques effectively because they had a good grounding in biology.

Retirement

Friends of mine in the department of microbiology knowing my penchant to work with my hands chipped in and bought me a wood turner's lath as a retirement gift. It was a good gift for me and after I retired I started making bowls, lamps, and various types of furniture. A bit later I started to do small sculptures. I don't think it was an accident that I started this by making animals, goats, rabbits, squirrels, and other wild and domestic animals. My pets and my veterinary training helped me in this activity. Later I started making human figures also (Figures 8 and 9). These were most often female figures (Figures10 and11). I always thought that in humans and also in many other species, think of the elephant seal for example, the female is by far the better looking one.

As my retirement continues, I have increased my activities in woodworking and sculpture as above described. These activities give me pleasure, fill my time and keep me in motion. This is vital as sitting inactive is not only unpleasant, but I believe is physically damaging. There is one activity however which played a large role in my life as a scientist and academician which I continue in my retirement. This activity is speculation about patterns which exist in the world of ours. It is so much a part of me that I am unable to stop it. I can't stop it probably because it gives me much pleasure. For me, at least, the search for patterns helps me to handle the vast amounts of data that accumulate as a result of scientific studies by many people of diverse aspects of our world.

I feel that persistence of this type of thought in my head and the importance this type of thinking had on my career justified my providing a short discussion of it in in this final section of this chapter.

I believe that a variety of universal patterns exist in biologic systems and that at least in the systems we refer to animals the pain and pleasure principle is almost universal. We all repeat doing things that yield pleasure and we avoid doing things that cause pain. I believe that this type of system developed quite early in evolution. I have watched motile bacteria and protozoa in a wet preparation on a microscope slide trying to escape when a noxious chemical is introduced on one side of the slide. If on the other hand, if the chemical that could serve as a food is introduced the organisms move toward it. This behavior with much variation is the basis of the senses of taste and smell and our reaction to pain and pleasure.

There is another pattern in nature that I have found to be very interesting, one which is very significant in biology but exists to some extent in chemistry and physics also. My thoughts about this pattern I can trace back to a course in virology I took as a graduate student. It was given by Salvador Luria then a professor at the

University of Illinois. He said that there two types of building blocks making up all of the structures in nature. One of these types is the common brick which can be assembled in almost any way and in which the information for the assembly is a plan produced by the assembler while the other type of particle was one in which the information for assembly is programmed into the particle. This latter type of block makes up the structure of most of the universe. Think about the subatomic particles of so much interest in contemporary physics. It appears they all have attributes that determine how, and with which other particles they can combine. In fact the possibilities seem to be quite limited at least in their ability to produce atoms. There are only about eighty eight atoms possible. These as seen by Dimitri Mendeleev, the Russian scientist, who developed the Periodic Table, also carry much information which determines how they can interact with each other and what can be built by those unions.

The union of atoms produces structures we call molecules and these molecules make up ourselves and the world around us. There is a vast amount of information contained in each molecule just as there is in each atom. This information which exists in each molecule determines with which molecules the molecule can combine and with and how the binding may occur and how one molecule can affect other molecules. Think for example of enzymes. They are capable of modifying other molecules in specific ways. One of the most interesting attributes of molecules is their ability to assemble other molecules that carry information required for assembly of complex cells, which can then rebuild themselves and thus maintain themselves and even reproduce themselves. DNA molecules for example carry information for their own duplication, as well as information for the production of molecules that spontaneously organize themselves into cellular structures that are organized at the simplest level that manifests properties that we call life. I believe that some of our DNA also carries the information that controls when individual genes produce their products and for how long they continue to produce them which provides vital information shaping the cellular structure.

And now I will say a few words about how the ideas I have just discussed affect evolution and, in particular, natural selection. As a student many years ago, I was taught that the variations that occurred in biological systems resulted from errors in duplication of the information-bearing molecules in cells that is the genes. Most of these changes were considered to have no effect or to be lethal, but some brought about new forms that were better adapted to the environment or permitted the organism to enter new environments or survive in a changed environment. As a result of what I was taught, I always thought that the possibilities of change were unlimited and that chance selection provided the survivors in the selecting environment.

With my growing awareness of the magnitude of information programmed in the molecules of which we are assembled I have begun to think that chance selection is very much limited by the limits imposed on the pools of variance available for selection by chance. These limitations are the result of the vast amounts of information built into each particle and determines which newly assembled molecules are available to participate in the selection process and even determines if such a molecule could exist in the first place. We must consider the limitations imposed on natural selection by the limitations imposed by information programmed into all of the components available for selection. All I can say now about the process is that if other planets are similar to ours, and are made of the same molecules and atoms, there must be quite a few similarities between their biological systems and those here.

Scientists can only find how the existing systems work. They can then sometimes use the information they obtain to manipulate how it occurs. At best however they can only do what occurs naturally and tweak the system to some degree. If genes did not move around with the aid of viruses in nature than scientists would not be able to do it in the laboratory.

If we ever completely understand how all the biological systems work we would still not know why they work, or what would be the purpose of their working, and whether there exists any moral significance to it at all. Morality is a human creation or as the Jewish Bible says humans ate the apple of knowledge of good and evil. It is an interesting result that they were expelled from the Garden of Eden as a result of their eating the Apple and trying to create moral systems. At any rate science cannot tell us why we were here only how it works. The creation of moral systems is the job of theologians.

In final summary I want to emphasize the large role played chance plays in what you do but you still have some control as a result of your decisions on how to handle what chance offers you. You can even to some degree shape what chance offers by what you do. You must have some plans and goals. You must place yourself where you will encounter appropriate chances. If you want an academic life you have to enter institutions where chances in that field are likely to occur. In other words you have to be active and be there and be capable of making decisions about how to handle the chances that arise, but now I think I have said enough. I have described how my career and to a lesser degree how my life has developed. I believe that the important message for someone just starting out in the difficult task of leaving childhood and entering adulthood is to try to make the best use of your assets and then to try to enjoy what you do and be as happy and as positive as you can.

Good luck I wish you success in your search for a good carrier and happy life.

Figure 1. Three Colombian ruins located near Papantla. The pyramid called the Tajin is on the left while the picture on the right is of some other ruins on the site.

Figure 2. Papantla town square on a Saturday night shows
local people in the square. Top picture. Lower picture
is a close-up of a Totanack woman in tribal dress

Figure 3. Typical small rural towns in the regions of Veracruz in which I worked. In the top picture my co-partner and I are standing in the street watching a mother turkey and her chicks. In the lower picture a group of people are gathered in the street of a somewhat larger village

Figure 4. District veterinarians examining a cow for signs of foot and mouth disease. Man on left is me.

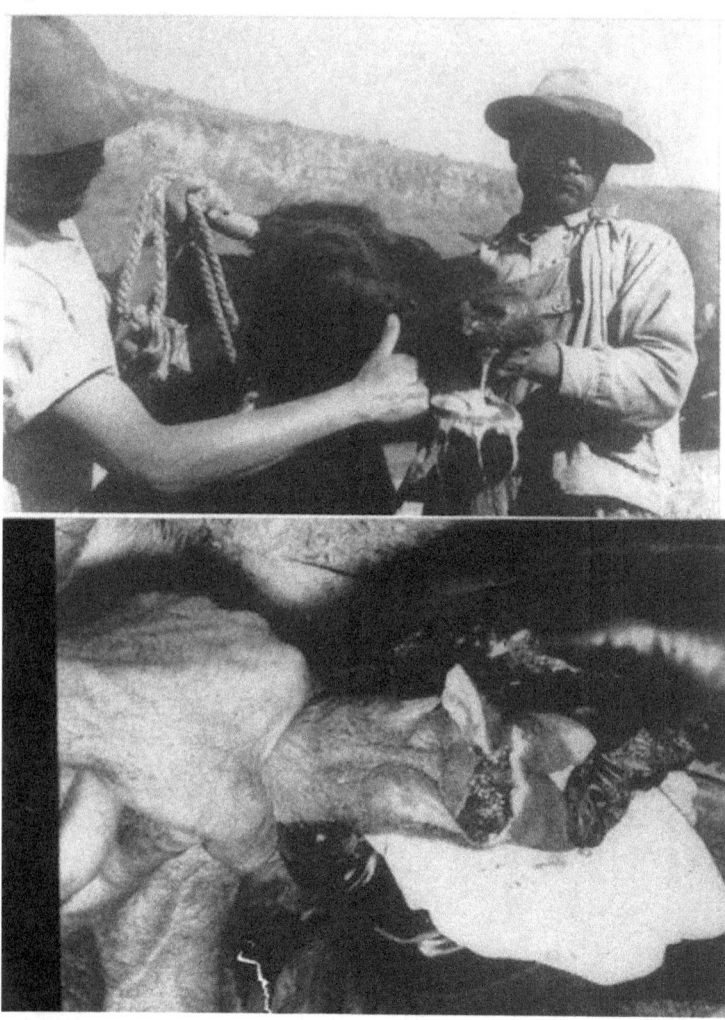

Figure 5. Foot and mouth disease lesions. The lesions occur in the mouth causing excess saliva (above) and on the feet causing blisters were the skin meets the hoof.

Figure 6. Burial of an infected herd. The animals were shot with 22 caliber long rifle bullets.

Figure 7. Heading out on horseback to inspect cattle on
ranches not assessable by wheeled vehicles (above).
In the lower picture two Mexican veterinarians
and I are on a road going to inspect a ranch

Figure 8. A squirrel scratching himself. Making figures was one of the projects that engage me since retirement.

Figure 9. More of my work. A mouse standing on a log, a squirrel, a sitting fox, a sheep and with its lamb.

Figure 10. A girl and her dog, a goat with her kid,
and a woman holding her hat on her head.

Figure 11. A ballet dancer tying her shoe.

CHAPTER 9.

Growing Up International

Jesse Kreier

Let's be clear from the start: I am no scientist.

As a child, I received the usual gifts: a little microscope, in its wooden box, accompanied by a set of slides; a telescope, on its tripod, for viewing the night sky; a chemistry set. It didn't take. I had no real affinity for these objects, no natural sense of what to do with them, and they were soon abandoned in a corner of my room, or dumped into the old black chest with my less favored toys.

From a science perspective, it was downhill from there. I was a good student, and most of my friends were science nerds. Jim, who wanted to go to medical school. John, whose father was an engineer and who aspired to the same. I remember them proudly showing me their most recent TI calculators, miracles of computing power at a time when the home computer was a distant dream and the slide rule was still the gold standard. But science was not my forte. I made it through ninth grade biology okay, but chemistry the following year was a struggle. Amazingly, that satisfied my high school science requirements, and I never looked back.

Neither am I an academic.

This one was a closer call. I did at one time assume I'd become a university professor. I have what my wife Susan rather unkindly refers to as the "lecture button", the proclivity to launch into a topic – provided it doesn't relate to the sciences! – And pronounce myself, until I notice that my victim's eyes have glazed over. But despite these professorial attributes, I did not ultimately pursue an academic career. Admitted into a doctoral program in Latin American economic history at Tulane, and having accepted, I got cold feet. The field was a narrow one, I was told. Did I want to be an eternal post-doctoral, waiting with impatience for a tenured professor to retire, or die, perhaps of hemlock administered by one of his already balding protégés? So I took the LSATs....

Yet despite first appearances, my formative years in an international and academically-oriented household shaped my life profoundly. My father Julius Kreier – one of the editors of this book – was a research scientist at Ohio State University. Much of his work was in the field of tropical diseases of humans and animals, and as a result his laboratory was a revolving door for students and post-docs from around the world. The nationality of his students varied over time, depending on the work he was doing, and on the vicissitudes of US foreign relations, which influenced the availability of funding. In the early years I remember many Indian students – relations were good and US AID was generous – but over the years I remember among my father's students and collaborators Iraqis, Sudanese, Mexicans, Brazilians, Israelis, Turks, French, Germans ...

From the earliest age, I was therefore surrounded by foreign students and academics. When the new students arrived, they often bunked in the spare bedroom while they looked for an apartment, bought a car and generally got oriented. In later years, I would sometimes serve as chauffeur and advisor as they settled in. But in the early years, I remember more the social events, usually dinners, we would share. Thanksgiving was particularly international. My parents would often invite new arrivals to experience this most quintessential of American holidays. The turkey usually went over well, but I recall then – as now, when we invite non-American friends and colleagues to our Swiss home for the feast – that stuffing was peculiarly inaccessible. Despite my insistence that it was the best part of the meal, small portions would be pushed about the plate and seconds rarely accepted.

In this environment, somewhat exotic in a one-time provincial town like Worthington, Ohio, it is perhaps not surprising that I became interested in things international. It was obvious to me from an early age that a grasp of languages could be useful, that people from different parts of the globe had deeply different perceptions of the world. I began to collect stamps, as my father brought me home beautifully colorful envelopes from a colleague in Moscow, covered with philatelic commemorations of the glories of Soviet sports, industry and culture. Like most kids, I took Spanish from seventh grade; but unlike most of my fellows, I took the exercise seriously, knowing that language was a tool that could be used, not just a list of verbs to conjugate and forget.

The revolving door of foreign visitors was supplemented by several overseas trips, linked to my father's work. The first big expedition, which I will never forget, came in 1968. My father had a conference in Teheran, and my mother, sister and I arranged to meet him afterwards in Tel Aviv and then travel back together via Rome and London, staying a week in each country. Wandering around east Jerusalem, only a year after the Six Day War, was an eye opener for a 10-year old boy who's most exotic foreign destination to date had been Expo 67 in Montreal. Beaches covered with pottery shards in Caesarea, the Coliseum, a train ride to Naples and on to Pompeii, the houses of Parliament and the Cutty Sark -- these things remain as sharp and clear in my head as this morning's breakfast.

So what did we talk about around the dinner table and on these trips? I'm sure there was ample discussion of science, but quite frankly none of it stuck with me. What I remember is continual discussions of history, of languages, of demography, archaeology, religion, and, of course, of international politics. Not surprisingly, my own interests began to run in these directions. My hobbies were coin and stamp collecting; I gathered my collections through a network of pen pals reaching from Spain to Korea. I added Russian to my language studies, devoured historical novels set in distant lands, stunned my history teacher by knowing where Tonga was located. Graduating high school, I chose to spend my summer in Mexico, studying Spanish – what choice more natural, given my father's two years in the jungles of Veracruz -- and enrolled in an international relations program at Johns Hopkins. My international career was launched.

While my professional goals shifted with the years – I imagined in turn being archaeologist, Foreign Service officer, history professor -- my international orientation remained a constant. Junior year in Italy, a 3-month love affair with Brazil that *almost* took me to graduate school in Latin American history, and finally a joint degree program in law and international relations at Georgetown. It was perhaps natural that I landed in a private international trade practice in Washington, that this practice took me to Geneva for trade negotiations, that I stumbled on a job there at the General Agreement on Tariffs and Trade – now the World Trade Organization -- and that I've spent the past 22 years living and working in this most international of environments. A few times a year, I find myself in Delhi or Tegucigalpa, lecturing government officials on the intricacies of anti-dumping investigations, or speaking to a visiting group of masters of law candidates on multilateral dispute settlement.

Was my career predestined? No, but who I am was clearly shaped by those Thanksgiving dinners of my childhood, watching perplexed new arrivals politely picking at the stuffing of a turkey dinner, and discussing the caste system in India or the domestication of the potato with my father

CHAPTER 10.

My Global Journey

Raul Machado Neto

I was born in 1949 in Piracicaba, in the State of São Paulo, Brazil, son of sweet and kind Helena Moreira Cesar Machado and Raul Machado Filho. My father was a traditional MD in our town and greatly influenced me with his attitude and scientific mind. In 1974 I married with the wise and generous Maria Luiza Barros Cavalcanti Machado and we have two admirable children, Carolina and José Eduardo. Carolina married special Rodrigo and now we have three wonderful grandchildren, Helena, João and Pedro. I am who I am today because of my family.

About Education:
BS in Agricultural Science (1970-1973) and MS in Animal Science (1974–1977) at Escola Superior de Agricultura Luiz de Queiroz, the College of Agriculture of Universidade de São Paulo, USP/ESALQ, Piracicaba, São Paulo, Brazil; PhD in Animal Physiology (1976-1980), University of Illinois at Urbana Campaign, USA. 1980; Postdoctoral Fellow at Agricultural and Food Research Council-AFRC, Institute for Animal Health, England, (1989-1990).

About Career:

Assistant Professor of Universidade de São Paulo/Escola Superior de Agricultura Luiz de Queiroz, USP/ESALQ USP/ESALQ (1974-1980); Doctor Assistant Professor (1980-1985); Associate Professor (1985); Full Professor (1997); Visiting Professor at University of Bristol and Unilever Research Center, England, 1982; Member of the Postgraduate Committee – Animal Science, ESALQ/USP, 1984-1989; Coordinator of Inter-American Bank of Development Program (IBD) for ESALQ/USP equipments, 1987-1989; Member of the University's Board Meeting, ESALQ/USP, 1989-2007; Head of the Zoology Department, ESALQ/USP, 1991-1996; Coordinator of the Electron Microscopy Research Support Nucleus (NAP/MEPA) of ESALQ/USP, 1991-1994; Member of the Coordination Committee for Specialized Institutes of USP, 1991-1994; President of Research Committee, ESALQ/USP, 1991-2007; Coordinator of CNPq Scholarships (PIBIC) for USP Undergraduate Research Students, 1995-2005; President of USP Undergraduate Research Program, 1995-2005; Assistant of Pro-Rectory of Research, USP, 1998-2005; Associate Provost for Research, 1995-2005; Coordinator of the USP Research Fund Committee, 1996-2005; Vice Dean of USP College of Agriculture – ESALQ, 2003-2007; President of USP Call Technology Academic Committee, 2005-2006; Visiting Distinguished Professor at Rutgers, 2007. Current Assistant of Pro-Rectory of USP Graduate College, Assistant of Research Pro-Rectory and Vice-President for USP International Relation Board.

About Honors and Professional Service:

Undergraduate Research Scholarship, FAPESP, 1972-1974; Research Scholarship from Coordination of Brazilian Postgraduate Programs (CAPES), USA, 1976-1980; Research Scholarship from British Council, England, 1983; Honored by the Undergraduate Students of College of Agriculture ESALQ/

USP, 1985 and 1987; Research Fellow of CNPq, 1985-present; best paper presented at National Conference of Animal Science (XXIV SBZ), nominated to receive two prizes, from EMBRAPA and AGROCERES, 1987; elected by the undergraduate students for one of 10 best teachers of ESALQ, 1994; Scientific Merit Medal of State of São Paulo Governor, 2001; Reviewer for the Coordination of Brazilian Postgraduate Programs (CAPES), 1997-present; Reviewer for the Brazilian National Research Council (CNPq), 1988-present; Reviewer for the São Paulo State Research Foundation (FAPESP), 1988-present; Reviewer for the Government's Education Ministry's (MEC), 1996-present; Associate Coordinator of Executive Committee for the State of São Paulo Higher Education Project, 2005-2007; Member of National Committee for Undergraduate Research Program (CNPq), 2007-present; Coordinator and founder of the Tripartite PhD – International Graduate Program in Molecular Cell Biology - Plant Science, USP/OSU/Rutgers; Co-Coordinator for the Project FIPSE/CAPES, USP/UFRGS and Rutgers/OHIO State University, 2007-2010; currently coordinator of the agreement between USP-The Ohio State University; currently coordinator of the agreement between USP-Rutgers University, Member Special FAPESP Committee to evaluate grants, Member Special CNPq Committee to evaluate grants.

About Undergraduate Research (Iniciação Científica/Science Initiation in Brazil) during my college days:

I began my research activity in 1971 as an undergraduate research student developing projects in the area of plant nematodes with Prof. Luiz Gonzaga E. Lordello. From 1972 to 1973, our project was supported by FAPESP, for which I qualified for a scholarship. During the same period I developed research activities in the biochemistry area at CENA under the supervision of Prof.

Otto Jesu Crocomo, a special adviser from whom I learned a lot. Another important contribution to my early days in science was Professor Henrique Vianna de Amorim, a man whose scientific entrepreneurship spirit I also admire.

In 1972 I met William Rod Sharp, a talented young professor from The Ohio State University, USA, who came to work with Prof. Crocomo. I was one of his monitors in the course "Application of Nuclear Energy in Plant Tissue Culture", an international course in the new area of tissue culture. The course was coordinated by Prof. Linda Caldas and Rod Sharp and supported by CNEN (National Nuclear Energy Commission) and The Organization of American States (OAS). Rod was a very special person and became my generous friend, my partner and my professor since then, for 40 years now.

About USP Undergraduate Research Program Under My Responsibility:

As the first President of USP Undergraduate Research Committee, my contributions included: organizing the USP symposium initiative and mobilizing and aggregating all our colleges in an event called SIICUSP. It became USP's most important undergraduate academic event with nearly five thousand presentations. Considering advisers and coauthors, this would involve more than twenty thousand USP participants. The result was an organized system of scholarship distribution/selection, and the creation of an identity for this activity at USP.

The internationalization of this activity started when Rod Sharp visited my office at ESALQ/USP and asked about two books, on my desk containing the abstract summaries for the undergraduate students participating in the USP Undergraduate Research Program (Science Initiation). I explained that the books contained the student research abstracts resulting from the undergraduate student's research projects to be presented in the forthcoming Undergraduate Research Symposium (SIICUSP). He was very impressed, and with his entrepreneurial and visionary spirit, asked if I could talk with his colleagues at Rutgers University about this initiative during my next visit. I accepted, suggesting this would be an opportunity to develop an exchange program of undergraduate research scholars from USP and Rutgers. We agreed and that is how we broadened the USP event expanded to a global event. Initially the event was called SICUSP, but it evolved to become SIICUSP in 2000, with an additional "I" in the acronym representing the internationalization of the event – Simpósio Internacional de Iniciacao Científica da USP (International Symposium of Undergraduate Research of USP.) Rutgers and OSU undergraduate scholars participate in SIICUSP during November and USP undergraduate research scholars participate in the Rutgers and OSU Undergraduate Research Scholar Symposia during May. OSU and Rutgers have both been successful in establishing five million dollar endowments in support of their symposia. The recent inclusion of other universities has resulted in further expansion of the initiative.

About the Electron Microscopy Nucleus at ESALQ:

In 1987 I was designated by the USP Rector Prof. José Goldenberg the coordinator of USP Inter-American Development Bank (IDB) for ESALQ-equipment financing. This grant was very important to the acquisition of state-of-the-art equipment for the USP research facilities, allowing us to perform world-class research. This initiative and the efforts of Prof. Goldenberg to upgrade the USP infrastructure significantly contributed to USP's reputation for global academic research and teaching. After two years of reviewing departmental requests for equipment purchases at ESALQ under my coordination, I proposed to all departments to abdicate the remaining individual grant requests to buy an electron microscope and create a center to serve the entire academic community. After personally communicating with and a long process of convincing each of the department heads, the remaining funds for two years of projects became available to buy the electron microscopy. The Rector, driven by higher academic spirit granted approval for two equipment purchases instead of one, and we acquired transmission and scanning electron microscopy equipment. This center is called NAP/MEPA, Nucleo de Apoio a Pesquisa em Microscopia Eletronica Aplicada a Agricultura – (Research Nucleus on Electron Microscopy Applied to Agriculture Research). Prof. Elliot Watanabe Kitajima, whom we hired, after a complex negotiation, from the University of Brasília, played a fundamental role in the organization and functioning of the Center. Under his coordination, the center dramatically enhanced research analytical capacity, becoming one of the most active centers in this area in the country. I have to highlight the relevant support of Prof. João Lúcio de Azevedo, dean of ESALQ at the time.

About the Tripartite PhD Program on Plant Molecular and Cell Biology:

In 1996, Rod Sharp and I, as Director Rutgers University Experimental Station and USP Associate Provost for Research, respectively, decided to involve our two institutions in biotechnology, a very hot topic at the time. We decided to organize a workshop of large dimensions at ESALQ in the area, along with other institutions of the US and Europe. The two-week event took place in December 1997. We dedicated an exhibit hall to accommodate booths and poster for the display of laboratories research information of both institutions. In an additional section with representatives of all participating institutions, we did a reflection about the future of academic relations. The event was a great success and this initiative continues today with the addition of our The

Ohio State University partner. Since the first event, the workshop has been conducted on a biannual basis alternating between institutions: 1° Workshop – 1997 USP/ESALQ; 2° Workshop – 2001 Rutgers; 3° Workshop – 2003 USP/ESALQ; 4° Workshop – 2004 Ohio State University; 5° Workshop – 2007 Rutgers; 6° Workshop – 2009 USP, and 7° Workshop – 2012 The Ohio State University.

In 2004, we organized a brainstorming conference to explore additional synergies between the institutions beyond workshops and undergraduate research scholar exchanges. To accommodate research faculty and administrator leadership schedules, of the three universities we met at the VIP American Airlines lounge at Newark International Airport (a creative idea arranged by Rod). It was the first time I had to go through airport security checkpoint to attend a meeting. After a long lunch discussion, I proposed that we do something quite different: create a tripartite graduate program, making use of the competence already shared in research initiatives by the three institutions. Thus, Rod and I began a long endeavor, still a work in progress, to create a unique joint PhD program.

The three research institutions formally agreed during September 2004 to develop a tripartite Ph.D. Graduate Program, initially in Plant Biology and Plant Biotechnology. During the 2007 Tripartite Workshop, the institutions reported plans for execution of the Ph.D. Graduate program, which is expected to expand in the future to encompass additional disciplines. The program is designed to provide adjunct faculty status appointments for faculty members in their respective graduate programs and the joint mentoring of graduate students. The mobility of students and faculties is an important program characteristic, creating a real and complete international academic environment. The program provides graduate students the opportunity to conduct research at USP, Rutgers and The Ohio State University. Participating graduate students have the opportunity to qualify for a Ph.D. degree from two or three of the partner institutions. Students enrolled in the program are expected to demonstrate language proficiency in both English and Portuguese. Faculties in tie humanities and linguistics departments of the three institutions have been engaged in the preparation of new language and culture courses, resulting in yet another academic collaboration. The program comprises national and international course modules, embracing the campuses of USP, Rutgers and OSU. Students are required to take basic courses at their home institution during the first year of the program. Only students who complete the national module are allowed to enroll in the international module.

The remarkable synergism and commitment of faculty and administrative leadership from these three institutions allowed the accommodation of undergraduate research scholars. In 2006 Rutgers, Ohio State, USP and UFRGS received grant awards from CAPES and FIPSE for a four-year exchange of undergraduate research scholars in plant biology and microbiology for one semester of academic studies, an internship and cultural immersion. The program is expected to provide a pipeline into the Tripartite Graduate Program.

Related to the "tripartite" initiative, Rod and I are pleased to acknowledge the great effort and competent support of the following professors from the three institutions: Helaine Carrer, Marcio Castro Filho, Ricardo Azevedo, Michael Lawton, Eric Lam, Gerben Zylstra, Jerry Kukor, Bob Tabita, Erich Grotewold, Andrea Dosself and Pat Osmer.

Today, the Brazilian end of the program is coordinated by the dedicated and enthusiastic Helaine Carrer, a key person who has worked with us from the beginning. The program has been approved by CAPES, Brazil's federal agency that regulates graduate programs, as the first in the country with an international characteristic. Because the program is a reference point of the national academic scenario, I have been regularly invited by CAPES president, Prof. Jorge Guimarães to attend meetings to discuss the program with other research institutions. Prof. Guimarães has enthusiastically supported the initiative and highlights the tripartite proposal as a template for new initiatives at other institutions. We are currently working on some specific and final adjustments required by the US. The program has enrolled seven students that are finishing the national module and starting the international module.

The magnitude of the collaboration between these three institutions is due to the academic excellence, institutional respect, and commitment by the respective administrations and faculties. The qualifications of these institutions and the shared expertise of their researchers was the motivation to do something more. This international synergism allowed the creation of an outstanding program of human resource training at the PhD level in a truly international academic environment. Looking at the future, the USP global collaborative degree granting approach provides an opportunity to integrate the recent Massive Open Online Courses (MOOCs) initiative.

About Other Areas of Internationalization:

As vice-dean of ESALQ, working with dean Parra, my competent friend of many years, I had great support in implementing several initiatives that contribute to the consolidation of the internationalization at ESALQ. Contributions from these international collaborations include establishment of the first double degrees in ESALQ, with the French Agriculture College, ParisTech, and with group FESIA, headquartered in Toulouse. Prof. Maria Lucia Carneiro, with her dedicated competence, has been instrumental in coordinating this initiative with French institutions. I also had the opportunity to take part in the process of establishing

Wageningen University representation in the ESALQ, the first initiative of its kind in our campus.

As a result of our collaborative relationships with Rutgers University and Ohio State University, we have been able to enhance our partnership to include chemistry, mathematics and humanities. In the area of bioenergy and biomass research, we are working to organize an Inter-American consortium including Rutgers University, The Ohio State University, Michigan State University, the University of Wisconsin and the University of Georgia, that will deal with mega-research projects for alternative energy.

With the recent increased visibility of USP resulting from increased improvement of our academic indices, we have been sought after by a large number of qualified institutions. In 2012, working with Prof. Marco Antonio Zago, USP provost for Research, we established, among several others, two very special new strategic partnerships with the University of Toronto and Princeton University. We have two main areas of mutual interest with Toronto, medical science and study of cities, and with Princeton, global health and anthropology. The process of establishing the partnership, was very enriching, resulting in friendships with the two presidents, David Naylor, from the University of Toronto and Shirley M. Tilghman, from Princeton University. It is worth noting that these two universities have very few international agreements. At Princeton, the partnership with USP is the only formal existing international agreement.

As professor of animal physiology, in my research activity, I have had several outstanding students, but three of them were very special, Rosana Bessi, Patricia Pauleti and Débora Moretti, not just as advisees, they become my partners and colleagues who shared with me the responsibilities in achieving the goals of my lab. The students are thanked for their intellect and continuous stimulation which is so important to the academy's teaching and research endeavors.

Finally, I have been privileged during life trajectory, to meet and share experiences with very special people of almost fifty years, Paulo Botelho, Osny Bacchi, wistful Raul d'Arce, Parra and Rod Sharp, all of them related to agriculture research and teaching.

At USP, we continue to pursue "consolidation of a world class international academic environment, where excellence prevails over borders".

CHAPTER 11.

My Unusual Pathway through Academic Life:
A Biographical Reflection

Mark T. Muller

An Overview.

Every scientific publication has an abstract that defines the body of the work and crystalizes the key findings. This is a good way to start. As a child, while living in the Orient, I developed an interest in science and specifically in microbiology. I questioned everything and dismantled just about anything that I could. Years later as a second year undergraduate, I finally got my chance to experience the thrill of learning, problem solving, discovery and research. Graduate school followed and by the age of 26, as a newly minted Ph.D. with a reasonably good publication record and I started Post-doctoral studies. At age 29, I joined the faculty at The Ohio State University, and in 1987 with several colleagues, we founded the Department of Molecular Genetics, the first in the country. A few years later, I started a small but profitable bio-technology company and 25 years later, I was recruited to the University of Central Florida in Orlando and retired from OSU. This is my story that is defined by influential colleagues, mentors, family and friends. Along the way, I developed new passions in aviation, music, photography and the freedom that comes with greater financial independence. A man's life is his message to the universe; I hope mine can inspire others in some small way.

The Early Years: Coming of Age and Developing an Interest in Science.

My father and I share the same name, Mark Twain Muller. My paternal grandparents were emigrants from Austria (Vienna) arriving in NY in or around 1910. My Grandmother (maiden name, Topper) arrived a few years after my Grandfather, I'm told. I have a number of relatives, according to my father, that were Jewish (including a great Uncle who was a rabbi). Evidently my Grandmother was a Samuel Clements fan, hence the namesake. After arriving in NY City, my Grandfather worked as a diesel mechanic and was one of the first Daimler engine experts in the US. They lived in Brooklyn, NY. The Muller family was small by early 20th century standards (from youngest to oldest, a daughter, Rita, Paul, Mark and William). My father was born in 1915 and William (Bill) Muller was born in 1911 (Bill was my favorite uncle and I got to know him well). My father graduated from Cornell in 1939 and promptly entered WWII as a signal officer (Fig 1). He was in the Pacific theater and worked tangentially with some famous people while stationed in Australia (Fig. 2), including Douglas MacArthur, John D. Bulkley and L. Ron Hubbard and a number of NY Times correspondents. Years later, I met Admiral Bulkley (nickname the "Sea-Wolf") who was awarded the Medal of Honor for his role in getting General MacArthur out of the Philippines after the Japanese invaded and conquered these islands. My parents met in Australia during the war, fell in love and were secretly married in Australia in 1945 and then 'officially' married back stateside in 1946. Soon thereafter, my siblings arrived (Conni, the oldest, Mary, next). I was born at Walter Reed Hospital, in Washington, D.C. in 1950 when the family was stationed in Baltimore, MD.

As a career military officer in the early 50's my Father was getting his Masters Degree from Boston College as well as teaching communications and military science at MIT. I don't recall too much about living in the Boston area, other than the house we occupied seemed huge to me, with a large basement where I recall getting a fishhook under my tender skin and occasional visits from my grandparents who lived in NY City.

My father fought in the Korean conflict and when I was about 3 years of age the family relocated to small town in east Texas (Pittsburg), my Mother's hometown. I recall more about this time of my life and the tragedy that shaped me from a young age. My mother came from a large family (7 boys, 2 girls) and most lived in the little east Texas town of Pittsburg. A couple of my uncles were prominent citizens. I spent some time with my Uncle Leo and Willie, who pioneered the sweet potato industry in Texas. It was a big operation and I grew accustomed to eating raw sweet potatoes for breakfast, while spending time at the warehouse. To this day, I only really eat sweet potatoes raw. I also witnessed day-to-day operation of a highly successful business venture, and the trappings of that success. I was imprinted at an early age by this experience, I believe. Life in east Texas in the early fifties was horribly marked by intolerance, bigotry sadly and segregation, something that I could not comprehend at such an early age. The town had two main streets, one for whites and behind another for blacks. All stores had front (white only) and rear entrances (blacks) to maintain separation. In my opinion, bigotry sadly is still pervasive in the rural areas of Texas..

The tragedy I spoke of arrived on a warm spring day in 1956, while I was playing at my Aunt's home outside of town. My Mother had left that morning early to drive to Dallas with Uncle Willie and his wife Girlie for a medical checkup (Willie was a diabetic). My Grandmother was babysitting for the day. A man walked up, while we were all sitting on the porch, and started acting strangely; almost as if he could not speak the words. What came out sent my Grandmother in a wailing, screaming, crying hysterical fit. My older sister, Conni, was crying and wailing. Mama (as everyone called my grandmother) told me in a calm voice that my Mother was in an accident. Actually it was a head-on collision, and my Uncle Willie was killed instantly and my Mother was not likely to survive the massive head trauma. (My Aunt Girlie also had life-threatening injuries including a nearly severed foot and many broken bones.) All were on the way to the ER. In those days, cars were like huge metal ramming machines and without seat belts or airbags or any safety features we take for granted. Occupants became human projectiles driven into whatever immovable object was in the path. My Mother was thrown clear and found 100 feet from the wreckage. There were a number of fatalities in the other car, driven by a young man (that car also had a number of children). I recall Willie's funeral. I had 'shadowed' him at the Sweet Potato House and really liked him; however, I really could not understand the loss. My Uncle Frank quietly stated that Willie was sleeping as he held me over the casket to view the body. I accepted his explanation, which seemed reasonable since Willie did appear to be asleep. Mom was in the hospital and when I saw her I was horrified. I simply did not recognize my Mother who had a swollen, stitched and blood caked face. This mangled body in the bed could not possibly be my Mother, I thought in horror. She looked more like a monster and it was frightening to the core. Her injuries were very severe and included a shattered femur, massive internal injuries, head trauma and being partially scalped. Mentally Mom was not fully functional, but awake and aware. My Father came home on emergency leave from active duty in Korea and we started the long process of rehabilitation and recovery. Some months went by and gradually she improved and after plastic surgery, a metal thigh pin implant and much rehab, she was released but wheelchair bound. Her mental state was still not back to normal and she was not capable of being a responsible mother. About six months after the accident, my Father was stationed in Okinawa. I was five years old. There was some disagreement in the family with my Dad suggesting that he go with the kids to Japan, and leave Mom to recuperate in Pittsburg. Mother refused and we all moved to a new existence on the tropical island of Okinawa. Prior to arriving in Japan, my Father spent time explaining Japanese culture and a little bit of the language and customs as I recall. It came in handy actually when the family boarded a taxi in Naha Airbase and the driver took off leaving my Dad on the curb. Everyone was screaming for him to stop, but the poor guy spoke no English, but I piped up and said "Papa-san, Papa-san" and he instantly knew the situation. It was at this age that I realized that speaking a new language had some real rewards and I have always enjoyed learning languages. While not fluent, I have working knowledge of German, Danish, Italian and Japanese.

Since my mother was not fit to take care of us, my Dad hired two young Japanese women, Kazuko and Sudoko, as nannies. Kazuko was tasked with dealing with me and she spoke little if any English. This turned out to be a good opportunity for me to learn some Japanese. I started to learn the language at a very early age, and got reasonably proficient according to my Dad. The bad news? Well, the Japanese language is not complicated but it is somewhat different between males, females and children. Essentially, in this context the language was different in tones, construct and what I call 'harshness' of the dialect. There are also multiple levels of 'politeness' or supplication. As an adult male, visiting Japan many years later, the language I recalled was largely inappropriate in polite society. This led to some interesting and awkward scenarios (when speaking Japanese in the lab I visited, a graduate student asked if I was gay, since I used female oriented language cues). On the other hand, when dealing with Tokyo businessmen (on business matters), I never let them know that I understood the gist of what was being said in various business meetings. This gave me an edge, since I could easily peg friend from foe. When they would refer to me with term 'gaijin' which is generally considered politically incorrect (GAI meaning outsider, JIN is person), it was easy to see who was on my side.

I recall a few other snippets of early events in Okinawa. I recall sitting with my Dad at the beach while he was painting a seascape. While exploring, I stumbled across a live land mine buried in beach sand and for once refrained from trying to take it apart and study its inner workings. I promptly reported the find to my Father who turned it in to the authorities. I also recall that my Father had many Japanese friends and associates and that he was acutely interested in Oriental culture.

We spent about 2-3 years in Okinawa and Mom finally recovered. The next move took us to Ft. Huachuca, Arizona in 1959. We lived on the base, which became my new world. Ft. Huachuca is near Sierra Vista and has an elevation of almost 5000ft. While hot and dry in the summer, we sometimes had snow in the winter months, a new thrill. There were miles of hiking trails, rattlesnakes, venomous Gila monsters, horny toads that defined a biological potpourri to be sampled and studied. My Dad had this tiny little Vespa scooter and he and I would trundle through the back roads on weekends, exploring and generally making random searches for interesting bits of culture, antiques and crafts. I'm convinced that several events at the time impacted my future proclivities. For example, the Vespa sojourns got me interested in motorcycling and I later learned that my paternal grandfather was an avid motorcyclist who owned a 1915 Indian side-car bike. As part of my Father's duty assignment, he worked on communications systems in unmanned drones and I watched a number of drone launches. I was fascinated by flying and took every opportunity to watch and learn about drones. Even at this young age, the thrill of full 3-D movement in space seemed so exciting. I acted on this fascination many years later, as described below and I believe that my aviation interest grew from watching drones fire off with the JATO (or Jet Associated Take Off) bottles. One Christmas, my parents got me a telescope. Not some small tripod window toy but a real nice 6 foot long refractor telescope with reasonable light gathering ability and floor-based tripod. I systematically studied the heavens, but soon realized it was far more fun to view the Sierra Vista Drive-in movies about 4-6 miles down in the valley. No sound of course, but I viewed first run movies and endured the eye strain that attended the process. Evidently, astronomy had little if any allure.

There were many fond memories, including working with my Dad, an avid ham radio operator (K5LOW), who taught me basic Morse code and the phonetic alphabet. We spent time on hunting, fishing and camping trips around Cochise country with my Father and his friends and we visited Bisbee, Tombstone, large copper mines, the OK Corral and many other sites. I also recall getting in major trouble for setting off a fire alarm in the school (remember, I was curious). Basically I was a good kid but very active with a reasonably active social life. I love to take chances and push the envelope. My life promptly changed from a very comfortable social life when my Dad got his next duty assignment. The family packed up and again we made a move abroad, this time to Taipei, Taiwan. The family drove from Arizona to San Diego to board a military ship called the USS Barrett. The whole family then endured a three week voyage to Taiwan. I don't recall my Dad being with us on the drive from Arizona to California but we had a fun and exciting time driving the '57 Chevy wagon across half of the country. My Mom had a wonderful sense of humor and we spent several days laughing and poking fun at each other.

The Military transport ship (USS Barrett, which was scrapped in 1973) was immense and insanely complex to my 10 year old brain. There were not so many families, and few kids my age. I recall a large number of the passengers were US or Chinese soldiers (mostly enlisted men). We had a small berth high up on the port side and porthole views of the water. My parents had close friends who had a son my age. Danny and I became fast friends and we got into considerable mischief. The ship was this enormous city with a heartbeat that was manifest as central activities including ping pong with the Chinese enlisted men, free movies, lots of food, arts and crafts each morning at 10am and so on. Still, Danny and I got bored and engaged in monkey business. My sister, Mary, gave me this plastic light bulb (a gag) that looked exactly like a real glass bulb but unbreakable. The ship had a central staircase winding up about 10-12 floors and open in the middle. I would stand on the top floor and drop the fake bulb down to Danny on the first floor. The MP's caught us and hauled us both away after closing down the staircase due to some 'idiots dropping light bulbs'. Lesson learned on this one, courtesy of my Dad's stern verbal thrashing (and confinement). On day three at sea, everyone got sick including yours truly. It was dreadful and you just had to ride it out for about 12 hours before you get your 'sea legs'. After that, it was no problem. Going across the international dateline was a both fun yet disgusting (blindfolded, hands placed into vats of spaghetti and jello, being hosed down with frigid seawater). We were 'pollywogs' and Neptune (my Dad) did the initiation rights.

In 1961, we arrived at the Port of Keelung and moved promptly to Tianmu at the base of Grass Mountain outside of Taipei. Many American service families and expatriates lived in Tianmu which was a lovely rural setting, surrounded by lush rice paddies. I was fascinated with reptiles, which were plentiful. I started a reptile club and we captured and kept a number of snakes, all fodder for my science fair projects. I asked the local army hospital if they would provide me with sterile agar plates for bacterial growth, as I was becoming more interested in culturing bacteria

from my snakes. Despite repeated requests, the military technicians refused, saying it was too dangerous to let a 10 year old play
with bugs (and they refused to offer oversight). Again, an early
influence was an acute interest in microbiology, although I was
unable to act on this desire. So I turned back to my snake 'zoo'.
Tropical Taiwan is a haven for venomous snakes. Vipers, bamboo
snakes and other colorful serpents were interesting subjects of
study. I was careful not to handle the real killers, for example
the '100 pacer' (100 steps and death) or the Habu but many others while venomous were not capable of killing with a single bite.
Fortunately, I never got bitten but I had close calls while hiking in
the hills around our home.

Based on my interest in language, I tried to learn Mandarin
and my Father was supportive since he spoke the language with
reasonable fluency. I totally faltered in learning the language
for a couple of reasons. First, I was not really immersed in the
Chinese culture and did not really use what I learned along the
way. Second, Mandarin is extremely difficult for English speakers, especially for a 12 year old. I really admired my Dad's ability
to communicate with the locals and in Mandarin. My Dad had
many Taiwanese friends and associates. He was acutely interested in oriental antiques (such as cloisonné vases, coins, jade
snuff bottles, and jewelry) and rather unique among American
personnel (many shunned the local indigenous people and had
no interest in language or culture). I had many interesting family dinners with other Chinese families, courtesy of my Dad. We
lived in a very nice home in Tianmu, with a large pond with fish,
a lush garden, a walled-in yard and servants. Charlie and Su-
yen (husband and wife) were Chinese expatriates who escaped
through Hong Kong in the early 1950's. Charlie was the cook
and Su-yen the maid. They had four sons (all boys) ranging in
age from 10 to 18 years of age. Jer-Pong was my age and I had a
built-in playmate who was salaried! He spoke no English, and as

I said, my Mandarin was terrible, but we managed to communicate with sign-language. He was a good baseball player and kept me in training for little league (I was a catcher). Most of the language I learned from Jer-Pong should not be repeated in polite company and I can to this day shock Chinese colleagues (and wives) with expletives. Chinese is an inflective language and otherwise rather difficult to learn. During this time, while attending Shi-Lin Middle school (part of the Taipei American School system) I discovered that I was particularly good at gymnastics; actually any sort of activity that involved innate knowledge of the position of one's body in aerial space. Several 'offshoots' of this ability made me good at peripheral activities, including pole vaulting, high jump, diving and of course tumbling. As a catcher in little league, I was enthusiastic and aggressive but not great. I loved being the center of attention however since every pitch was watched as the ball terminated in my well-oiled mitt. My enthusiasm was significantly attenuated when a tipped fastball made it somehow into the side of my facemask, leading to a massive shiner. The physics of impact is a good teacher it seems. I switched my attention to soccer and enjoyed playing goalie, for the same reasons I liked playing catcher I suppose.

My time in Taiwan taught me so many things about myself, largely through trial and error. One influence, I believe positive, was the fact that all the years growing up in Asia, there was no television so very little time was devoted to this particular distraction. I developed an appreciation for listening to the radio (local US Army media). Like many in that generation, I loved rock and roll (early-mid 60's). Another defining event was being beaten up by a Chinese man while riding a bus to the American Compound in downtown Taipei. In retrospect, I believe this person was mentally ill. After ranting in broken English about atrocities in 'Red China' he turned his invective to me and began a tirade about America. This rant was terminated by a rather painful pummeling and I was unable to defend or protect myself against this attack (he was a full grown man in serious anger mode and I was an 80 pound 11 year old). The bus driver pulled him off me, thankfully. No major injury but I was terrified.

My friend from the USS Barrett (Danny Robbins) lived in Hsin-Shu, about a 3-4 hour drive down island. Hsin-Shu is now a suburb of Taipei, but back then it had a small Army compound with about 50-100 families, a nice pool and officers club. Danny was a rather sadistic young man and while I enjoyed catching and studying frogs, reptiles as noted, he took great pleasure in exploring such life-forms from the inside out. While cruel, it was nonetheless morbidly interesting and satisfied my enquiring mind. I really wanted to know how things worked mechanistically and visual inspection worked well.

Dan's father, Lt. Col. Robbins (aka Robbie), was with MAAG (Military American Advisory Group) and we sometimes got to accompany him on field exercises (as observers). Robbie was a hardcore career military man; all business. At the time, I recognized that career experts were a wonderful resource to learn from and I was full of questions regarding organizational details, 'what if' scenarios and just general insight. My father, being career military (but not regular army) clearly enjoyed his work and I considered that such a career path may work for me as well. Regrettably, my questions to Robbie were usually answered in a way that involved ridicule. "How could you be so ignorant" he would say with a scowl. I quickly learned three things. First, don't even think about asking questions to people with closed minds. Second, I vowed never to put another down for asking a simple question. Third, it seemed that a military career would never work for me. I would say he was a negative role model. A real negative inspiration! The rocky relationship persisted and I especially noticed how that Robbie was rather harsh with his son, Danny. I believe this is reason Danny took his own life years later at the age of 21 (Dan was their only child). The parents became bitter and actually blamed me, rock music or whatever was convenient, never considering that high, unachievable expectations are a crippling and demoralizing attitude. I heard he left a suicide note but the contents were never revealed. I will never forget his mother throwing herself on the casket and wailing, while Robbie stood stoically by with no emotion. Actually, Danny was supposed to be in my wedding as a groomsman; thus my comment to his father that I hoped he and Mrs. Robbins would try to attend my wedding, was not well received. Not surprising given the circumstances. Oddly, the Robbins family and my parents remained good friends for many years; however, they never ever enquired about me or my career (but asked regularly about my sisters). Bitterness and despair terminated the life of these two individuals and they passed on.

In 1963 my father was assigned to Washington, D.C. (my birthplace) and we packed up and moved again. I started the 7th grade at Gunston Junior High in Arlington, VA. We lived on Columbia Pike, a pretty busy road, in an apartment complex. I completed grades 7-9 at Gunston and during these three years learned a bit more about myself as I matured toward puberty. I loved music and especially The Beatles (didn't everyone?). I was a pretty quick study on the guitar, but I was cash-strapped and could not buy my dream guitar (Gibson SG) and amplifier (Gibson Eagle) and my parents were no help. I was fortunate, however, and managed to secure (in 8th grade) rather lucrative paper route delivering the Washington Post to several nearby apartment buildings. I had roughly 200+ customers and garnered an income of about $150/ month including tips. This doubled over Christmas (bonus tips). I learned the value of working in a business and the reward structure that comes with being clever and working hard. The Gibson/ amp combination soon followed, but I got buyer's remorse, because I actually decided on a Gretsch Country Gentleman that

Harrison played (gold hardware, sophisticated electronics, a visual treat). This was almost $800, an ungodly amount of money, so I stuck with the SG. In the 8th grade, I was reasonably accomplished on lead guitar, so I put together a band called "The Dimensions" which included lead guitar (me), a bass (Rob Hughes), a lead singer (Mervin Hall), a rhythm guitar (Rob McHeleny) and drummer (Gary Padgett). Our repertoire included Beatles, Stones, popular rock tunes from one-hit wonders and a lot of 3 chord progressions that everyone loved to jam and improvise over. Probably the weakest musician in the group was Rob Hughes, my best friend. He was such a popular and good looking guy (and a jock), I convinced all band members that we needed his help to get gigs (but he really did not contribute musically). We played small venues, private parties, school dances and so on and only did cover songs. We never composed our own music, however, and this clearly limited my advancement in music. Although I did not realize this limitation, I thoroughly enjoyed the popularity, notoriety and financial returns of playing semi-professionally. In the 9th grade, the band significantly improved when Lenny Aaronson moved into the Riverside Apartments (by the way, Ponce Cruse aka 'hints from Heloise' also lived in Riverside and was a close friend). I knew Lenny in Taiwan (his Father worked for Voice of America) and we quickly re-established our friendship. Len was a wonderfully gifted bass player (and as good as I was on guitar, perhaps a shade better). I invited Len to substitute for Rob and I came to realize just how important a bass player is in a small group! So Len came on board. Something else I learned about myself from this experience was that I am very uncomfortable making decisions that hurt friends (Rob took it very hard). To this day I tend to be conciliatory and I place collegiality as a valued virtue. The bass defines the 'pocket' that glues everything together rhythmically. Most people (including me) may not realize just how critical a good bass player is to the overall connection in the group. The band prospered and improved with Len holding it all together at the low end. We made decent money and invested in better equipment including matching outfits ("Beatle boots" and equipment upgrades) which made for a more professional package.

A major influence on me at the time was my track coach in 9th grade. James Van Dyke played college basketball and was a spitting image of the 'Marlboro Man'. I was also on the gymnastics team specializing in tumbling, trampoline, horse and horizontal bar; however, rings were not possible due to my poor upper body strength. I excelled in track and field (like my Uncle Bill and to a lesser extent my own Dad). There was no pole vault in Junior High school, which I was reasonably good at, so I focused on the high jump. At the time I recall being about 5'5" and setting the school record in 9th grade by clearing 5'9". I performed best with a crowd, probably because I could muster a rush of adrenaline and focus my effort. Mr. Van Dyke really encouraged me and often gave me a ride home after practice, since he lived in the next apartment complex. I looked up to him (as did many) and he was consistently positive and well disciplined. He was also a tough taskmaster and would take a huge paddle to guys who stepped out of line. I recall that he told my Mother that I was cocky and arrogant, but he viewed that as a positive attribute. I am not sure my Mom liked or agreed with this assessment. In summary, these years in Washington DC area were primarily dedicated to music, sports and socializing, all of which are linked closely to mating behavior. Academic pursuit was pretty low down on the priority spectrum, however, I made excellent grades and was not challenged in any classes. I suppose there was no room in my life at the time, although I still enjoyed reading science fiction and learning about scientific developments of the time. My Father was very active with early computer technology at the Pentagon and I recall seeing the CRT displays in his office. This was another source of fascination since the project involved AI (artificial intelligence) and my Dad was way ahead of his time because of his job at the Pentagon. My Father told me that the very early AI programs were tested in a variety of ways. One question that was asked was "are you happy with your job?" and the program responded with "who wants to know?" Basic AI routines understood well the chain of command!

In 1966, my Father made a decision that impacted the family directly. For his career advancement, he agreed to a tour in Vietnam and left us for a year. He informed me of my new role as 'man of the house'. With his tour in the Vietnam conflict, my Father would be on track for General Officer (he as a Colonel at the time). Father never spoke much about his experience in Vietnam, but based on his biography, he was General Westmoreland's Chief of Communications. As such he had a high security clearance and was not allowed to fly in helicopters or aircraft with less than 4 engines. In any case, the year he was away came and went quickly. A new duty assignment followed his return to the US and we once again moved. This time we relocated to Hampton Roads, Virginia (Ft. Monroe). I spent the first semester of the 10ᵗʰ grade at Kecoughtan High School, named after the Algonquin Indians who settled at Hampton Roads in the early 1600s'. I spent only one semester at Kecoughtan High, but many years later I collaborated with a fellow faculty member at OSU who was in my class. We did not know each other at the time (she was a cheer leader and I was a transient nobody). Still it was a surprising coincidence and my colleague showed me the yearbook and my 10ᵗʰ grade picture, which I had never seen.

The military brass gave new orders to my Father for another year in Vietnam, ostensibly to be promoted to General. I am not sure of the details, but within a few months of moving to Hampton, my Father retired from the Army as a full Colonel, after almost 30 years of service. I recall getting into a fistfight with the son of a Master Sargent at Fort Monroe, but I cannot recall why. I took a beating that required 4 stiches in my face and my Father got a 'juvenile delinquent' report over the incident. He always joked that this JD report prompted his retirement; at least I think he was kidding.

We moved to Austin, Texas where my Father took a staff position at the University of Texas in electrical engineering and worked on computer aided instruction (CAI). A Google search reveals a number of publications from my Father and his team at UT Austin on CAI and they attracted considerable federal grant support for the project. I was intensely interested in the project and learned much from working with faculty on the project and in particular interacting with my Father who took me to the CAI lab on weekends. This influenced my thinking about a future in science. I liked the problem solving environment, I enjoyed interacting with highly intelligent researchers and perhaps most importantly, I saw people intellectually engaged in an academic environment that was not 'work' or a job. I realized that if you love what you do, the reward is simply showing up (Woody Allen claims that 95% of life is simply 'showing up' anyway). This was the first and single most influential event in my young life. I saw a viable future path and was intrigued.

In high school in Austin, I re-established myself in music and got into a local rock group. Being a bit older and more experienced, I was highly selective about getting the best mix of talent and we were reasonably good, but not great. We played a lot of gigs, usually fraternity parties, local dance and high school events around Austin and most weekends were taken up with this activity. I tried out for sports (track) but realized that there were many more athletes at the high school level who were much better and competition was fierce. I was simply not big and strong enough to compete and win medals, so I quit the team. For example, my best pole vault was about 10" feet 11" inches and while other team mates were easily clearing 12' on the first vault. Contributing to my exit from track was an ankle injury stemming from an aborted vault where I came down in the foam feet first and twisted my foot (you are supposed to land on your back in the netted foam). I chipped a bone in my ankle and was out for the season. I still recall the intense pain of the injury and I have an oversized left ankle to this day.

Inspirational events, both positive and negative, during my Austin high school days are noteworthy. In the latter category, I witnessed a bad outcome of a University spin-off company gone terribly wrong. My Father got involved with 4-5 other U.T. engineering faculty on a failed business venture. The business was a printing company located in San Antonio and had some links to CAI and publishing somehow (details unknown to me). All partners put up cash to buy the company and my Dad worked long hours each weekend to grow the business. In retrospect, I believe that the company 'cooked the books' to make the business appear highly profitable. On top of this misinformation, the partners trusted a manager and former equity partner to oversee the business. Long story short, the business was mismanaged and to make matters worse, the partners were pointing fingers at each other and no one was accountable. There was no conductor with an equity stake who could take the time to run the company on-site full time. The company was mismanaged by the administrative staff (former owners) and within 6-12 months the company went under. All of the faculty partners filed bankruptcy and did not have to re-pay loans (although they had ruined credit ratings). My Father refused on principle and repaid the bank his share on the equity in monthly installments over many years. Lessons learned from this experience were manifold. First, it is important to control the business if possible. Limit the number of chiefs and keep it small initially. Second, never put your own money on the line. Use OPM models (other people's money) or grant funds and bring expertise and scientific acumen to the table. Third, control the equity and use that equity as leverage to motivate (and control) the stakeholders. Fourth, recognize that a start-up must evolve and the business model originally proposed can be re-directed as opportunities arise. Fifth, risk taking must be carefully managed and when money is 'on the table' be highly suspicious of input that can lead to misdirection. Finally, as a result of this experience, I realized why people don't like to start businesses. This was a painful and highly destructive experience that my Father had to suffer through over a number of years. He was a highly honorable person who repaid every penny to the bank, when all other partners bailed out with bankruptcy filings.

I made mention of the positive influences during this period of my life. I was strongly inspired by my Father, who set an amazing example of honesty, integrity and honor. There were others along the way, including parents of my high school friends, who were successful entrepreneurs. My best friend, Joe Bond who lived across the street, moved to Austin my senior year. Joe's Father worked at IBM and was a very impressive guy who influenced my thinking about business. Mr. Bond started a splinter company with links to IBM that was highly successful. I still recall his advice fondly. "When you start a business as a young man, and you lose the shirt off your back, all you lose is your shirt." This is sage advice. As you age and responsibilities increase (kids, family, debt, etc.) the risk to benefit ratios change considerably. Seize the opportunity when all you have is the shirt. In fact, I discussed some ideas for business start-ups with my Father, while in High School; however, he vigorously discouraged me and stressed that I should get an education first and foremost. In high school, I also worked at a Department Store called 'Gulf Mart' in Austin (in the sports department). The manager, whose name I do not recall, made a compelling statement with such conviction that I still recall it verbatim. "Go to school, and stay in school as long as possible". I'm thankful that I followed this advice. While I did not act on Mr. Bond's edict, I took a somewhat derivative route by combining my academic pathway with an entrepreneurial one, as described below.

Choices of Undergraduate and Graduate Education.

Academically, I was mostly an A student with B level performance in areas of low interest. I did well on testing platforms but was never really challenged and other activities held my attention. During my senior year, I made the rational decision to focus on studies and become engaged in curricular and extra- curricular activity. I started to look into different colleges and universities and my Dad encouraged me to apply to his alma mater (Cornell). However, I was reasonably sure that I would not get into an Ivy League program (grades) despite doing well on the SAT exam.

Choice of a University was molded by a number of internal and external factors. I was very concerned about getting drafted and slotted into the Vietnam conflict as a foot grunt enlisted man in a rifle company. This was a major issue driven by positive survival above all else. So, if I'm going in, I wanted to be an officer. In addition, I was leaning toward engineering (specifically electrical). I decided on Texas A&M which is a 2+ hr. drive from Austin because it had ROTC and a strong engineering major. I also considered a major in microbiology because I was also interested in quantitative biology, derived from my numerous science fair projects. To sort this out and make an informed decision, I decided to attend summer school at A&M prior to starting my freshman year. In this way, I could sample some courses (and my performance), while learning in more detail opportunities available in graduate or professional school. I quickly realized that math and calculus (differential equations) were not my strength. Although I could do well in such courses, it did not come easy. A decision to major in microbiology was made at the end of my summer term; however, my next step (graduate or professional school) was far off and I was not focused on this point. Unfortunately, I was not able to get good counseling at the time, so I was in limbo anyway. I finished out my freshman year in the A&M Corps of Cadets as "Fish Muller" in Medical Company E-1. In this first year I met many interesting and lifelong friends. My parents were very proud of my academic performance, as I made the Dean's list and managed to survive the Corps (hazing, crap-out sessions, white rats, the Fighting Texas Aggie Bonfire, etc.). The Bonfire, which preceded the fall classic 'Turkey Day" football game with Texas University (aka University of Texas at Austin), was a great experience and I saw a fantastic entrepreneurial opportunity, which I will describe later. In sum, there were no real academic influences worth noting in my freshman year; however, being in the ROTC, I learned military discipline and organization. The Corps imposes its collective will on its members, which includes attributes like the value of hard work, discipline, honor and respect. Such attributes are important to a 19 year old kid like me and despite the hardships, it was extremely important in my maturation.

In the summer of 1970, I moved back to Austin to live at home. I got a job as an orderly at a local hospital (Brackenridge). That summer I fell in love and met my future wife, Annette, who was enrolled as a nursing student, taking courses at U.T. and performing clinical duties. I had to work hard to win her over, since she was involved with a rival suitor in Vietnam. Lucky for me, I won out and we have been together for almost 44 years. She has been a huge and ongoing inspiration with skill sets that complement some huge voids associated with my character. She is a genius with people has excellent social skills and is also extremely positive and optimistic in all matters. Thus, she is well liked by any who meet her. Sadly, I am just the opposite, so we make a good team. In some situations, I may not feel like interacting socially, she will come to my aid and act as the interface.

In the fall of 1970, I started back at A&M, now as a "Piss-head" in the Corps (2nd year). Things got more interesting academically when I took my first genetics course and started the chemistry series. I especially like genetics and got to know one of the instructors, Dr. Clint McGill, an assistant professor in the Genetics Department. His wife, Jane, was also on faculty, but I'm not sure if she was tenure track. Jane taught in the Department of Biochemistry as I recall. Clint worked on *Aspergillus nidulans* and I approached him to inquire about undergraduate research. I asked if there might be someone to work with on a project, but unfortunately, his lab was full. He directed me to Dr. Gerard A. O'Donovan (Gerry) who was in the Department of Biochemistry and looking for students. Jerry was a fiery Irishman who could really spin a yarn. His stories were legendary and he was one of the most entertaining individuals I've ever met. He was a former Soccer player (he claimed) and actually taught the A&M punter how to kick field goals. Jerry claimed he could nail the uprights at 40 yards. It made for great story telling, but I have my doubts about the veracity of his claims. Gerry's lab worked on bacterial pyrimidine metabolism and he and Jan Neuhard (University of Copenhagen) published a very nice J. Bacteriology review on the subject in 1970, and I read and re-read this review so many times I practically knew it by rote. That semester I read James D. Watson's book entitled the "Molecular Biology of the Gene" and I was hooked. For the first time, life was not a mystery but founded in solid chemical principles and I was ecstatic. I read his book at least three times and had many discussions with Gerry, Clint and other lab members. In the spring semester, I started auditing graduate courses in biochemistry and genetics, in addition to my other courses. This was not necessarily a good thing since I devoted too much time to the lab and faltered on organic chemistry (making my first B- in a science course). I started thinking about a future in academic science at the time and received encouragement from the McGills and Gerry O'Donovan (along with some of his post-docs). I finished my sophomore year and was excited about the future. I was fascinated by molecular biology and microbial genetics. I struggled with the primary literature so most of my reading centered on textbook sources and reviews. I developed strong expertise in nucleotide metabolism, at least at a theoretical level and even at that early time of my career, I saw the potential for applications in cancer (at the time, 5 Fluorouracil was just starting clinical use). In retrospect, it was unfortunate that I did not start working in the lab on a specific project; however, my duties in the Corps, a heavy course load and evening audits of graduate courses left no time to get going in the lab.

Due to my poor performance in Organic chemistry, I decided to take my last chemistry course over the summer term (Quantitative Analysis) at U.T. in Austin. This would free up some time in the normal academic year for other activities related to undergraduate research. I spent time hanging around the UT campus, well spring of liberal anti-war sentiment at the time and interacted with my Dad whose office was near the chemistry labs. I did not take a job that summer and focused on reading and my studies. The Quant instructor was a chemistry graduate student and we had many lively discussions about science. I did well in the class and enjoyed a carefree time spent with two things I loved most, academics and my soon to be wife. Things were definitely looking up, except for a major issue that loomed before anybody of draft age: Vietnam. My UT summer experience made me re-consider my stance on the war, which as a Fighten' Texas Aggie, had been rather hawkish. I began to realize the futility of the war and how irrational it was to most young Americans who would fight, die or be injured. I met several "SDS" advocates (Students for a Democratic Society) and they made persuasive arguments against the Vietnam conflict. These guys were serious archradicals and I recognized this as well. On the other hand I recognized flawed logic being perpetrated on the public and when events affect you personally, you will think long and hard. The prospect of having my education derailed by the draft hung over me like an ugly black cloud. I could never openly discuss this with my Dad, a retired and decorated career man who served this country in WWII, Korea and Vietnam. Some things are best kept to oneself. My mother, in contrast, clearly did not want her only son going off to war, but she kept quiet about such matters.

Starting back at A&M in the fall, as a 3rd year student, I declared my major officially as microbiology and began to think about getting into a research project in earnest. I did not sign the ROTC 'contract' to enlist as an officer, but instead continued through the program as a "D&C" or Drills and Ceremony cadet. I began working with Jerry O'Donovan's group and learning microbial genetics with a focus on *Pseudomonas aeroginosa*. I always liked working with Pseudomonads, simply because they always smelled nice (sort of grape like). To this day, I can tell *P. aeroginosa* spoilage a mile away using my nose as a diagnostic tool.

Late in the semester, I recall that President Nixon established what I call a 'quantitative metric' on the draft lottery. The lottery system, based on birthdates, established a priority list based on random drawings of the date. For example, if your date of birth was June 12th, and that date was assigned a lottery number (say 1), all men born on this date were inducted first. Clearly, the lower the number the more likely you would be called up. The process was not necessarily random and there were differences between draft boards. For this reason, I stayed in ROTC since my number was 127, which is well within striking distance for Vietnam! In 1971, it was decreed (I think by Nixon) that all draft boards would cutoff at 125. This meant that I would not likely be inducted; thus, in December of 1971, I dropped my student deferment and became 1A because I was surely not going to be drafted. In January of 1972, I then would be '2nd priority' meaning that I would not be drafted unless everyone from 1 to 356 got drafted first. I was safe! At the end of my junior semester, I dropped out of the Corps of Cadets, handed in the uniform, grew my hair and a beard and led the liberal life of a 'non-reg' (slang for an aggie outside the Corps). This opened up more opportunities for spending time in the lab.

In early 1972, I met several individuals who influenced me significantly. Dr. O'Donovan continued to be a jovial figurehead and a source for great entertainment, but offered little research direction (his initials were prominently displayed on all lab items as "G.O.D.") He was typically 'hands-off' with graduate student and post-docs, so I decided to work more closely with lab members. Gerry O'Donovan had two lab groups, one in genetics and one in biochemistry but most of my time was spent in the genetics group (with Clint and Jane McGill and graduate students, John Womack, Cassandra or Cassie Smith). These individuals provided research training microbial genetics and worked with me on the project, which involved analysis of pyrimidine metabolism (specifically salvage pathways) in *P. aeroginosa*. During this time, I met a Danish couple, who were visiting post-docs from Jan Neuhard's lab at the University of Copenhagen (Soren and Marianne Norby). Soren (actually the 'o' has a slash making his name sound like "Siren" to English speakers) was an MD researcher while his wife a Ph.D. Marianne was rather cold and distant with classic Scandinavian features. She was probably the most beautiful female I had ever seen, with large exotic and intense blue-green eyes and a lithe, athletic physique. She made an impact on all around her, yet was largely unaware of the affectations of natural feminine beauty. Too bad her personality did not match her looks. She barely acknowledged me and was fiercely arrogant. Her arrogance was not a result of her looks but rather derived from an intellectual haughtiness borne from a keen analytical mind and acute intelligence. It was intimidating and I stayed well away. Soren was just the opposite, however. Soren looked like Steve McQueen and many people made note of this while he was living in the U.S. His persona was engaging and he was extremely articulate, speaking essentially unaccented English. On occasion, he would recite poetry or launch into a Shakespearean sonnet with a proper British accent of course. Everyone liked him and he was very helpful to me. Soren was more of a classical geneticist working with Drosophila (fruit flies) as a model system to probe nucleotide metabolism in a tractable eukaryotic system. He was also well read and broad based in his frame of reference. For example, he had an interest in comparative linguistics and when I expressed an interest in learning languages, he worked with me to learn some Danish phrases. Being exposed to such interesting and intelligent young scientists (both were post-docs) convinced me that academic science was an ideal pathway for me. I continued to struggle with the lab chief (aka "G.O.D.") largely because he did not provide any reasonable direction or objectives, but was more intent in 'holding an audience' with any who would listen. Soren and Marianna also complained about Gerry O'Donovan. Evidently, Gerry had made wide sweeping claims about the availability of resources and equipment in order to lure them into the lab. They were most unhappy about the misdirection. For my own work, I decided to focus on a key enzyme, ATCase (aspartate transcarbamylase) a key early step in pyrimidine synthesis. I obtained a diploid strain to over-produce the enzyme for biochemical studies.

Another influential person I met during that semester was Rodney Kellems, then a Ph.D. student at the University of British Columbia in Vancouver. Rodney worked with Tony Warren at UBC (Dept. of Microbiology) and was collaborating with Gerry on the Pseudomonas project. Rodney was also outgoing and extremely knowledgeable. By outgoing I mean, in your face, up close and non-stop, but I liked that. We worked closely together and despite his short visit (1 month) I learned so much about the biological system (Pseudomonas). He strongly encouraged me to consider UBC for graduate school, which was sage advice.

At some point early in the spring of 1971, I asked G.O.D. if there was any chance to go to Jan's lab and work over the summer at the Enzyme Division at the University of Copenhagen B in Denmark. As usual Gerry was very positive and encouraging saying that he had NATO Funding for this exact purpose. He said he would arrange my visit to Denmark on the NATO Fellowship. All I had to do was purchase a ticket and upon arrival, contact Soren (no email back then, so communication was slow). In summary, I trusted that all the arrangements would be made, based on reassurance from my boss.

I was thrilled to be working abroad and excited about the prospect of learning Danish and becoming culturally immersed in everyday life in Copenhagen. I arrived in early June with a backpack and a gut string classical guitar that was my 'road guitar'. I was given the lab number to call on arrival. Much to my total surprise and dismay, no one knew that I was coming. Gerry had totally misled me on the whole affair! Gerry clearly screwed everything up and a number of people (me included) were upset. Soren and Marianne were very kind and I could tell felt sorry for my plight. There were some perplexed and probably angry phone calls from Jan's lab (Gerry's collaborator) and I don't know the details; however, somehow things got settled and I was provided a dorm room in "Nordisk Collegium" near the campus (after a temporary residence in an vacant apartment of one of the lab staff). Soren and Marianne were managing the Collegium facility, probably like dorm supervisors, in exchange for a nice spacious apartment (they were in the process of buying home in a place called Birkerod (slash o), a suburb in Copenhagen. This was a nice fit for me and I was grateful to have my own place that was rent free.

That summer, I worked closely with Soren and Jan (and Elsebeth, Jan's Technician). Some of the highlights were as follows. Learning to speak Danish (at a "Folk University") made little sense for a couple of reasons. First, only about 6 million people worldwide speak Danish and second, nearly all Danes speak English. I enjoyed mastering the language AND going to bars and using my language skills. Most were quite surprised! Other notable events included learning to get to know the lab staff who became close personal friends, including Just Justesen who later visited me in the US. While practicing Danish, I asked Jan (the lab chief) a question which I frequently employed in my evening exploits in local Copenhagen bars. The direct translation is "are you happily married?" I found this phrase useful for obvious reasons but Soren also pointed out that the Danish word for marriage (gift) also means 'poison'. Thus, when you ask "Er du lykklelig gift?" You are also asking if you are happily poisoned. Those Danes have a real sense of humor! This humor was lost on Jan. He was outraged that I would ask such a question, even in jest. I later learned he was having an affair with his technician (Elsebet) who he later married (following the divorce from his attorney wife, which must have been interesting). Clearly, I hit a nerve. Lesson learned: laboratory romance is an ongoing phenomenon in Danish academic settings.

My unannounced arrival in Copenhagen landed G.O.D. in serious hot water with his colleagues and collaborators. I think it was the first time he really got caught in this fantasy world of storytelling that he created. Gerry turned his vindictiveness toward me and I was caught between the two warring factions. This was my introduction to University politics on an international level. I recognized that truth was on my side and Gerry clearly knew he was culpable; still I had to stand my ground yet not totally alienate my boss in the US. The Danes were impressed with me and I accomplished a lot that summer, but I was not able to get published, unfortunately. The training and real world experience was price-less and despite the backlash due to irresponsible behaviors, I do not regret the experience.

One experience that I will always cherish was an invitation to visit Soren and Marianne in Jutland at their summer home on the sea. I borrowed a small motorbike from Just Justesen, strapped on a backpack and took the ferry from Sjaelland (the island where Copenhagen is located) to Jutland (a peninsula that connects to Germany). This involved many hours of driving 50km/hr. on side roads and shoulders across the whole country of Denmark. I got by easily with my newly acquired language skills as long as I was in Sjaelland; however, Jutland was a totally different experience. I could not understand the Jutland accent. They seem to swallow whole sentences and spit back some strange guttural explosion of words that sounded like they were gargling fish. They, on the other hand, had no problem understanding me and often commented "you are from the city". I slept in youth hostels for the most part. Just Justesen had given me the address for a farm his family owned in Jutland and told me I could sleep in the barn (the place was abandoned) so I spent one sort of spooky night in the countryside with no one for miles around.

I arrived at Soren's summer home and slept on the couch in their small bungalow. Soren and I played Frisbee, rode bikes through the countryside and fished. He had a small rowboat and a gill net and we would go out and get our day's catch for dinner. One time, I got terribly seasick due to the large rollers, so I was reluctant to go back. Soren had a great sense of humor, usually play on word humor, so I spent a lot of time laughing. (Example: while pointing out cows, he asked which cows were females and quipped... it should be utterly obvious). Marianne was quiet and aloof for the most part; however, she spent the entire week topless (and recall she was a perfect 10). Soren suggested I wear blinders; more cow humor, I guess. We had a visitor, Elsebeth Lund, from

Aarhus University. She was a post-doc who spent several nights. I also found Elsebeth to be cold and distant, like Marianne. I figured out that this must a Danish trait shared with female scientists, and accepted it for face value. Soren told me that Elsebeth was a bit of a home wrecker and was known to sleep with senior faculty; more intralab romance drama. The Danes are pretty tolerant and tend to be open minded about these things. I returned to Copenhagen and completed my work in the lab and packed up for home. The week or so before leaving Gerry O'Donovan paid a visit. In the end, he made good on his promise to provide the NATO Fellowship support and I was reimbursed for most of my expenses over the summer, so things turned out well indeed.

I recall returning to Austin and seeing my fiancée (Annette) who had since graduated and was a practicing R.N. making very good money. She had her own apartment in North Austin, a nice new car and was very happy to see me. This was one of the happiest times in my life, being young, in love and experiencing a new beginning. The future looked very good indeed and I was thoroughly invigorated.

I started my senior year with all batteries charged up. I saw my pathway as an academic scientist. I next needed to find a graduate program. My positive experience with Rodney (from UBC) influenced my decision and I applied to UBC Department of Microbiology in Vancouver, British Columbia. I got very strong letters of recommendation from Gerry, Rodney, Soren and Jan Neuhard (I learned later) and was accepted with full support into the program on a "Ph.D. Direct" path, which meant that I would bypass the Masters degree. The Ph.D. direct path was contingent upon passing a rather rigorous general and written exam and required a perfect 4.0 performance in all graduate courses. I graduated with honors from Texas A & M in early May of 1973.

I would like to digress a bit from my academic history and discuss a business venture that I started as a junior at A&M. I learned from this experience that creating a business is exciting and that a for-profit outcome is unbelievably seductive. I seriously considered getting out of science and honing my business skills as an entrepreneur and risk taker; however, my Dad talked me out of it (and I am glad he did). I started a photography business that filled a needed niche and was totally a cash activity. I have always like photography and with my Dad, we built a home darkroom in Austin that was set up for B&W processing. I convinced my Dad that we should invest in professional grade equipment (high end enlargers, contact printer, high capacity drum dryer, print washers and of course a panoply of medium and large format cameras). It was done right and had reasonably high throughput. It took very little to convince him to invest. He collected antique cameras and was a total camera-phile, so it was an easy sell. As an aside, my Father's antique camera collection was donated to the University of Texas and is still on display in Austin (the "Mark T. Muller Camera Collection"). There are hundreds of cameras dating back to the 1860's along with daguerreotypes, tintypes, ambrotypes, carte de vis, and Civil War images in wonderfully carved union cases. I recall him saying that the world's oldest known photograph is in the collection, but I have not verified this.

In the fall of 1971, 72, during construction of the Texas Aggie Bonfire, I recognized that many in the Corps of Cadets would pay for 8x10 B&W pictures of their company activities, peers and themselves working building and organizing bonfire activities. Nobody was providing this service and since I was now an upperclassman, I had easy access to all bonfire activities. I bulk loaded 35mm film and made 40 frame cassettes, carefully labeled and cross-referenced with detailed record keeping. I used a state of art Nikon 35 (with multiple lenses) to capture thousands of images of individuals, groups of working aggies, whole companies and of course the bonfire itself. These images were all contact printed and I handed them out to the company sergeants in each dorm in the quadrangle and the orders rolled in. I charged $1 for an 8x10" print (only 1 size was offered) which was very fair and provided me a tidy profit margin. That first year, I cleared more than $5000 in two weeks of work. This was a huge sum of money back then and I put it to good use (saving for the future, paying back my Dad, buying a rather pricey ring for Annette). The next year was a repeat, but I engaged a sales force and paid them to help peddle the images and for darkroom help, I hired one of my best friends (Bill Gillespie, aka 'Skyboy' since he lived his life in a weed infused fog for most of his undergraduate years at A&M). Bill was no dummy and got an engineering degree from A&M and is an oil executive! I recall charging a bit more per picture that year, and reaping an even larger amount of money. It was easy money, it was my hobby and it was great fun. These funds were put in a savings account. During my Senior year, I also did several commercial photography jobs including shooting portraits and weddings. I hated wedding photography since it was high stress and sometimes the personalities were difficult. There was very little profit in weddings in any case, so this was a low priority activity.

After graduation, I started flying lessons. I have always been fascinated with flight and aviation and signed up for the ground school at Bergstrom Air Force Base (in Austin) and enlisted a flight instructor (CFI or certified flight instructor) named Wendell Fuqua to train me in a Piper 140 (a low wing trainer). It was hot that summer (as always in Austin) and the thermals (turbulence) were brutal but I enjoyed the flying nonetheless. Since my Dad was retired military, and technically I was still a dependent, I was allowed to use the Air Force Base facilities for all of my training. I passed the medical and an easy FAA written exam and did my solo. I needed some cross country time to complete the program (along with night flying); however, I simply ran out of time. I was to be married in late July followed by our honeymoon in Vancouver (to find a place to live for graduate school).

In Vancouver, I met several faculty including Tony Warren, Bob Miller, Jack Campbell, Barry McBride, Julia Levy to name a few. All were extremely nice and acutely intelligent. They were genuinely interested in me as a first year Ph.D. student who might be interested in joining their respective research programs. I returned to Austin and in August (1973), packed up a U-Haul trailer with my new wife and we drove our Dodge van from Texas to B.C. to begin graduate school. The cross country trip gave me time to think and reflect about my life and the future. I was awash with enthusiasm for the future.

My choice for graduate school was strongly influenced by the people I worked with as an undergraduate; however, I was careful to validate that choice. My 'due diligence' involved looking carefully at the UBC research base and curriculum. I evaluated funding potential (stipend support) but never considered how much I would be making as a Ph.D. student. It was important to have some sort of ongoing support, but actual amount seemed irrelevant at the time. Part of the reasoning was that my wife could work as an R.N. and would make very good money, so I was less concerned about such matters. UBC is an exceptionally strong Canadian university. The Department of Microbiology was also a strong academic program, perhaps even the best in Canada. UBC is a currently a premier center for biomedical research in Canada. I have no regrets on this choice.

Choosing a Research Lab: Graduate School.

I started graduate school at UBC in September of '73. The UBC Ph.D. program is loosely patterned after the British graduate education paradigm (hence the name 'British' Columbia). In a nutshell, the program is characterized by minimal course requirement and the overarching view that the Ph.D. is a research terminal degree and students should start working on projects with little or no coursework after that first year. To allow nascent

Ph.D. students to make informed decisions about laboratory choice, the first year also involved 10 week rotations in at least three different areas within the broad field of Microbiology. I did four rotations in a phage/microbial genetics lab (Bob Miller), in a bacterial physiology group (Barry McBride), an immunology lab (Julia Levy) and a molecular virology lab (Jim Hudson). I learned so much from this experience and developed a broad appreciation for model system science, available research techniques and hypothesis driven science. Most of the courses were seminar style courses where the students presented papers to groups of faculty who posed questions and probed the student's basic understanding of the primary literature. I recall my graduate class, where Bob Miller assigned 7 papers for the next class period (the class had 8 students). We were told to be prepared to present all 7 papers in the next meeting and students will be selected at random to present one of the papers. This was pretty intense and the pace was maintained throughout the semester. I also took a course on "Nucleic Acids" that was taught by Gordon Tenor and Michael Smith.

Michael Smith was a very influential scientist in Canada and I would like to discuss his influence in my own career. Many years after graduate school (1991-92), I did a sabbatical in Mike's lab in the Department of Biochemistry at UBC. Like me, Mike was a morning person, and I loved coming into the lab and having coffee with him early in the day. The next year, 1993, Mike won the Nobel Prize in Chemistry for his work on oligo directed site mutagenesis. That year was an unusual Nobel year because two 'technique-Nobels' were awarded (Kerry Mullis, shared the prize for PCR). Mike gave me the following sage advice to those who aspire to win the Nobel Prize. Wear Birkenstock sandals and sleep naked. Mike's sense of humor was legendary. I also met Mike in the late 70's when he gave a seminar at McArdle Laboratory for Cancer Research (University of Wisconsin, where I was a

Post-doctoral Fellow with Dr. Gerry Mueller). Mike had done a sabbatical in Switzerland (to learn molecular cloning which was just taking off) and was invited to Madison to give a seminar. He told the audience that when he returned to UBC, the Dean asked him to give a seminar on his accomplishments in the area of re-combinant DNA. Mike then proceeded to report on his sabbatical 'progress'. His project was to craft recombinant DNA strategies to turn cow manure into chocolate (he was in a Swiss lab after all) and he was delighted to report that he made great advances on the project. A major advance had been made in fact. They had already successfully used recombinant DNA to get the color just right and now were diligently working on taste, texture and aroma!

My graduate school research rotations were challenging and intellectually rewarding. In each lab, my curiosity and creativity was stimulated and my scientific maturity advanced. I was fasci-nated with viruses and their simplistic life-plan yet complex pi-rating of the host biochemistry. Molecular virology became my passion and I enjoyed the trappings of the field; specifically, working with animal cell cultures, molecular aspects of DNA and RNA, analysis of novel viral proteins and thinking about the work from a molecular perspective. The decision to join Dr. Hudson's group was made. I liked Jim because of he was very soft spoken and had an easy going personality. On the other hand, Jim's easy going character meant less direction on the research, which pre-sented a problem at first. Fortunately however, there were several good people in the lab that helped offset the problem. Also, I tend to be a self-starter and independence came natural. One of my class mates, Vikram Misra, was a very positive influence on my life. Vikram was a strapping 6' tall guy from Jodhpur, India, who oozed charisma and we became fast friends. We shared a house for a while in 1974 with Vikram and his wife, Lyn, living downstairs and my wife and I upstairs. At the end of the rotation year, Vikram also elected to join the Hudson lab, which was fantastic. Vikram worked on transcriptional control in lytic MCMV (murine cytomegalovirus or MCV, mouse cytomegalovirus as the Hudson lab called it) while I worked on the impact of the host cell cycle on viral replication. We were a good collaborative team and complemented each other. This worked out to be a synergy that led to a productive outcome. I established a nice project looking at the delay in replication of MCMV in G1 arrested cells and using this as a possible model for herpes virus latency. It was an exciting project and my advisor, Jim Hudson had little if any role in deciding my research direction. Jim was better at organizing data and writing up papers but otherwise was a pretty loose lab chief. I learned the importance of self-reliance and independence in the lab. My Ph.D. committee included Hudson (chair), Gordon Tenor (Biochemistry), Bob Miller (a T-phage expert), Gerry Weeks (Microbiology) and Ray Reeves (from the Zoology Department and a chromatin expert). When my thesis was completed, an external committee member from the University of Calgary was added. All were very good scientifically and brought their unique expertise to the committee. Ray Reeves was instrumental in providing advice for a post-doctoral position a few years later. I got very little input from Jim on post-doc positions as I recall; however, he was very supportive in my decision.

My next hurdle was the qualifying exam (or general exam), a three day written portion and a 2-3 hr. oral exam on general knowledge related to your research area. Late in year 2 of my graduate studies, both Vikram and I took the general exam to officially enter the Ph.D. track. I worked every day at a cubicle in the library, designing the future project and establishing a frame of reference in the literature. I did this over a 1 month period prior to the exam. Vikram went first (about a week before me). Both of us did well on the written and the project-based oral loomed before us. Much to my surprise and dismay, Vikram did not pass; however, he was given an opportunity to retake the oral portion. Since he was a more seasoned graduate student and scientifically more mature, I felt my chances on the oral portion were remote at best. This forced me to work doubly hard at the preparation stage. I passed with surprising ease and got glowing recommendations and positive comments from all of the committee members. I learned something about myself in the process, specifically, that I can rise to an occasion given enough motivational time. Vikram eventually took the oral again and passed. We both set out working on our respective projects in earnest and in year 3 we made significant progress. I eventually published 8 papers from my Ph.D. thesis (not counting the thesis itself). Some of these were collaborative manuscripts. One paper that was very nicely done and was a chapter in my thesis was the demonstration that MCMV did not encode its own viral thymidine kinase (TK). In science, it is very difficult to prove a negative; however, I engineered a foolproof approach using a TK minus HSV-1 mutant from Saul Kit (I recall having to drive down to the Univ. of Washington, in Seattle, to avoid a customs hang-up). TK is an important 'Achilles heel' in herpes virus since drugs like Valtrex and acyclovir target the viral enzyme selectively. However, it was important to determine whether the MCMV genome encoded this enzyme, which I showed using a halogenated pyrimidine analog (IUdR). IUdr (and related BUdR) both strongly inhibit HSV-1 replication and the TK mutant was an important control to show that in our hands, these mutants were wholly resistant to the drug. MCMV behaved exactly as the TK minus mutant, being resistant; therefore, the virus must not encode for TK. This paper was accepted immediately and without revision and is well cited.

I went through some personal problems after our first son was born that nearly de-railed my marriage and career. My wife Annette suffered post-partum depression for some months and I felt the marriage was doomed. I eventually suggested a trial separation (which in my mind meant divorce). Contributing to this was my desire to be free to do what I wanted, a terribly self-serving attitude driven by hormones and a narcissistic attitude. She moved to Austin with my first born (Soren, named for Soren Norby now on faculty at the University of Copenhagen, Department of Genetics). Coincidentally, my Ph.D. advisor went on sabbatical for a year and I was largely unaccountable for any research progress since no committee meetings were held. The day of reckoning came after his return and the committee really unloaded on me for lack of productivity. After much discussion, Annette and I decided to give our marriage another try and after a 6 month separation we were re-united. Both of us adjusted our attitudes with Annette becoming fiercely independent and less 'clingy' and me settling down and nosing into the grindstone while being an attentive father and husband. I am so glad that we both gave the marriage a second try.

I worked exceptionally long hours in the final year of my thesis. I developed, in collaboration with Vikram, computer software to analyze RNA expression complexity based on "R_0T Curves" and regression analyses. I loved working with DNA /DNA and DNA/RNA hybridization kinetics and the data analyses that followed. I designed and implemented machine based programs to analyze the data (on an HP 'desktop') and we recruited a software consultant to run the mainframe analysis of complex regression analysis.

Nearing completion of my Ph.D. research, I began a quest for a good post-doctoral lab in the U.S. I had the opportunity to attend a symposium in early 1977 and heard a fascinating seminar by Gerald C. Mueller, M.D., Ph.D. from The McArdle Laboratory for Cancer Research at the University of Wisconsin. He gave a wonderful talk on eukaryotic DNA replication. I was advised by several people that one should do a post-doc in an area that is distinctly different from one's thesis topic. The idea was to be employable and enhance one's resume to fit different teaching needs in the job market. This is very good advice in point of fact as I had job offers in both areas of expertise (in my case, virology and cancer). I made a formal application to Dr. Mueller's lab and was quickly accepted into The McArdle Laboratory for Cancer Research to begin post-doctoral studies in the fall of '77; I was 26 years old. I found out later that Dr. Mueller had made several phone calls about me and spoke with my committee members, including Ray Reeves. The thesis defense went well; however, there were a couple of committee members who were quite formal about the writing style. Gerry Weeks, a recent addition to the committee strongly objected to my 'loose' writing style, which others did not seem to mind (for example referring to protein 'cognates' as two different DNA binding proteins that bind the same DNA sequence). I made the edits of course, all the while grumbling under my breath. Gordon Tenor made a rather humorous comment I thought (which was valid). I commented in the presentation and thesis about virus yields being a 'log' apart. I knew it was jargon, but he asked me if I meant 'presto logs'? More corrections but it was an easy fix. Recall that this was in the days before word processors. I wrote my thesis out longhand and because my penmanship was terrible, I had my wife recopy it into beautiful script and then the Division secretary (aka Rosie) typed it up. I will forever be in debt to my wife for this herculean effort and also of course Rosie. I finished up my Ph.D. about a year before Vikram, which worked out very well since he and I had collaborated on a few papers. Vikram stayed in the Hudson lab to oversee and complete these manuscripts and perform the necessary changes that reviewers always impose (in some cases, adding experimental data). This was a great synergy and we both benefited. I will always be indebted to Vikram for following up on the publications and seeing that I was represented. He is a very good friend. Vikram completed his Ph.D. and went to Roger Hand' s lab in Montreal. Vikram is currently Chair of the Department of Microbiology at the University of Saskatchewan in Saskatoon, Canada. He has had a distinguished career and has remained with his wife Lyn.

Post-Doctoral Studies.

After completing the editorial changes, getting papers submitted, attending an international meeting in Calgary (and presenting) and saying our goodbyes, we packed up our household goods in a U-Haul and set out across the country for the mid-west. Fortunately, I was able to sell the small sailboat-trailer combination that I had acquired two years before when my wife was employed as an R.N. working at VGH (Vancouver General Hospital). We were flat broke and this small sum allowed us to defray the cost of moving back to the States. The sailboat had a small cabin and we did a number of very fun trips around English Bay (avoiding the cargo ships was daunting), up Howe Sound and to the Gulf Islands while living in B.C. Vikram and his wife Lyn also had a sloop and there is more safety in numbers, so we ventured far and wide on our explorations of the Gulf Islands.

Driving over the mountains pulling a trailer with an aging van led to some problems that had to be corrected enroute (radiator issues). We did an overnight stay in Wall, South Dakota since we saw signs hundreds of miles away drawing us in! We got up early the next morning and enjoyed a 57 cent breakfast, one of big attractions at Wall Drug Store. Along the way, we paid a visit to my Aunt Rita in Minnesota and spent time with my paternal grandmother. My grandmother looked like a plump version of a little muffin, with wiry white hair pointed in all directions looking much like a version of Einstein in drag. She was in fact a little old Jewish lady in her nineties in a nursing home who looked at me and recognized me as her son (my Father). My name was always 'Markie' so when she called me 'Mark', it was clear. Granny spoke with a heavy German (Austrian actually) accent and when I spoke to her in German she became visibly upset because I butchered the language so badly; ok… English only! Granny lived to be 100 but was in her own world the last years of life. My Father said that he believed that she was older than her stated age, because she kept her actual birth year a secret (there were no records from Vienna in any case). It was good to see her one last time, even if she was in a fog.

After a long trip (40 hrs. of windshield time) we arrived in Madison, unpacked our belongings in the storage basement of our apartment (we did not move in) and then got in the van and drove to Austin to visit my folks. More drive time with a 2 year old (Soren) and a van without air conditioning! Good training for purgatory. I tried to keep everyone's spirits up but we were miserable moving into hot weather in August. We finally got to Austin and spent some needed vacation wrapped in the cloak of luxury that was my parent's very nice home in an upscale suburb of Austin called Northwest Hills. The baby was whisked away by my Mother and Annette and I were free to reconnect socially with old friends, classmates and others. We spent a very pleasant time for about 2 weeks. We were back on the road to Madison to start my post-doc in G.C. Mueller's lab at McArdle in Madison. This was in September of 1977.

I started working in the lab on a project trying to understand the role of cytosolic support factors in DNA replication using permeabilized cells. We devised a clever strategy to analyze events in vitro. DNA replication origins do not initiate in permeabilized cells; thus, we were looking only at completing already initiated DNA chains (pre-existing *in vivo* 'extension'). To discriminate, we used BrDU (bromodeoxyuridine) to tag the replication chains in the intact cell, then we isolated and prepared the permeabilized cells and 'completed' the chains of DNA synthesis. Here is the cool part: BrDU substituted DNA is highly sensitive to long wave UV irradiation and becomes fragmented. The 'new' DNA that was made in permeable cells was released and we analyzed this part without complications of the bound DNA from intact cells. I was the first one in the lab to run 'submarine' agarose gels for this analysis. Back then, DNA cloning was in its infancy and gel separations were the hot new ticket in technology (of course routine these days). We published a very nice paper on this work (after two years of post-doctoral research). I came to realize that my mentor was only interested in publishing very high profile papers and he was not inclined to allow post-docs to put out anything less. While this is laudable, it can get out of hand since post-docs need publications to get jobs and if they don't strike gold on a project, they get nothing. Hard work and long hours in the lab are no assurance; it is necessary but not sufficient. I was in a big group with many post-docs and all were competing for that next big Science, Cell or Nature paper. Welcome to the world of high profile, high impact research, I told myself. For the first time, it became obvious that working extra hard to gain an edge was not going to work (it was necessary but not sufficient). After my second post-doctoral year, I informed Dr. Mueller that I had sufficient data to publish another paper (or two even); however, he resisted. I considered changing labs and going into Janet Mertz's group. Janet had done some exceptionally nice work on *X. laevis* microinjection with John Gurdon (Nobel Laureate). When Dr. Mueller got wind of this, he changed his demeanor and was a bit more encouraging since he did not want me leaving the group (in retrospect, I should have left). I discussed matters with Bill Sugden, a younger faculty member at McArdle (working with EBV and a rising star from Harvard). Bill saw my point of view and felt that Dr. Mueller was acting a bit irresponsibly, but to keep the peace, he suggested staying in the lab. I stayed on but my ambition was blunted by the experience and I came to grips with the prospect of only getting a single publication from my post-doc experience.

During my time at McArdle, I got more experience and maturity from a vast number of individuals, many of whom I am still in close contact to this day. Probably the most influential was Howard Temin who won the Nobel Prize (shared with David Baltimore) for discovery of reverse transcriptase in retroviruses. Dr. Temin was very well known of course and respected. He worked at the bench (seven days a week). He would come to the lab early and examine the cell cultures using an inverted phase microscope, looking at various retroviral mutants and cytopathology. He had an intense demeanor and was highly focused. Several years later, I invited him to Ohio State and he gave a high profile seminar which was one of the best attended talks ever given. During this visit, he was interviewed on a radio and a TV program. Dr. Temin also oversaw the tumor virology training grant with many talented trainees. We got along well and I respected him. He died of lung cancer at a relatively early age after I left McArdle.

One of brightest people at McArdle (beside Dr. Mueller who was a true genius) was Bill Sudgen. Bill and Gerry Mueller did not get along well for some reason (probably due to age related divergent opinions about how to conduct research). Bill was the most creative and amazing intellect I have ever known. We got along well and he was receptive, kind and giving of his time; he read my proposals, gave insight on experiments and was spot-on with his advice. Bill is still active in science but at a greatly slowed pace it seems. Many years later (mid 90's) he was called in as a consultant on a Venture with Senmed Medical, a VC firm out of Cincinnati, Ohio. There were other influences at McArdle. Richard Burgess, a mid-career associate professor, was arrogant but an excellent biochemist with wonderful insight on protein structure and function (he had worked extensively with E. coli RNA polymerase and sigma factor). There were a couple of graduate students that I respected. One was Jeronimo Blanco, from Barcelona, working on a Ph.D. (in the Mueller lab). Jeronimo worked 12-16 hr. days on technology to analyze protein-DNA binding in HeLa cells. He did crazy experiments, like hooking up Sucrose gradients to electrophoresis systems to resolve chromatin structures. This was also messy business as you can imagine, with sticky sugar solutions all over the bench (and lab). I became close personal friends with Jeronimo and his wife Jeannie (an American) and after leaving McArdle he took a faculty position in Spain. I visited him once back in the 1990's and was impressed with the quality of life and his academic environment. I overlapped briefly with a graduate student (Dan Schoenberg) who received his Ph.D. working with Dr. Mueller. Dan and I had many discussions over Dr. Mueller's resistance to publish good science (not great science but nonetheless solid). I met Dan years later when he joined the Medical Biochemistry Department at OSU (mid 1990s) and established a successful RNA processing program. Dan was a bright guy who produced a quality Ph.D. thesis that really never saw the light of day, publication-wise, thanks to Dr. Mueller's lust for great papers at the expense of the careers of his PhD and Post-doctoral fellows. I also met and became close to a husband and wife team from Japan. Both were M.D./Ph.D. researchers that I collaborated with for some years on topoisomerase projects. Ken and Kimiko Tsutsui were both on faculty in the medical school at Okayama University (well south of Tokyo) and I worked with them in Japan (summer of '83) and they both worked in my lab on projects. Ken was pathologically quiet, unless he was lubricated with his favorite beverage (single malt scotch). Ken is unusual for a Japanese guy because he speaks essentially unaccented-English which occurs on rare occasions when he actually speaks. Ken's father was the Dean at Okayama Medical School and his mother spoke fluent Esperanto (I never met anyone who learned the 'universal' language). I liked Ken immensely. He was very intelligent and had unique insight on all scientific matters. His reticence probably did not help him career-wise. For both Ken and Kim, science was their life. They never had kids and were totally dedicated to research. Ken was also an accomplished jazz pianist and had a special room in their home for his 1885 Steinway (dissembled in New York City, and re-assembled in Okayama). His music room was accessible through a trap door from the first floor master closet. Go figure! Both Ken and Kim were major influences in my life and are good friends to this day. My last visit was in 2010 where we made a sojourn over to Naoshima Island (the art island) which was an amazing experience (Fig. 5).

I would like to make few brief comments about Japanese academic science and my peer group operating that environment. It seems that Japan acquired the German (or European) structure for academic hierarchical organization. Most likely this comes from WWII. While this system functions reasonably well in Germany, it is not ideal for a homogeneous society like Japan. Essentially there is one full professor who runs a large research enterprise (equivalent to a small Department in the US) while there are several underling associates and assistant professors. The full professorship is loosely based on seniority and political capital and this person (usually a male) totally dominates all resource allocation in the group. This imposes strict funding penalties on young investigators who don't "toe the line". If the lab chief is brilliant, creative, well balanced and a good manager, the system can work, but I believe this is generally <u>not</u> the case. Advancement based solely on age is really not a good practice because not everybody gets 'old and wise' (some just get old). Some senior investigators are intellectually tired, moribund or scientifically out of touch. To put such individuals into a power position controlling younger researchers can effectively stifle a generation of researchers. I spent many hours bemoaning this problem and being critical of the system when Ken, Kim and I were young assistants. We all recognized the major defects; however, nothing could be done. The Japanese have an expression. "If the nail sticks up above the board, it must be hammered flat." For Ken and Kim, they were in a somewhat oppressive lab environment (lab chief was a guy name Professor T. Oda) and held hostage. The basic problem is that young investigators tend to be more productive and creative than their senior colleagues. They must pay homage to an individual who has no skin in the game, only the power to control others. This is stifling in science and there is an acute need for sweeping reform. Such reforms can only spring from senior faculty who control policy. Of course that would mean imposing changes that work against their own self interests. Once in power, the full professor has paid his dues so change is unlikely. Those discussions long ago with Ken and Kim, bemoaning the Japanese system, are long forgotten. They also paid their dues. Adding to the problem is the mandatory retirement at age 65. Usually, the lab chief is not appointed until late in the career (mid-late 50's I recall). Thus, there is little time to change the system, even if the senior faculty were so motivated.

Kim Tustsui was very successful and managed to advance to full professor status. I think that she is one of the very few full professors in Japan that is a woman and at the time of her official promotion, I think she was one of the first. Like Ken, she paid her dues, enduring long clinical duties as a psychiatrist at distal hospitals. She published with Ken for the most part. Kim's CV was probably very similar to Ken's record of publication and they both are full professors and share a large lab with associate and assistant professors under foot. The Japanese hierarchy of academic science will carry on it seems.

My First Academic Position

Sometime in late 1979 (Fall), I started on the job circuit looking for a tenure track assistant professor position. I had a good project (analysis of nucleosome assembly factors in cancer cells) and a reasonable overall publication record (despite only one paper as a post-doc in two years of work). Fortunately for me, there were plenty of viable positions for someone with a molecular virology background, an area where I was well established. Reading through each issue of Science magazine, I looked for ads for assistant professors in either virology or molecular biology. I interviewed at a number of state universities (and one in Canada) and got multiple offers. I only considered tenure track appointments. My selection was based on academic environment, research infrastructure, collegiality, teaching demands and lab start up package. Salary was not a serious consideration. Ohio State University offered me an attractive package; however, the salary was the lowest of all of my offers (my 9 month salary was $21,000 starting in September of 1980). I really enjoyed my visit and seminar in the Department of Microbiology. Everyone was enthusiastic and sincerely interested in my research. I recall a lively meeting with the Dean (Pat Dugan) and the Chair (Bob Pfister) and being impressed with both. The Dean informed me that they were quite impressed with my credentials and he wanted to 'sweeten' the offer with some additional attractions. Dugan informed me that being on faculty, I could get OSU season football tickets! I thanked him and said it was very nice to offer me tickets; to which he said that I would have the 'privilege of buying tickets' and they would not be gratis. I am not a football fan, so this whole concept was lost on me. My wife however, loves college football and was thrilled to get tickets. I also recall meeting Julius Kreier, who was on the search committee (composed of John Reeve, Neil Baker and Julius). I liked Julius immediately and after joining the faculty at OSU, we became close personal friends. Julius is well known in the malaria field. He has a Willie Nelson phenotype and a limp that imposed a bird-like dynamic to his gait; however, he was still quite agile and nimble. Julius was very bright and had a reputation for being a bit quirky but I really liked this about him. I recall seeing a 'rodent zoo' in his -20 freezer, courtesy of

Dave, a graduate student working on rat malaria (*P. berghei*). After euthanizing adult rats, Dave had frozen the animals in doll furniture sitting in various poses around a tiny round table. Some of the rats were smoking tiny cigars while others appeared to be holding small playing cards. All were frozen in place, waiting for necropsy. While strange, it was very entertaining! I suppose that some students get bored after long hours in the lab!

1980-1986: Getting Tenure at OSU.

One of the great things about OSU Microbiology was that I had a teaching free year to establish my research program. I was replacing another assistant professor who was denied tenure due to low research productivity. This outgoing faculty member was very bitter of course and I was able to avoid him most of the first year when we overlapped. He felt he got a raw deal from senior colleagues. Evidently he was partially appointed by the Cancer Center in the OSU Medical School (with very attractive start up package and minimal teaching loads); thus, when he did not deliver on publications, he got denied tenure. This person left science and completed law school and is now a highly successful patent attorney in Washington, D.C. I worked with him many years later when he prosecuted a patent I developed on gene delivery; small world it seems!

I worked extremely long hours in the lab, in addition to writing NIH grants, papers, dealing with graduate students and teaching. My service work was minimal and the Chairman shielded me from committee assignments. In those early years, I arrived at the lab around 8am, worked until 6pm, went home to dinner and returned by 7-8pm and worked until 11pm. I did this 6 days a week leaving Sundays to spend time with my family and new baby (my second son, Branden). I built the lab up with some highly capable people, including my first graduate student (Douglas Trask) who received his Ph.D. in 1986 and then went to medical school and a successful career as a physician-scientist. After I secured my first NIH grant (a cancer grant on topoisomerase, an important anticancer target protein), I opened up a second project area on Herpes Immediate Early gene regulation and was fortunate to get another concurrent RO1 grant. With multiple RO1 grants, I expanded the lab with post-docs and technicians; however, I was having problems attracting Ph.D. students from the Microbiology Department. I joined another graduate program, the MCDB or Molecular Cellular Developmental Biology program. This interdisciplinary program was more focused on eukaryotic models and was an excellent fit for my research interests. I was actively involved in the MCDB program, from an administrative standpoint (running the seminar program, committees, etc.). A couple of senior faculty were administering the program, including Phil Perlman, Tom Byers and George Marzluf, all very good scientists whom I respected. The only problem with the program was the lack of training grant support; however, this was remedied eventually. In addition to the MCDB group, I was invited to join the OSU Comprehensive Cancer Center, which grew into a very large and active program as an NCI designated cancer center. In the early 80's, the Cancer Center was rather loosely knit and not well organized. The OSU-CCC did however provide some support for my start-up activities and provided access to clinicians interested in basic science (but there were very few of these opportunities).

In the early 1980's biotechnology startups were ubiquitous and the investor community was ablaze with activity. Fueling the process was the enormous perceived potential for recombinant DNA in general (and gene cloning in particular) for improving the human condition. Early patents by Boyer and Cohen showed that restriction and ligation of DNA could create new DNA 'chimerias'. Thus, molecular (not reproductive) cloning was big news and investors were salivating. The very early startups like Genentech in the mid 1970's, established the biotechnology industry. With cloning and overexpression approaches, Genentech had its first FDA approved product (human insulin) in the early 1980s. Everyone recognized the potential in human medicine, veterinary medicine and agriculture. For the first time in history, biology was transformed into a predictable and quantitative hard science. Model system science was all the rage and a historic reset of phage systems biology (dating back to the 40's) was being re-established in more complex systems, such as yeast, human cells in culture, plants, worms, and flies to name a few. The driver for all this work was the lowly bacterium, E. coli, and its engineered plasmid (an extrachromosomal element which served as a vector for cloning). In addition, phage and cosmid based gene library technology was well established (a gene library is similar to a conventional library but instead of a collection of books, it has a diverse collection of gene fragments). As the field advanced, major revolutionary technologies punctuated the meteoric rise of biotech. These included the development of transgenic technologies in mice, the polymerase chain reaction (PCR), site directed mutagenesis, DNA sequencing, over-expression systems (in diverse model organisms), plant transformation with agrobacteria, gene and expression libraries and ultimately the human genome project. All of the above were made possible by the basic research on restriction enzymes followed by rational extension by Berg, Mertz, Cohen, Boyer, Axel and many others. As noted above, I knew a number of these folks. Janet Mertz working in Paul Berg's lab made the first recombinant DNAs. Michael Smith at UBC (my instructor in a Nucleic Acids) later developed oligo directed site-specific mutagenesis in the 80's and was awarded a Nobel Prize for this effort. My sabbatical with Mike (1991-92) was the year before he won the Nobel Prize. This was indeed an exciting time.

As a young faculty member at OSU in the early 80's, I was not directly involved in biotechnology business activities. I was building a research program and trying to get tenure; however, I did get an occasional glimpse of the business end of science. I recall giving a presentation to Isaac and David Blech, out of NYC, two shakers and movers in startup funding (organized by Rod Sharp who founded DNA Plant Technology, another successful startup). The business of biotechnology involves commercializing exciting new ideas and products, attracting investors and creating spin-off companies. Everything about the process was appealing. In addition, my experience with teaching undergraduates gave me an ability to convey complex scientific ideas to business professionals and investors who were very bright people but lacked scientific backgrounds. It is important to reach such people effectively and carefully titrate the amount of information to be concise yet simplistic.

I came up for promotion and tenure in 1985. I was well funded and had published well-cited papers. While it seemed highly likely that I would be promoted, I needed backup plans; therefore, I put my C.V. out for consideration at other institutions (confidentially). I interviewed at several different places and received offers. I delayed my decision on accepting any offers until after the promotion and tenure review, which came back positive. One of the offers (U.C. Davis) was attractive. The research infrastructure was excellent and the academic environment was very rich with many excellent colleagues. I was tempted. I was a little unhappy about not having access to graduate students in my home department at OSU (Microbiology). Relocating my lab was going to disrupt my research program and I would lose at least one year of being productive; this was a huge negative and seriously eroded my enthusiasm. In the end, I stayed at OSU. This was the right decision.

The Creation of a New Department of Molecular Genetics.

About this time, the Department was undergoing some leadership changes. John Reeve, who had the lab next to mine, became the Department Chair, replacing Bob Pfister. I respected

John and we got along well, so I was very encouraged about the new leadership. The future of the Department was in good hands and a new beginning was taking place.

In early 1987, Phil Perlman, the Chair of the Department of Genetics, asked if I might be interested in joining a new program. Phil asked me to help him create a Department of Molecular Genetics, the first in the country. The idea of creating something new was intoxicating and Phil was enormously enthusiastic. He could really excite a crowd and he was extremely careful about who should join this new program. The new Molecular Genetics Department would focus on eukaryotic models. It was immediately obvious that my research 'fit' in this academic program would be ideal. Moreover, I should be able to attract students from a pool of Ph.D. applicants with interests similar to my own. There was another aspect of the new program that I found attractive. I would be able to select my colleagues. Usually, the reverse is true (they select you).

Switching academic departments is risky business and can alienate your colleagues. There is considerable territorialism in academic units since each department is competing for a fixed sum of the available budget. When a well-funded faculty member leaves, that department is diminished in its grant portfolio and teaching support. I met with the Micro Chair and explained my reasons for leaving Microbiology. He understood my rationale (I was a poor fit in a prokaryotic based research unit) and seemed supportive; however, other faculty were not as sanguine. I told the Department that I would try to help with teaching duties until a replacement could be found for my main course (Virology) and that I would be happy to retain a courtesy appointment if they were agreeable (they were not). In the end, I decided that a clean break was the best exit strategy. Fortunately, people tend to forget, and over time things got back to normal. I never regretted joining the new department and I relished in the creative process.

Starting a Company: 1990

In the late 1980s I noticed that some faculty were very active in consulting and outside entrepreneurial activity. One faculty member in the Biochemistry Department was quite active in this arena and my 'hallway' discussions with him were very stimulating. My work with topoisomerases was really getting traction and after publishing papers showing that these proteins were high value drug targets in cancer, I started getting calls from different pharmaceutical companies to come and give seminars or consult. Indeed, I was traveling quite frequently and the compensation was a very nice perk. I liked the idea of outside consulting and recognized the potential for a profitable start-up company, solely based on the kind of consultation I was providing. My thinking followed along these lines:

1. There is a demand for my expertise, technology transfer and reagents in the field of anticancer drug discovery.
2. I was spending a fair amount of my time visiting various companies and consulting in the area.
3. Why not develop a Research Diagnostic company to meet the demand as a way to reach many other academic and industry labs in the U.S. and abroad?

While this was a great concept for a small startup business, there were serious rate limiting steps to consider (along with some risks). I am not risk averse by nature, so I decided to figure out a pathway through the maize. It might sound trivial now, but to learn as much as possible about start-ups, I enrolled in an evening course at a local community college on "how to start a business". This gave me a basic foundation within a short time and I found it useful to know the nuts and bolts of running a generic business. I also did a fair amount of reading on the subject of gorilla or bootstrap marketing and advertising. These resources were not sophisticated but helped me with the basics. A major influence in my life at the time was my Uncle (William Muller), my Dad's older brother, a Ph.D. in Chemical Engineering who had started several successful companies and was well-versed in the IP arena.

There are many reasons why people do not start companies but emotional and psychological risk factors are clearly high on the list. On top of the emotional issues, there were serious legal concerns. For example, at the time the State of Ohio had antiquated laws making it illegal to own more than 5% equity in a company, as a state employee. This was an old law to keep politicians honest (or so I was told). Clearly, risk factor number one was jail time! This puts an overriding damper on matters! Here are less weighty concerns that I had to consider (in no particular order).

1. Seed funding. No matter how small the start-up or venture, you need cash. I could not approach venture capitalists or investors for a number of reasons (lack of time, no networking, a poorly defined business model and no business plan, among others). So my idea was to start an escrow account (with a new LLC) to hold my consulting funds. These funds were signed over to escrow over a period of about 1 year. It was not a fortune but it was sufficient to defray some basic costs (incorporating, paying attorneys, CPA, etc.).

2. IP and OSU Conflicts. As a University faculty member, everything I do in science, related to my research, is owned as IP by OSU. Things can be murky here, but OSU could claim ownership (or worse) terminate my employment due to conflict of interest issues. On the other hand, there was no formal policy in place, so they might be open to some creative solutions. While I was nervous about this potential complication, I figured there must be a rational strategy to make it all work. This involved a dual pronged approach: First, implicate everyone above me in the chain of command (full disclosure) and get them on board; second, file a complete invention disclosure to the University officials. First, I met with Dean Jensen (Biological Sciences Dean) who was largely asleep during the meeting; however, Associate Dean Gary Floyd was fully engaged. I explained what I planned to do, specifically, create a company spin-off based on my research and use my own lab facilities to do so. Gary Floyd was a capable administrator with a good head. I was surprised when he said (without hesitation!) "go for it". This was 1989 and no faculty (to my knowledge anyway) were bootstrapping startups. There was no playbook on how to proceed, so the Associate Dean decided to give it a shot. One other reason he was supportive was my prior disclosure of the technology to the Patents and Licensing Committee at OSU (*viz*, technology that founded the company concept). In 1988, I disclosed the technology and the concept for a company to the Committee, loosely composed of a faculty members from chemistry, engineering, physics and biochemistry. The IP disclosure was reasonably well detailed and I argued strongly that OSU should file at least provisional patents on these ideas. The Physics guy said bluntly "everyone knows there is no real money in biology". Interesting perspective! Thank you very much. I got a formal letter from Ed Jennings (OSU President) releasing all disclosed IP back to me, with no strings. Such "no strings'" releases are a thing of the past because, these days, the University 'release' actually has a 10% (or similar) option that benefits the institution. This is really absurd in my opinion for one reason. The faculty member has to pay for the IP and commercialization and take on all risk. For the University to then turn around and make claim on successful commercialization seems outrageous because they gave up their first refusal rights (sort of like the "cake and eat it" concept). My current employer, the University of Central Florida has adopted this policy. Perhaps a class action litigation process will be required to stop this institutional behavior (which stifles the entrepreneurial spirit).

The letter was very nice and while at first I was upset, I later realized this was a great outcome; very nice indeed. Since all technological aspects of the company start up were released back to me, I could approach the Dean of Biological Sciences and legitimately ask for approval without any conflicts. I still have the letter of release from Jennings. The IP associated with the company was to be considered 'background IP'. I used the concept of background IP to negotiate a new faculty position at the University of Central Florida in Orlando (College of Medicine, my current employer). This shields me from Licensing and Patent "tech transfer" predators who are suspicious of faculty who create new University spin-offs. I simply do not understand why any University should look suspiciously at faculty who start companies; however, the current 'compliance' climate in U.S. science is largely all stick and no carrot. Major state universities are risk averse as a result.

3. Bricks and Mortar. At the time (1990) OSU did not have a wet lab incubator; at least not as a formal, functional facility (although there were some unofficial pockets here and there around the campus). I was in no position to create production, shipping or business facilities for the company I had in mind. I approached Gary Floyd (now the Dean of the College of Biological Sciences), and bluntly asked if I could use my lab as 'seed space' to ramp up the business. I explained that there were no conflicts (*vis a vis* IP or commitment) and assured him that any sales activity (shipping of products for example and a business office for billing and tracking purposes) would take place off site. (I created a separate business office in rented space and all shipping billing and accounting activity took place at this facility.) I still needed a production facility. Since the kits and reagents I planned to market as "research diagnostics" were in my research area (topoisomerase biochemistry) I explained a bootstrap model for production of products. In a nutshell, I told him that the Company (TopoGEN, Inc.) would defray all costs associated with materials for both inventory AND my research. For example, a typical enzyme preparation would yield one million units, which the company would defray *in toto*. Half of the preparation would then go into inventory (500K units) and the other half would go to my research (which would help me get future grants and maintain my research). In this way, I created a win-win situation for OSU and for TopoGEN. I did not think he would approve but he did so enthusiastically. I was thrilled and he was excited to see a new growth enterprise and was keen to see how things worked. Nobody was thinking about such entrepreneurial ventures at the time. I had crafted an innovative strategy that got people's attention. All the negotiation took place over a period of about 3 months and the next phase was ready to begin: Putting together a technology transfer company that would market a number of molecular biology kits and reagents.

4. Products and inventory. Having reasonable experience with designing undergraduate laboratories for teaching purposes, gave me reasonable insight into design and implementation of 'kitable' technology. I recall having a meeting the Arthur Kornberg, a distinguished Nobel Laureate (discovered DNA polymerase) who was giving a seminar at OSU. I mentioned in passing the start-up concept and he wanted to know more about my idea. We spent most of the hour in my office discussing entrepreneurial inroads in science. He must have been impressed because he made note of our discussion in his seminar later that day. He explained that providing products to promote science was innovative and honorable and he commended me for my efforts.

I spent about 6 months, never on University time but in the evening and weekends, designing the products we would sell. Recall that I was consulting with many different companies on how to do these sorts of assays and utilize the enzymes for drug discovery, so I knew there was a demand and based on the published literature, which was showing exponential increases in topoisomerase related papers, I had a very good idea of what would sell. All commercial products were based on some aspect of topoisomerase biochemistry and/or molecular biology. Many of these same products are sold even today, and I have opened up other niche markets. Basically, I put together kits for drug discovery using DNA based assays, such as relaxation of plasmid, cleavage assays of DNA targets with high affinity topoisomerase binding or recognition elements, decatenation assays and so on. None of the kits were sold with enzyme, so I created a suite of enzyme products that has grown over the years.

We also made and sold DNA substrates, antibodies to key topo enzymes and what I call "support" products (reagents to replenish kits). In addition, we started to collect and market as many different topo active drugs as possible and we became a repository for these compounds for researchers. Note that any investigator could approach our chemical suppliers and get drugs or other companies and get DNA for that matter; however, the catch was that we certified the drugs, DNA and related reagents as being pre-tested and guaranteed to work for the topo research in question. This certification was akin to a 'technology lead' or trade secret. Recall that we really had not patent protection, so we relied on trade secrets and careful marketing to get sales and repeat sales of our products. Moreover, the margins are huge on these products (we can produce $100K worth of enzyme for less than $2500, including tech time). Any time an end user was not happy with a product, we simply replaced it gratis and gave the customer even more than they ordered. This yields many satisfied and repeat customers. Often we find that the end users can be unsophisticated, sloppy or just inexperienced; however, they rarely admit fault. By simply replacing the product for free, they will eventually realize that they themselves are often at fault. Clearly this is a niche market and some of the 'market forces' that drive retail simply do not apply. Another advantage to a 'niche' business is technical support. Other companies can purchase bulk quantities of our enzymes (topoisomerase, Gyrases, etc.) and sell them at highly competitive prices. The problem is when the end user has technical issues with the product. The warehouse concept of reagent sales, cannot offer detailed support to clients. The company, started by experts in the field, can offer very strong technical support. When dealing with technology transfer, this is a very important advantage.

A Successful University Spin-off.

TopoGEN was officially incorporated in 1991. I was President and CEO and the major stockholder. I relied on my accumulated consulting funds to launch TopoGEN, Inc. These funds were used to defray the set up costs, including attorney's fees, miscellaneous filling fees, paying for a very creative and excellent CPA (Thomas W. Brandkamp and Associates). The key to creating a good income streaming enterprise is to minimize overhead. Don't create infrastructure unless unavoidable. Use all available resources, but do so with full disclosure and implicate everyone by being totally honest and above board. That said, I tend to use the 'thin end of the wedge model' whereby I start small and expand. This means that I adhere to a policy of "asking for forgiveness rather than for permission". I structure things so that I can always apologize for any miscommunication or missteps and immediately back away. Mea culpa. This is risky, but I like and manage risk well.

I needed technical help with the company and that meant man (or woman) power. I asked my Ph.D. students if they would like to earn extra funds by helping the Company on weekends. I paid the students for their time on an attractive hourly rate and I never demanded their participation. It was voluntary and there was no requirement for students to perform Company duties. I also made sure that their research activities would not suffer by monitoring their project progress. In most cases, Company duties overlapped with research directives. For example, at one time I had three Ph.D. student working on various topoisomerase problems in cancer. There was an ongoing demand for resources (plasmids, kDNA, enzymes, antisera, etc.) and as noted above, the preparation of these key reagents was paid for by TopoGEN and the students had access to Company products gratis. These reagents were pre-tested and well controlled and materially contributed to students' progress toward their Ph.D. This lead to publications and made me more competitive for future funding.

One other safeguard I built into the Company/OSU relationship. I told the business office at TopoGEN to never process purchase orders from OSU researchers. Sales back to OSU were absolutely forbidden. I was to be told who wanted our products if such orders were sent. I then contacted that faculty member and explained that we could not sell them any TopoGEN product; however, I informed these faculty that they could have any product free of charge. The only thing I asked for in return was a nice letter (or email) stipulating that all products were supplied by TopoGEN without charge (or that the Company had donated the products). In all cases these faculty were very grateful and also acknowledged how important TopoGEN was in the success of their research. I have a stack of such letters of gratitude from numerous colleagues in the OSU College of Medicine. I call this good, cheap insurance, when someone comes sniffing around thinking that I was exploiting my position as a University faculty member by doing business back with OSU and profiteering on the backs of colleagues who don't have tenure. It paid off in fact. Some years later, a colleague of mine started a company with family members; however his company was a middleman type of day to day business that sold plastic-ware and disposables (tips, pipettes, tubes, racks, etc.). He was essentially a warehouse that advertised these things that he obtained from a variety of vendors.

His company (run by his wife and adult children) would mark up and distribute the products. This was a sales operation of a very different type. The problem was that he was selling back to OSU faculty labs and making a profit. People in his own department were buying these items, and he was in a position to purchase these products using his own NIH grant funds, clearly a conflict of interest. The local newspaper got wind of his operation and a reporter called the faculty member, ostensibly to interview him on his "exciting new research findings". The interview quickly turned ugly when the reporter wanted statements regarding his company dealings with OSU. In other words, a trap was set and sprung on the poor guy and they even got a photographer to document with guilty looking photographs. It was major embarrassment to OSU and the faculty member, who was reprimanded and forced to dissociate from all business relationships and the company. I was also contacted by the same reporter, who asked to 'interview' me about my latest research findings. He showed up in my office with a photographer. We spoke briefly about my science and the topic turned to TopoGEN and he asked how it is that a faculty member could be profiteering off OSU colleagues. He had done his homework and knew about TopoGEN. He requested a full explanation of my relationship with OSU and the Company. I explained that my company was a business, separated from OSU and that the University had released all IP back to me to develop as I see fit. I further explained that TopoGEN was a manufacturing enterprise that made products for science and our customers were worldwide. Leaning forward on my desk he asked pointedly to see records of sales to OSU faculty. I informed him that TopoGEN conducted no business with OSU investigators. I further explained to the reporter that any researcher wishing to purchase TopoGEN products would be denied; however, all Company products were provided free of charge to any OSU lab. As proof, I pulled out the stack of emails and letters from OSU

Cancer Researchers expressing their gratitude for the Company largesse. The reporter spent a few minutes thumbing through the support letters in silence. He then stood up, told the photographer that pictures were not necessary and that there was 'no story here'. He did not thank me or shake my hand on leaving. I never saw anything in the local newspaper and there was no follow-up. Nonetheless, the whole experience left me shaken. After all, the State of Ohio had not yet repealed the law stipulating limited equity ownership for state employees; thus, I was still vulnerable. That law has since been changed, thankfully.

The OSU Edison Foundation: A New Research Park and Incubator.

Word got around fast that I had created a successful for-profit company. TopoGEN was in the black and a strong income streaming enterprise. Eventually, OSU administrators were asking questions about faculty entrepreneurism. Why were there so few spin-offs or start-up companies emanating from OSU faculty? What was missing? First and foremost, it was recognized that ownership in a corporation was a 4th degree felony in the State of Ohio. As noted, this law has since been repealed. Another problem was the lack of incubator wet lab space near the campus. I was asked by the VP of Research, to sit on a blue-ribbon committee to create from scratch a new research park and spin-off incubator. The committee operated under the aegis of the VP for Research. It took many months of meetings to flesh out the details but the general idea was as follows. The University purchased (or was gifted?) the old Simmons mattress factory on Kinnear Road (near campus). The building was at more than 100,000 square feet and over several years was renovated into a first rate facility. We organized the space for wet labs, central equipment, engineering, chemistry, bioscience labs and so on. In addition, we built infrastructure for secretarial support, reception staff, mailroom and janitorial services. Labs were well organized and reasonably priced at subsidized rates. There was a business board of professional and local business leaders (including some faculty from the Fisher Business College at OSU). Client companies were given modules of space of 600 square feet and each company had to be reviewed by the board of directors. Criteria were established to ensure that growth-oriented, high technology would populate the incubator.

TopoGEN, Inc. was quickly approved as a tenant in the new incubator. The facilities were very nice. We incorporated our shipping, production and business office in the space and I gained access to free consulting input on business activity and new ventures that I was working on with colleagues. Twice each year, I made presentations to the Edison Business Board and garnered input and suggestions for new products and business opportunities.

One new joint venture was started by myself and two other individuals. The spin off attracted a prominent Venture capital firm and we managed to secure some SBIR funding to develop the company. TopoGEN sponsored the grant and paid for a business plan development class to create a full-blown plan. There were some "teething" problems as the company founders and I attempted to figure out equity splits for the new entity. The situation was acute because there was money (in the form of equity) on the table and each founder had his own disparate vision for what was fair. For my part, I had assumed all along that equal partitioning was appropriate. Imagine my surprise when I was informed by the two other partners that I would actually get less than 10% of the equity. There are no ground rules for this situation. One of the partners was completely implacable and insisted that he would be diminished if I got more than this arbitrary percentage. When the Venture firm saw this behavior on behalf of one partner, they immediately declined future funding. No VC investor wants to inherit litigious partners. Several years later, the company did get going somewhat and was acquired by another company (and I received equity through that acquisition). The other founders migrated into the new company. Roughly 4-5 years later, that parent company went under and there was no return on equity. I learned a few powerful and valuable lessons. First, don't work with greedy people who value their efforts above others. Instead, work with well-adjusted people who understand that if the business is successful, there will be plenty for everyone. Do write an MOU (memorandum of understanding) that spells out all details of the relationship before there is a funding event. Make sure everyone signs off and has a chance to edit the document; get signatures and keep copies. Be sure everyone is comfortable with full disclosure to University officials, if appropriate. Maintain open lines of communication and engage everyone in the discussion process. Whenever possible, use the OPM model (other people's money) as seed funds to add early value to the company. OPM can be in the form of NIH-SBIR funds, angel investors, venture input or any other sources (crowd funding through the internet is a new and attractive model). Be creative and tap all available resources.

TopoGEN, Inc. was my first company, as noted, and for many years it has provided me a pool of funds to support activities that I was passionate about. This company is still going today and revenues are stable. Over the years, I have made improvements that keep abreast of the internet and e-commerce marketplace. I designed and implemented the first generation Website for the company (www.topogen.com) in 1998. In the early 2000's e-commerce was put in place with on-line ordering and feedback. Our distributor network is now worldwide. We have more than 20 distributors around the globe including in Asia (three in Japan, two in Korea, one in China), South America, Israel, Europe, Mexico, Singapore and Scandinavia. The Company has its own IT staff, a business office (and MBA who direct all business and sales) a technical staff, a business board and a SAB (Scientific

Advisory Board) that I currently chair. TopoGEN owns and oc-
cupies a 3600 square foot building in Daytona Beach. This facility
houses a business office, production lab, cell culture facility, QC
clean room facility, business office, conference rooms and wet lab
capability. The income stream is based from three sources: product sales, grants and contract research (and related subcontracts
for drug testing). In addition, the Company has recently acquired
new IP assignments to promote growth. Since TopoGEN is able to
sponsor NIH research grants, such as "R" grants (R01, R21, R43)
we can partner with other researchers and form alliances with
these investigators to develop new technology, new business ideas
and products for funding.

Many may consider that TopoGEN, which was founded on a
niche R&D market, may be self-limited and not a strong growth
model as a business. Moreover, the company was founded based
on 'technology leads' and trade secrets rather than an IP portfolio. Despite such criticism, the Company continues to be profitable. I have met a few University administrators over the years who
don't understand (nor support) my business approach. I shake off
negativity and keep pushing. As the Chinese proverb says, 'the tiger does not lose sleep over the opinions of sheep'. A positive attitude is critical especially in the face of hard economic times. Due
to the recent downturn in the economy, where the US Congress
has been cutting the NIH budget, sales activity dipped by 10-20%;
however, screening contracts continued, most likely because our
big pharma customers cut their R&D staff and outsourced. It also
worth noting that a 'trade secret' based company is, in some ways,
superior to a fully IP driven enterprise for a couple key reasons.
First, trade secret-based companies do not experience the 'patent
time-line'; thus, it is possible to keep things going for an infinite
time. In the situation with TopoGEN, we have been profitable for
>20 years (i.e., long past a patent lifespan). A trade-secret based
operation is more efficient and less costly because of ongoing

expenses associated with IP filings (government fees, attorney fees, etc.). An IP portfolio can get costly, especially when you consider that not every patent is a winner. Finally, the TopoGEN model is a rapid or at least expedited pathway to profitability, where income is rapidly generated early on since it is based on sales of a product that can be made in large quantities with enormous margins. Also noteworthy is that once a startup is profitable, the company can still acquire technology and conventional IP.

I have been involved in a number of other startups, some of which are still incubating. These are all cancer based enterprises in various stages of development. My business model, which is a bootstrapping approach to growth of a startup, is applicable in all cases.

A Passion for Aviation (1996-Present).

I started flight training in 1973 and my aviation interests were on hold due to many other obligations (family, expenses, career, business, consulting, etc.). Once established as senior faculty member and having a critical mass of researchers in my lab, I decided to get back to flying. My goal was to regain currency and get my private pilot's credentials, obtain an instrument rating (for IFR, 'instrument flight rules' flights) and acquire a suitable airplane to meet my business needs. Airplanes are strongly mission specific, but I had in mind the type of flying that would help the Company and provide a traveling machine for personal use.

Getting my Private Pilot's License. There are a couple pathways that have evolved, under FAA oversight, that can be used to obtain the Private (Visual or VFR) rating. The Private Pilots certificate is a VFR only rating, which allows the pilot to fly an aircraft when visibility is good (typically >3-5 miles). This rating prohibits flying in zero or low visibility and in clouds. The VFR rating is pretty limited, but still a necessary first step. This rating can be obtained by training with a certified flight instructor (or

CFI) within a flight school program or as a standalone individual with or without a school affiliation. There are also higher education based training programs (OSU has a flight school) that are called "141 Programs", which are more structured with formal curricula, integrated ground school and 'stage checks' by other check pilots to ensure that students are being properly trained. The 141 programs are rigorous and, as a result, more expensive. As an educator, I opted for the 141 approach and OSU has a very respectable aviation training program and a dedicated airport (Don Scott Field, KOSU).

As a consummate 'morning person', I decided to do my training at least 3 times/week at 6:30am. This had several advantages. Mornings are almost always calm and cool. Low turbulence, minimal winds and comfortable temperatures would allow me to focus on technique (stick/rudder, landing etc.) in a consistent external environment. Also, the OSU airport was located just north of what was my home in Upper Arlington and I could drive over in a 10-12 min, complete the flights, debrief and get to my lab by 9am (and still be at my office well before most other faculty show up). I also signed up for ground school that met two days/week in the evenings at the OSU airport. In addition, the 141 program involves pre and post flight discussions with a formal syllabus, which I found immensely helpful. As an OSU Faculty member, I also received a hefty discount on wet rental of the aircraft and instructor flight fees. Any faculty member who has an interest in flight training should take advantage of the 141 Aviation Program.

My flight training progressed well and the other areas, airwork, ground school, radio communications, ATC interface and numerous stage checks, part of the 141 program) progressed smoothly. My biggest hurdle was landing. Maneuvering the airplane accurately within a few feet of the ground is a challenge. You need to control your airspeed within a few mph window, you must maintain a proper landing configuration/attitude and most critically, manipulate the controls to land straight ahead without 'side loading' the landing gear. Part of the landing process is a something called the 'flare', where the nose is gently raised to bleed off airspeed and allow the airplane to 'settle' on the runway and stop flying. In zero wind or with a direct headwind, it's relatively easy to get; however, throw in a slight cross wind component and the devils come out. My biggest obstacle was getting the airplane alignment straight with my tracking over the runway, especially in cross winds. If you touch down while 'cocked' left or right of your forward vector, the resulting side loading can blow out tires, damage the gear, or worse (dip a wing, strike the prop, etc.). Also, landing an airplane involves the basic physics of energy management. On my second or third stage check during my training, the chief pilot (a fellow faculty member in the Aviation Department) did something to solve my problem of keeping the airplane straight on landing. He took a 3 foot piece of blue painter's tape and placed it on the cowl directly in front of the left (pilot) seat. This provided a good reference point to 'feel' the airplane going in a straight line for landing. After a few perfect touch and goes, he moved the tape to the middle of the cowl (between the seats). This gave me the correct sight picture and solved the problem quickly and effectively. The tape came off after a few more landings. Ah, the value of a great instructor!

My first solo flight, staying in the pattern doing touch and go landings, was watched by my instructor and wife standing by nervously on the ramp. It all went well. I started going to the practice area solo and working on flight standards necessary to pass the check-ride with an FAA designated examiner. I made a perfect score on the FAA written and was nearly ready. One last thing was required. I had to plan and execute a long cross country flight of at least 250 nautical miles and two different airports. I chose to fly from Columbus to Toledo to Muncie Indiana and back to OSU airport. I selected a nice summer day for the flight. It all went well until the last leg from Muncie to Columbus. While in Muncie, I had a chicken salad sandwich, which I believe was tainted with Staph. Staphylococcus produces a toxin and in minute amounts makes one violently ill. The organism need not actually be present, only the toxin. This type of food poisoning is characterized by a sudden onset (within one hour). About an hour and a half after departure, it hit me. I was at 3500 feet when the vomiting started. These trainers have no autopilot, so I panicked because I would not be able to control the airplane if I got the dry heaves. In addition, I was not talking to ATC, so no one could offer assistance. I vomited into my (hastily emptied) nylon flight bag. Going into a sort of survival mode, I knew that altitude was my friend, so I trimmed the airplane into a climb and just tried to keep the wings level while being sick. Thankfully, at about 6000 feet, the vomiting stopped (I was empty) and I began to improve. I leveled back to a VFR altitude (5500) and eventually landed safely back at OSU. I quickly cleaned out the airplane as best I could, throwing my flight bag and its contents in the trash. I left the doors open on the ramp to help air things out but the stench in the airplane was overpowering. I really felt sorry for the next student/instructor who flew that airplane (they took it immediately after I landed). About two weeks later, I passed the VFR check ride with flying colors. The process requires a minimum of 40 flight hour (dual plus solo combined) and it took me about 50 hours (average is 60 hours). I was finally "free to move about the country" as the commercial for Southwest Airlines proclaims!

I totally embraced my new found flying skills. I purchased my first airplane, a very nice Cessna 182Q model. A four-place single engine that could cruise at 140 mph; however, it haul a huge load (I called it my suburban of the sky). It was not terribly fast by aviation standards but very stable and a good platform for IFR training. The instrument rating (or IFR Rating) is required to fully utilize the airplane for business travel. Many people think that IFR simply means flying by instruments in clouds; however, it is that plus much more. It's certifies the pilot to use the instrument/ATC system world-wide. In addition, flight standards are more rigorous because ATC may issue an altitude restriction (or compass heading) and the pilot must comply within tight parameters. For example, altitudes must be held +/- 100 and climbs or descents are expected to be 500 fpm (feet per minute) unless otherwise stipulated by ATC. Deviation from these parameters will get you in trouble with the FAA.

After getting the instrument rating (far more difficult than the VFR rating), I used the airplane frequently (going to meetings, visiting clients, giving seminars and so on). Commercial flying on jets became less frequent, as I developed my instrument flying skills. At the time, most of my flying was up and down the eastern seaboard and only rarely involved long east to west trips. This is important because weather patterns move west to east and long flights across the country are likely to involve fronts, storms and low visibility. In contrast, north/south travels were less weather dependent. Note that propeller aircraft typically fly less than 18000 feet (and I usually fly <12500 since above this altitude, supplemental oxygen is required). Commercial flights are in the flight levels above FL180 and usually from FL300-400 (30-40 thousand feet). In other words, my airplane is firmly embedded in the weather while jets are usually above the weather. I have learned to become an amateur meteorologist as a result of my IFR training. I watch cloud shapes, temperature due point spreads, visibility and (importantly) I pay close attention to visible moisture (clouds) and ambient temperature at altitude. All airplanes have an accurate external temperature probe so if the temperature hovers near freezing or below and you are in clouds, ice will form. If the airplane ices up, it can stop flying. I have no de-ice ability and am forbidden from flying into "known icing" conditions.

TopoGEN continued to thrive and by getting reliable staff, I could delegate routine operations and my time was spent running my research program, attending meetings and doing the things that I enjoyed. The airplane allowed me to travel with little or no advanced planning and my passion for aviation and the financial freedom to explore other avenues of creative enjoyment started in the mid-late 1990s. I did two things that greatly improved my life. First, I purchased a 2nd home in a 'Fly-in' community in Daytona Beach (Spruce Creek, the 'premier' fly in community). Spruce Creek is an amazing place and I started meeting and interacting with some famous folks (John Travolta had a home in Spruce Creek as well as some NASCAR drivers, like Mark Martin and Mike Skinner). I made many good friends and keep in contact to this day with these folks. The second thing I did was to purchase a place north of Columbus called the "Pond Farm". I would like to digress a bit off topic to discuss a very special country experience living in Amish country.

A Molecular Biologist Living with the Amish?

I purchased a rural property with outbuildings for the specific purpose of performing animal work for TopoGEN product development. At the time, there were rather few outsourcing entities for antibody production and I wanted to add topoisomerase antibodies to our inventory. A working farm with outbuildings and animal 'infrastructure' fit the bill perfectly.

The Pond Farm was a historical property of about 130+ acres of rolling farm land with a small lake and a wonderful original Victorian style farmhouse built in 1870. There were numerous outbuildings (large barns) and a modern pole barn that I converted into a makeshift lab for animal work. I never owned such a large property with a lake (full of large-mouth bass), hiking trails, large pastures of cleared land, an old growth forest and nobody around except the Amish (on all sides) who I came to admire and respect. These people were an inspiration to me and I think the feeling was mutual. My father-in-law had a great expression. "To have good neighbors, you have to be a great neighbor." I went out of my way to get to know these fine people, and to learn their customs, culture and language (we were called "English"). The Pond Farm was only one hour north of the outer belt of the Columbus metroplex! Amazing indeed!

Just a bit of Ohio history first. The Pond Farm is well known in Morrow County. We were only the 2nd owners since 1820 (give or take a few years). The property sits on the north edge of the Greenville Treaty Line, which bisected the state of Ohio east to west. In the mid to late 1700s, Tecumseh and the 100 tribes were provided reservation land that was north of this line and there were numerous native settlements on our land. This was set aside land ordained for the Indians and all went well until about 1795 when the newly formed Continental Congress went into the red. A major problem was that our fledging government could not support the large (and aging) veteran population who fought in the revolution. To stem the tide of red ink, the US Congress decided to give land to surviving veterans, as payment for their services. Abel Pond and his first wife were bequeathed one section of land (640 acres) just on the north side of the Greenville Line. Of course, the native Indians were dealt a bad hand on this deal since the US gave their land away. I am not sure about all the details that followed but I'm sure it was not a healthy situation. The Government handed the deeds to the veterans and simply stipulated to the recipients, we are rooting for you! Abel Pond took over the land in the early 1800's and it was deeded officially sometime around 1820, probably after the Indians were pushed out.

I found the family burial plot within the original 640 acres of land but it is no longer part of my property. On inspecting the small cemetery, I located his grave marker and his first and second wife buried at 90 degrees on each side. Talking to the Pond heirs, I learned his first wife died in childbirth and he remarried and outlived the second wife. The children and other assorted relatives (grandchildren) are all interred in the small cemetery that was not being maintained. My son, Branden, and several friends and my wife and I did what we could to clear and clean the cemetery; however, it was too much. I later found out that the US government is obligated to maintain any cemetery containing the remains of US veterans and I was able to coax Morrow County to come in with a crew and perform a first rate restoration. There was an enormous sense of accomplishment that came with this activity and I was very pleased with the outcome.

Andrew Miller, my closest neighbor, was a rotund, loquacious Amish gentleman with a wife and 8 children (two girls and six boys). He was a small man with a quick wit and a toothy smile in serious need of cosmetic dental work. Like most of the other Amish, he was bearded (married) and wore glasses. I got to know Andy and his sons rather well and developed a close relationship with their families in the neighborhood. Andy and Clara, at last count, had 58 grand kids and probably others have been added since my last visit in 2004. Clara knew all the names and birthdays by heart (not so for Andy however). She commented to my wife Annette once that it was her great disappointment in life that she had only eight offspring! This is not unusual, since there was another Amish family down the road with 18 kids. Two baseball teams! (One set of twins if memory serves.)

Andy was an amazing businessman who ran a canvas tarp making operation (covers for semi-trailers, boats, buggies, porch surrounds and the like). The Amish can use any technology that existed in 1880 (my interpretation of their lifestyle). His workshop had an old time central pulley system drive belt to power all of the sewing machines for these industrial tarps. They cannot own or use any vehicle that has an inflatable tire for example. They are not connected to the grid and do not use electricity (solar or other) although they use batteries, lanterns, and a liquid propane source for light and in some rare cases they operate small internal combustion engines (some use of modern conveniences is permitted by the church elders and permitted use can vary somewhat). The rules of the Church district (Ordnung) must be observed by all members and the rules not only specifies modern technology prohibitions, but also rules on dress (no zippers) and social behaviors. The Amish travel by horse and buggy and do not own phones (neither cell nor landline), hot water heaters, central heat or power tools of any sort. I once looked at an Amish home under construction and noticed that the bathtub had only one water line (cold) as the feed. Ohio has cold winters! They bathe once a week (Saturday) and on Friday, the aroma is a mix of horse musk, body odor and chicken crap. They go to school until the eighth grade (age 14) in one-room schools and do not learn English until they are about 10 years old. They own guns and hunt and fish and live off the land as much as possible. They have no refrigeration. They do have access to and utilize modern medicine, however. They also rely on old fashion types of medical intervention (chiropractors, healers, and natural products, etc.). A typical community will involve 50-100 families, all in the same church group, who pay into a sort of self-insurance program for health care. The church elders make the rules as a collective regarding all matters, as I noted above. There is no formal Church building; rather, they hold services at various homes in their community every two weeks. The location of the services is easy to spot because there is a special horse drawn wagon (painted dull grey) that holds all the pews. In addition, there will be 50 or more horse and buggy outfits tied up outside.

The Amish eschew legal pathways and avoid the court system whenever possible. As far as I can tell, they practice a kind of submissive or "leave it be" mental perspective as the religion places high value on "Demut" or humility and "Gelassenheit" (calmness, placidity). Individualism is generally frowned on in the culture and self-promotion is taboo. This explains much about the Amish. They will not engage in lawsuits, file criminal charges, challenge any party legally or use or game the legal system at any level. In one rather famous case, a drunk driver ran into a group of Amish kids from one single family who were walking home from school. Tragically, all five children were killed and the driver skipped the scene. When apprehended, the parents did not even want to press charges; of course, the law intervenes here and justice was swiftly meted out. There was no litigation of course. The community gathered around the family and there was collective grieving, but life went on.

I got to know one of Andy's sons rather well. Wayne was 28 years old and, like his Father, of small stature but a powerful upper body frame. He and his wife had 6 kids and lived down the road about half a mile west of us. Both Wayne and his wife were wonderful people with soft European features and clear, fair complexions. Wayne had brown eyes and his wife had blue-green eyes and blond hair. I hired Wayne to help me build a porch on the old farmhouse. I was very happy with the result and spent many blissful hours on that porch watching horse and buggy traffic travel to and fro over, up and down the Greenville Treaty line. Like all Amish males, he was a good carpenter and was very good with his hands. They had a farm and raised their own crops and their kids were like little dolls, so incredibly cute and shy. The kids loved me to the bone because I always brought treats and ice cream in summer, which was enjoyed by all. Lacking refrigeration, an ice cream bar to an Amish 6 year old is like heaven. I spoke German to the kids and they would always giggle and squirm about. Wayne told me that my "German" was the same language as in their bible and only used in religious services. The Amish German was also somewhat hybridized with English (or Dutch?). For example, "What is your name?" would be "Wie heissen sie?"

in hoch deutsch (formal tone) but the Amish would say "Was ist du namen?". Clearly the latter is very close to English and would be considered "Pennsylvanian Dutch". As far as I could tell, the Amish are tri-lingual (English, Hoch Deutsch, and Pennsylvania Dutch) with most family language around the dinner table utilizing Pennsylvania Dutch, which is actually the Swiss dialect of German.

On more than one occasion, I had to get up in the middle of the night to drive one of the kids to the Amish "Doctor" in the next county or to the hospital in Mt. Gilead nearby. In addition, I installed a land line phone in my workshop and gave Andy and sons a key for them to use the telephone for their business. (Wayne worked in the tarp shop.) The Amish are totally honest in all matters and the family kept meticulous records of all long distance calls and reimbursed me each month to the penny. I tried to refuse such paltry payments but Andy insisted, so each month I might get a personal check for "three dollars and fourteen cents". They don't own credit cards or get mortgages and never work on business credit lines.

To the Amish, I was viewed as an outsider of course, but also they considered me to be a strange duck, even for an "English". I had motorcycles, nice cars, a collection of tractors, mowers, implements and tools and generally was a serious technocrat; someone who relishes in the use of modern technology. I was accepted and well liked and the Amish were very concerned about watching my home at all times since I only came on weekends. Andy's presence at all times, using the telephone for example, also helped ease my mind about security. Burglaries and break-ins are a fact of life, sadly, even in these very rural areas (we were burglarized twice).

In about 2001, I purchased an antique airplane (an Aeronca Chief) that was a conventional gear (tailwheel), two-seater single engine aircraft. It was a beautifully restored fabric covered airframe (ceconite). The restoration was done by two airline pilots who were brothers living in a small rural town near Pittsburgh. The Chief had no electrical system (I had to 'hand prop' to start the darn thing), a 65 hp Continental engine and it cruised at 85 mph. This is a very light ship and I had to get a special rating for the tailwheel, which was NOT easy (it took about 10 hours of instruction). But once you learn to fly a taildragger, your flying skills improve enormously since you learn fine rudder control. Landing the Chief was like wrestling a pissed off alligator with the tail whipping about while rolling out. And you never stop flying the aircraft until you come to a dead stop. The problem here is that a conventional gear aircraft has its center of gravity further aft (compared to a nosewheel airplane) and on roll out you have to literally dance on the rudder pedals (i.e. making constant adjustments ever so slight) to keep the thing from swapping ends. The Chief is also 'short coupled' meaning that the tailwheel is relatively close to the mains which makes it directionally highly unstable and susceptible to ground looping. I ground looped the Chief twice: Once while learning to transition into the tail wheel while attempting a 'wheel landing' (as opposed to a 3 point touchdown). This cost the insurance company about $6000 in wing repair. Another time I ground looped on purpose to keep from hitting a tree, after losing control on a very gusty cross wind landing that did no damage.

I wanted to keep the Chief at the farm property and I decided to build a turf runway on high ground (east-west orientation) on the extreme north side of the property. I had to bring in fill dirt to make a 2200 foot landing strip and clear trees on one end for an open approach through the very thick forest. The project took a total of three years to get a manicured turf runway, build a 50x50 foot hangar, and put up a 18 foot tall deck as a "control tower". The raised 12 x 12 deck was on high ground and a provided gorgeous view of the entire farm and the whole valley below. Most of the project was paid for by the hardwood I harvested from 3 acres of cherry, oak, and ash trees that were cleared on the east side of the runway. I kept my Chief at the Pond Farm Airport and had many happy days of buzzing the countryside, going for nice Sunday morning breakfasts, and landing at different country grass strips owned by a few Morrow County residents. On a few occasions in the winter, I flew out of a snow covered field and landed back in the snow which was risky but fun. I almost hit three deer that were eating grass on the runway at sunset. Evidently, deer have no predators in the sky so they generally don't look up. They are very hard to spot in the waning hours of daylight and a quick go-round saved me from destroying the Chief along with Bambi and her friends. Learning to fly out of a short, unimproved strip taught me to be a better pilot as well. I learned energy management and how to really nail the approach speeds and to 'hang on the prop' for tight landings in the Cessna 182. I enjoyed runway maintenance (mowing) and had the proper equipment to get a finished cut on the runway. The runway looked like a golf green after a few years and I was very proud of the result. I held numerous fly-ins with fellow pilots and EAA members (Experimental Aviation Association) in the area.

During construction of the runway, I asked Wayne and Vern (Andy's sons) to help me clear trees. I offered an attractive hourly wage for their services which was money well spent because I was deathly afraid of dropping multi-ton trees and having an bad outcome. Wayne and Vern assured me that they had the necessary expertise and experience and saw no problem with the job. One fine Saturday morning, I was wiring up the new hangar with fluorescent lights while the Amish boys were felling trees when Vern ran up and breathlessly explained that I needed to come quick as Wayne was hurt 'real bad'. One look at Vern's face told the horrible tale of what was to come. The tree had suffered a lightning strike at some point and as a result had twisted unpredictably in the felling process. It released from the stump and came directly into Wayne, who was handcuffed by the heft of the chainsaw. He dived away and the tree came down squarely on his foot, crushing the ankle bone. His foot was horribly mangled but held in place by the leather boot. Wayne was clearly in agony and I carried him out of the forest on my shoulders, with the help of a neighbor who spelled me off on the long trek back to the farmhouse. My wife, an RN, informed me that this was a serious injury and Wayne must not remove his boot. We iced the injury and took him to the E. R. in Mt. Gilead (20 miles away). I later found out that Wayne was life-flighted back to Columbus for Orthopedic surgeries (multiple) to repair his ankle. My heart was broken over this incident. Wayne had 6 kids to feed, crops to harvest and manual work to support his family. He was now crippled and I was at least partially at fault.

The recovery process was slow and the Amish families in the Church group went into a rotation of sorts to support his family. They harvested and plowed his fields, managed the livestock and horses, did his work in shifts at Andy's tarp facility and delivered dinner and lunch everyday so his wife (with kids in tow) could be at the hospital during the lengthy recovery. My wife and I visited him as well and did what we could to provide for his family and help out on weekends when we were in the country. We gave the kids wrapped Christmas gifts and I spent time with the kids, bringing goodies and toys whenever I stopped by. Wayne was back home within few weeks, on crutches of course. About two months after the incident, I was paid a visit by Andy and the church elders. They very politely asked me how I was going to pay for brother Wayne's medical expenses which were approaching $30,000 and would continue to accrue. I explained to them that Wayne was working on my land and thus was in the 'English world' when the accident occurred. The way we handle such matters, I explained, was through my homeowner's insurance. I further informed them that I was not going to simply write a check for the medical bills because I have insurance that I pay to protect me from such catastrophic events. They collectively glared at me, spoke to each other in the native dialect (which was incomprehensible) and shook my hand and left. Before leaving, I told them that I would investigate how to handle the medical bills with my agent. Andy simply shrugged and bid me a very polite farewell. I figured that my good, working relationship with the Amish was finished and that I would forever be despised. The insurance company contacted me with a claim number and said that the injured party needs to simply submit up to $5000 in medical expenses and they would reimburse, no questions asked. It's a formal process I explained to Andy, so please follow through. Moreover, they informed me that it would be possible, even very likely, to get the whole bill settled if they would go to an attorney and arrange for a settlement. Not only did the Amish not go to the legal system, but they even refused to file a claim number for the $5K. The incident was never mentioned again and there were no vindictive directed my way. It was as if nothing ever happened. Wayne eventually recovered but will walk with a serious limp for the rest of his life. He remains a good friend to this day and I occasionally visit and check on him and his family.

I miss the pond farm enormously. I will never again own such a unique property, I realize, nor will I have such interesting and diverse neighbors and friends. In 2003-4 I was recruited to join a new medical school in Orlando (see below) and the while the value of real estate was strong in Central Ohio (2003), I sold the property by dividing it into multiple large parcels. I sold the house and 10 acres, the remaining crop and pasture land (120 give or take), but kept the lake plus 5 acres for myself. I later sold the lake to another family in the area, so I am fully divested of the Pond Farm and I did well financially. The timing of the sale could not have been better, since property values dipped terribly in the crash of 2008. I have no regrets. The Pond Farm also contributed to TopoGEN because I used the facilities to raise New Zealand White rabbits for antibody production. As noted above, these antibodies were sold as inventory and I relied on the Amish to maintain my rabbit population. How many molecular biologists have hired Amish technicians I wonder?

Life after OSU.

In 2003, I contacted a former OSU colleague who had been recruited to join the University of Central Florida, in Orlando, to direct a biomedical sciences program. Pappachan Kolattukudy was recruited to OSU in the mid-1980s to oversee and grow a new "Biotechnology Center". I met Pappachan (or PK as everyone calls him) in 1986 soon after he was appointed as the new Biotech Center Director at OSU. PK was a dynamic and powerful intellectual force who was responsible for many (>75) Ph.D. level hires into the Center and across the campus over the years. He has a reputation for being hard to work with but that is because of high standards he imposes on everyone in his orbit. I got along well with PK; however, the Department Chair (Phil Perlman) did not. Phil and PK were at odds and eventually, joint hires between Biotechnology and Molecular Genetics completely ceased. This was regrettable and hurt both programs. Due to the blood-letting, I did not interact with PK or the Biotechnology Center faculty (another regrettable outcome). Later on, as the OSU medical school started recruiting more powerful leaders, PK was marginalized I suspect because of his leadership style and strong personality. Eventually, the Biotechnology Center was splintered and I believe PK gained control of a new "Neurobiology" group affiliated with the OSU College of Medicine. This arrangement did not work out and PK was back on the job circuit, eventually landing as the Dean and Director of the Biomedical Sciences as the University of Central Florida in 2003 and later as Chair of a large undergraduate program in Molecular Biology and Microbiology (in other words, he wore multiple hats). He was a savvy negotiator and received a very lucrative start up package that included a large number of faculty hires.

In 2003, I was getting concerned about the future of the OSU College of Biological Sciences because we were a small College led by a weak Dean and it seemed likely that we would be merged with a larger, more powerful College (Math and Physical Sciences). I had nearly 25 years of service and was feeling like I was in a rut. My R01 had just been funded and I thought it was a good time to look around a bit. I interviewed at some places for a chair position, but my heart just was not in it. In fact, I hated the idea of becoming an administrator. I had a home in Daytona Beach (Spruce Creek) an hour's drive north of UCF (in Northeast Orlando), so I contacted PK to see if there would be any interest in my joining the faculty in the Department of Molecular Biology and Microbiology. In the fall of 2003, I gave a seminar at UCF and began the process of negotiating to join as a senior faculty member. On my second visit, I met deans, VP, provost and of course John Hitt, the UCF President. President Hitt was especially enthusiastic about my entrepreneurial activities and strongly encouraged me to come to UCF and bring TopoGEN in tow. He assured me that accommodations would be made to bring the company into the academic fold with the UCF Research Park facility. I negotiated a decent start up package and salary. I identified acceptable lab space and facilities and all was going according to plan. This was not going to be a lateral move.

Suddenly, things stopped moving forward, for reasons I cannot explain. The communications with PK simply stopped coming and I assumed that they (UCF) got cold feet or perhaps had found a better candidate. I still did not have a letter of offer. At the time I was on the Scientific Advisory Board for a company called "Genomics USA" or GUSA, an academic startup company from the University of Arizona (a microarray fabrication company using new IP for producing the arrays). At one of the board meetings, I suggested that GUSA should investigate Central Florida as a place to base their operations because the State was interested in subsidizing Biotech companies who come to the sunshine state and the availability of a skilled labor force at UCF. Plus, commercial real estate was widely available (and affordable). Within a few weeks, I contacted the Economic Development Board (EDB) in Daytona and asked if they would be willing to consider GUSA as a tenant in the new "ATC" (Advanced Tech Center) near Beachside. Their enthusiasm was palpable. We arranged a dog and pony show with GUSA and various business leaders, bankers and investors in Daytona and Central Florida area. The day before the meeting, the EDB director told me that he had invited 'several prominent UCF staff' to attend the meeting. This was a bit surprising, since I could not see the relevance of including UCF people in a venture so far from campus in Orlando. The meeting was well-attended and I acted as the master of ceremonies for the board presentations. Much to my surprise, the VP of Research from UCF was in attendance as was PK. The meeting was well received with many questions and general enthusiasm. After the meeting, PK apologized profusely about not getting back to me and vowed that things would move ahead. The very next day I had a letter of offer in hand, clearly laying out all I had asked for and then some. Lady luck smiled on me that day! I went back to Columbus but kept the job offer a secret for several months, while I was actively investigating all legal matters (regarding the Company and any IP or other conflicts). I began to feel bad about keeping things from my Department since my absence would create a vacuum in teaching. Moreover, I felt advanced notice was appropriate in this case and it just seemed the right thing to do. I regret that decision. In situations like these, where one is leaving the university, it is a good idea to play your cards close.

I informed the chair that I had an offer about one year before my planned move to UCF. The chair was a close personal friend (Lee Johnson, also a fellow pilot) and is as respectable and forthright as they come. Lee did not want to see me leave the Department and approached the Dean of Biological Sciences for a counter offer. Essentially, the Dean said no way could the College match such a strong offer. Moreover, the Dean informed my chair that if I stayed on at OSU that following academic year, I would be strapped with a heavy teaching assignment in general biology the next Fall. Thus, by trying to be a good citizen and help, our Dean came after me with impunity. I don't know what I did to deserve this response. The Dean seemed to want me out. I had just been awarded a 5 year R01 the previous year, so this decision could not be due to 'dead wood' status of a senior faculty member. The next thing that happened was the Dean called for a detailed audit or some kind of accounting of all my lab expenditures over the years that might have come from Departmental funds. I suppose there was some coordinated digging up of "debts now due" to the College and I was to held accountable. Basically, the College was trying to rape the NIH grant for funds that I ostensibly owed for past support. The College claimed I owed something like $85K in miscellaneous supply and equipment expenses. Like many research active faculty, I occasionally would be in between grants, and the Department had a liberal policy of interim support. The College now wanted payback for all accumulated expenses (I was never informed of such a payback policy). Lee informed me by official email that the funds would have to come from operating expenses of my current R01 grant prior to transferring any funds to my new position in Florida. I had to inform these good folks that I could not simply transfer funds from the grant, other than the indirect costs which were allowable payments to OSU. The grant was set up to do a body of work and NIH would take a very dim view of diversion of funds in this manner. The Dean was most upset. Finally, Lee suggested that I pay with some release time in the coming year to help offset. This would net the Department about 1/3 of the red ink which in retrospect seemed fair; however, there were clearly some lingering hurt feelings. I appreciate Lee's handling of the situation and all turned out well, thanks to his skillful and creative problem solving. In contrast, the Dean was a bumbling idiot, in my opinion. She got greedy and saw an opportunity to pillage the grant and in the process penalize me for finding a better deal. This was but one example of the Dean screwing things up and creating grief. A few years later she was fired and the College of Biological Sciences no longer exists, thanks to her 'leadership'. I was not the only senior faculty member the Dean ran off. One of the most productive colleagues in our Department, with a recently awarded program project grant, left the College and joined the OSU Medical School. The Dean was wholly responsible for this as well (it happened soon after my departure from OSU).

In spite of the bad turn of events, my final year at OSU was nonetheless a happy one. The Departmental staff did a superb job of making up a commemorative picture book and throwing a very nice retirement party. They really took the high road on this and I will be forever in debt to Lee Johnson and Jessica Siegman for a fabulous and well organized send off. Many of my friends, colleagues and past students were in attendance. It was punctuated by a number of moving and in some cases humorous speeches.

I did not move TopoGEN to Florida right away and this turned out to be a very wise decision. Back at UCF, the Dean (PK) asked if I would consider moving the company into University lab space, presumably to garner some PR about Biotechnology company startups in Central Florida. I immediately saw compliance problems by moving a for-profit enterprise into UCF designated space. The legality of this arrangement was a serious concern for me, but not others at UCF. I thanked PK and respectfully declined his offer. I thought to myself "there is no way on earth that I am going to link TopoGEN with UCF lab space". To do so would invite serious trouble (destroy the Company for instance if there was a falling out with UCF administrators). In retrospect, it is my view that Universities may view a faculty start-up company as a resource to garner new funding (as grants or contracts) that will result in a flow of indirect costs back to University budgets. This 'cash-cow' indirect funding model is short sighted and misses the more important contribution of general economic development. Moreover, it is very naïve, if not patently spurious, to think that a company would simply give up the indirect costs to the University, unless of course, the University holds the IP (which is sometimes the case) and the company and University are co-funding the project (through STTR NIH programs). In my case, TopoGEN is able to sponsor research funded by NIH; thus, the indirect costs legitimately belong to the company.

Given the situation, I took my time and built suitable R&D facilities for TopoGEN in Florida, which was still housed in the Edison Incubator on Kinnear road, west of campus. It took a couple of years to identify and create space for TopoGEN in Central Florida, but I should tell you of a clever solution to the problem (described next).

Combining Aviation Resources and TopoGEN .

The solution to my dilemma of building a respectable R&D production and shipping lab for the company came to me soon after I purchased a commercial airplane hangar in Spruce Creek Fly-in where we lived. Aviation was an essential element of my business model and it made sense to utilize corporate resources to create a modern lab facility.

I was the proud owner of a new airplane called a Columbia 300 (Fig 3). I purchased this high performance fiberglass state of art aircraft in 2002 from the factory in Bend, Oregon. This was a different breed of airplane and exceptionally fast and efficient. Because it is fiberglass, it had to be kept out of the hot Florida sun; thus, the need for a hangar. As an aside, I had never purchased a factory new airplane before, especially one like this. The Columbia is totally amazing. While the Cessna 182 had a cruise speed of 130-140 mph, the Columbia clocked in at 225 mph (roughly 3-4 miles/min). I have seen ground speeds in excess of 300 mph (with a good tailwind at altitude). The airplane was not only fast but it was quiet, tight and had jet like performance with high wing loading and a side stick (not a yoke like the Cessna). Moreover, the avionics were fantastic, with moving maps, autopilot, altitude pre-select, fuel computer, pressurized doors and, as the advertisement stated, "the interior is one well fed cow". Because it is fast, I paid a bit extra and got 'speed brakes' which are spoilers to help manage energy and scrubbing off speed. Near the ground, the Columbia has the performance profile of a light jet, which also have spoilers to slow down. Because of speed brakes, ATC would slot me in with the faster jet traffic in busy airspace like Dallas, Chicago, Orlando, etc. Not only is the Columbia fast, but it has amazing endurance (i.e. it carries a lot of fuel, Fig. 4). Fuel is safety and as my IFR instructor was fond of saying "the only time you have too much gas is when you are on fire". I once flew from Central Colorado (200 miles west of Denver) to Orlando, Florida non-stop (>1600 statute miles). Of course that is a fair amount of sitting still in one place (8+ hours). The winds were neutral on that flight by the way, so I did not get much of a push.

The commercial hangar is 60 x 70 feet and big enough for 4 aircraft the size of the Columbia. I had an architect draw up a design for a two story lab facility in the rear of the hangar. The lab is almost 1500 square feet and hosts a production lab for biochemistry, a cell culture lab, a business office, two conference rooms, desks for staff and a shipping facility. Even with the build out, there is plenty of room in the hangar for my Columbia and another smaller aircraft that leases space (at one time, I also had a helicopter in the hangar as well). Once the hangar build-out was completed and equipment purchased and up and running, only then did I move TopoGEN to Florida. The Company is fully independent from the University and will forever stay that way. As noted, TopoGEN can sponsor NIH funded research in the same way that UCF, OSU, MIT or any University for that matter. I suppose this makes the company a competitor to the University and that can create consternation between the Company and the University, especially when the academic side is predatory about harvesting indirect costs from the Federal government.

Moving on to Florida

In the summer of 2004, I vacated my lab in Room 224 of the Biological Sciences Building at 484 W. 12th Ave and the moving vans packed it all up for UCF lab space. I was very careful to play it by the book with inventory tracking and reporting of all equipment items. I informed my post-docs, technician and Ph.D. students several months back about the move to Florida. I extolled the quality of life in the Sunshine State. I explained that the dreary, overcast Columbus winters are a thing of the past. No more snow, no more shoveling snow, you can sit on the beach in mid-winter! Hooray! Everybody get on board. Some of my staff came with me enthusiastically. I flew several folks down in the Columbia to let them see for themselves. Everything went great and the lab staff moved in July, 2004 to set up shop and start buying the equipment. Now the bad news: For the first time in 50 years, Orlando got hit by a hurricane the very next month (August). In fact, Central Florida got nailed directly or indirectly by four named hurricanes in August, 2004. When a place like Orlando has no major wind or hurricane for over 50 years, the trees get very big and mature. They also fall on stuff, like cars, houses and even people. A large oak tree landed on my post-docs car (obliterated). He approached me and basically said "I was safe in Ohio"! Of course I felt bad but I don't control weather.

Before the first hurricane (Charlie) hit us, I flew the Columbia to Texas, to get it out of harm's way. In some Central Florida airports, hangar doors were blown in by the strong winds and airplanes were damaged or destroyed. The day after the hurricane moved out, I flew into Sanford Airport (South of Daytona) to fuel up. On the ramp, I always tie down with the lines provided; however, in this case the tie down rope had a metal chunk of airplane attached. Asking line service what happened they reported that the airplane was blown off its moorings by Charlie and was gone. Gone where I asked? Dunno he said. They later found the aircraft (a Piper Cherokee) in a retention pond several miles away. Little airplanes start flying at 40-50 mph, thus 100+ mph winds are at 'cruise speed' and the rest is pure physics.

I set my lab up and got new equipment in and functioning. I settled in and re-established my research program. My first year was without teaching, which was wonderful; however, the first year after the move was spent tooling up the new lab and I was not very productive. Research productivity is very difficult even when things working perfectly. With a relocation of lab space, the process is dealt a death blow.

Operating an Established Biotechnology Company in a New Academic Environment Requires a Pioneering Spirit: Some Recent History.

From 2004-2007, I worked as a full professor in the Biomedical Sciences Center in the Department of Molecular Biology and Microbiology. I was teaching graduate and undergraduate courses, running my research group and generally doing the same things I did while at OSU. PK, as Dean and Center Director clearly had his favorite faculty who were largely untenured and therefore easily controlled. I was left alone for the most part, although the administration's philosophy for senior faculty like myself was unambiguous. I was given a very nice start up but after that, there was no support of any kind (no annual equipment upgrades, service contracts, peripheral support for research, etc.). To make matters worse, our program had no speaker program for prominent scientists to come and give seminars. Consequently, I began to feel scientifically malnourished. This culture was very different from that of OSU and unfortunately there was little that could be done about it. As far as I could tell, all senior faculty got the same deal, except for the Director (PK) who had his hands in the startup kitty and was free to support his own needs. I began to sense that senior faculty were either a threat or a competitor, so I stayed well away and did my own deal.

From the moment I arrived, PK was making plans to create a new 5 story research building, fully flush with state of art equipment. Also, based on a lucrative start up package negotiated with the VP of Research, PK was to fill about >30 faculty lines in the new building. I was hands-on involved with the building planning and design from the start (as part of a faculty committee). Soon, it was revealed by PK that he was making plans to build a new Medical School (2005) and he was pushing hard to get the administration and the State of Florida onboard. I have to give him credit; it was a wonderful vision. PK is a builder and I have always been impressed by his vigor, insight and ability to motivate people above him to join his cause. He worked tirelessly on this and went through open heart surgery in the middle of the action! This slowed but did not stop the progression.

By way of political background, it is sad to note that Florida is not an 'education' state and University budgets reflect a highly conservative (now tea party) legislature. The aging population represented by retirees from places like NY, NJ, Michigan, Ohio (the Northeastern demographic) explains our current political climate. These fixed incomers do not want taxes and do not wish to support education of any sort. They already educated their own kids, so they will not pay a penny more! I could make the same argument that my kids are already grown, so why pay taxes for schools? I personally support education because I don't want to live in a country populated by uneducated citizens.

There was much resistance at the Florida State level to add another medical school. John Hitt, the UCF President, desperately needed to prime the pump somehow. A company called Tavistock stepped up to the plate and donated 50 acres of prime Lake Nona land (southeast of Orlando and about 20 miles South of the UCF main campus) along with something like $10M. The idea was that Joe Lewis (CEO of Tavistock and a very wealthy guy from the UK) wanted to develop the Lake Nona area of Orlando. The plan was to create a 'medical city' just five miles east of the International Airport (MCO). The Medical City would have it all eventually (MD Anderson, Mayo, a State Medical School, Children's Hospital, a new VA, Dental School, and a large number of primary care facilities for specific diseases). The Florida Legislature could not say no and the process began in 2006. To his credit, President Hitt attracted substantial donor funds to the fledgling concept.

In 2007, largely as a result of leadership by PK and the President (and many others) the new UCF College of Medicine (or COM) was created. A COM Dean was identified after a short search and the arduous process of creating something from nothing was started for the first class (which was in 2009). It was most critical to secure LCME (Liaison Committee on Medical Education) approval to operate. The ranks of administrators expanded quickly but little effort went into organizing conventional departmental structures at first. Most of the early College of Medicine (COM) hires were in medical education, which was sensible. Additionally, the COM did not want a tenure track system, but instead placed any and all recruited faculty on contracts and consequently, these positions had no job security. No active researcher will be attracted to this sort of academic structure. On the other hand, all researchers (new and pre-existing) were recruited in the newly formed "Burnett School of Biomedical Sciences" (BSBS) as tenured or tenure track faculty.

In order to 'migrate' the tenured or tenure track BSBS research faculty into the new UCF medical school, some changes were in order. Because the new COM was recruiting contract faculty, tenure and tenure track faculty had to be given some kind of arrangement or grandfathered in as tenure track lines. Main campus UCF faculty work under a collective bargaining agreement involving the union; however, medical school faculty in Florida are not unionized. The pre-existing faculty had to be moved 'out of unit'. To address this, an MOU (memorandum of understanding) was designed to protect tenure and operational features of the Department, including budgets and future tenure track positions yet to be filled. This MOU was an operating document between the COM Dean and the research faculty and signed off by administrators (VP for Research, the Center Director and Chair) and was only in effect for a few years. Soon after UCF hired a new provost, the MOU was dissolved. My colleagues were concerned about this naturally (since our tenure was no longer protected and we were no longer represented by the union). Soon thereafter, PK was relieved of all leadership duties the COM Dean put an interim director of the BSBS research program. The COM Dean is a highly capable individual with regard to PR and schmoozing with potential donors or making grandiose plans to play to the local press. I give her credit for this insight, and our program is moving forward with excellent momentum, even in the post-sequester, shrinking of the NIH budget environment.

After the new interim director came onboard, the COM became aware of the fact that I had an established and profitable

Company operating independently from UCF. I had previously disclosed this of course, but eventually it was a topic of discussion. Adding to complexity of the situation was my lapse in NIH funding (never a good thing). The COM administrators decided that entrepreneurial faculty must be scrubbed robustly for conflict of interest issues. To be fair, this came from new federal guidelines on compliance and is the closest thing to Mr. Orwell's 1984 that I have ever experienced. I was hauled up in front of COM attorneys and miscellaneous 'deanlets' and forced to explain all details of dealings with my Company. It was like a firing squad tribunal and I felt victimized. When I explained to them that the Company had nothing to do with UCF (and never would have a relationship), and that no IP was involved, they changed the accusation. They then said I must be guilty of other crimes against humanity such as "conflict of commitment" because how could I run the Company and be a productive faculty member? Furthermore, one individual claimed he heard some 'rumors' about me, and that unnamed parties had come forward stating that I was using UCF resources and had been for many years. Why someone would fabricate such a yarn is unclear but petty jealousy is a safe bet or possibly it was an incorrect assumption stemming from the early UCF invitation to move the Company to my lab. In a fledgling academic environment like UCF, some of my colleagues, I believe, are insecure. I never noticed this at OSU; however, this place is still young and in a growth mode. The administration appeared to be questioning my integrity, saying in effect that TopoGEN did not exist and it was all made up somehow. The only way to deal with such hostile accusatory behavior is with the truth. I invited any who might be interested to visit the lab (located in Daytona Beach). One individual demanded to see photo-evidence instead. So I sent them pictures of the lab with staff members in various states of official activity. In the end, I think I won them over but it was highly stressful. If I was bringing in a few million dollars of

NIH grant money, I believe there would have been no real issue with compliance, but this is pure speculation. A well-known quote on funding goes like this: "if you have sufficient grant funding, you do what you want, if not, you do what you're told". The indirect cost recovery model from Federal dollars can be a pernicious metric to live by.

Sadly, the Orwellian compliance situation in medical schools can be used to punish faculty who don't bring in large grants (by aggressively cracking down on conflict of interest approvals and using intimidation tactics on senior faculty). These compliance laws are designed to make sure that clinicians engaged in drug trials are not being bribed by a drug company (with free trips, meals, cushy consulting deals, etc.). I witnessed this sort of thing at OSU with physicians engaged in clinical trials. An easy way to deal with this is through full disclosure and mitigation if required. My situation is very different but still falls under that purview it seems. Moreover, administrators were afraid to sign off on a document that mitigates perceived conflicts because they fear for their own skin. Basic risk aversion behavior, I suppose. I could go on, but in closing, I do not understand why these administrators cannot adopt a more positive and broad perspective on things. Active and vigorous faculty do many different things to contribute to the general mission of the University. Besides research, teaching and service, some of us are passionate about economic development. After all, academia should be a contributor to the economy and not just a consumer. As the new UCF provost once said about the faculty "no conflict....no interest". At least someone really gets it. The administration must recognize that any time a faculty member is successful, be it in outreach, economic impact or entrepreneurial activity, the university also wins.

Things are getting better. The new UCF College of Medicine is growing and we are recruiting competent and experienced academic leaders who should understand the real and manifold missions of the medical school and university. I saw this progression and evolution in the OSU School of Medicine when really gifted and bright leadership was recruited in the Cancer Center in late 1990s. The program ramped up considerably and became a state and national treasure trove of talented folks. These 'growing pains' are a natural process with a new academic program, and I am eternally optimistic about the future of the medical school and the entire 'medical city concept' that is actively growing at Lake Nona in SE Orlando.

The current status of the operation is impressive (after only 5-6 years). The COM has been granted LCME approval and has graduated its first medical class. UCF itself has nearly 60,000 students with 177 bachelor's and master's degree programs, >30 doctoral program and an annual operating budget of $1.4 billion. Total research funding stands at $129 million with endowments of >$125 million. My specific program, the Burnett School for Biomedical Sciences directs three undergraduate degree programs (Biomedical Sciences, Biotechnology, and Medical Laboratory Sciences) and has a Ph.D. program. Our program has 41 faculty that teach in COM, undergraduate, graduate programs, and do basic and translational research. COM has a new 200,000 sq. ft. research facility located at Lake Nona (20 min from Campus) and I now have a new lab in this very well equipped space.

My Current Life in the New UCF College of Medicine.

The UCF COM is unusual because it contains multiple undergraduate. There are few other similar programs in the country. My teaching involves a high enrollment UG course on "Introduction to Biotechnology" (several hundred per term) and graduate courses. As COM matures, I believe that the UG program will revert back to main campus and live in the College of Sciences. Despite some recent compliance issues, I am very happy with the situation in which I find myself. I run a funded lab still (although one never has enough research support).

I spend a crazy inordinate amount of time slaving away writing grants. With our government deficits, sequestration and a hawk-like congress bent on defense spending in various parts of the world, NIH budgets shrink by default. Our Department of Defense consumes resources faster than a sailor on a weekend furlough.

In addition to tight budgets, NIH Review procedures have changed and not for the better, I would argue. In the past, NIH panels would write full reviews on the science being proposed and generally I received useful insight from the 'pink sheet' reviews. Reviewers had to spend time reading and thinking about the research being proposed, including reading over the PI's past publications that form the foundation of the proposal. Thanks to new NIH leadership, along with a high profile panel of selected top researchers, the review process has been reduced to bullet points and a summary paragraph of discussion. Metrics are applied for several broad categories with the "Approach" being the most critical. This creates a situation where reviewers can cheat in a sense. They can simply decide to not spend time on the proposal and because of the truncated nature of the review, a few simple negative bullet points, and the proposal is doomed. Fortunately, there are others on the panel who review the grant, so there are some safeguards. Why reviewers would shirk their responsibilities may be related to job demands and over-commitment issues. I have also found a few NIH program officials are really lousy at what they do and misrepresent events, either intentionally or due to incompetence. I do not believe this to be a widespread problem at NIH but I have noticed it with NCI, since I usually deal with this particular institute.

Our government has its priorities in a constant state of irrational dithering. Ideologically driven members of Congress are a serious problem. I love airplanes and aviation and I support this activity but there are limits. According to what I've read, a single F-35 fighter jet rings in around a $250 million and as far as I can tell, they still don't work. In 2014 the Department of Defense (DOD) proposes to spend over $8B for this single aircraft. This is about 1/3 of the whole NIH budget. I believe that the DOD has invested well over $120B on this program; roughly more than 4 times our national research budget for all diseases (cancer, cardio, aids and infectious diseases). Can this be justified?

My life is comfortable and I am blessed with a great support network in my family, friends and social environment. I rarely fly commercial airlines when moving about the country. We have a lovely home in Central Colorado, on the Arkansas River, where I can mow the grass and then catch speckled brown trout off my boat ramp. My wife Annette cannot tolerate the summer in Florida, so around May 15, I fly her out to our Colorado home until sometime in the Fall. I own a second hangar at a regional airport very near our home in Buena Vista, CO where I keep my Columbia over the summer months (in the Winter, she is in Florida). I can easily beat the airlines from Orlando to Central Colorado, even though Jets are 2-3 times faster than my Columbia. This is because I can fly directly to my home in the mountains from Florida, whereas commercial flights land in Denver and I have to drive for 3 more hours to get home.

I like my UCF colleagues here and everyone seems to get along for the most part; although, I don't like the petty jealousy that I noted above. My compliance problems with the medical school have been resolved to a large extent. TopoGEN continues to thrive, even in a hardscrabble funding environment. Eventually, I will sell TopoGEN when I retire, which will be in the next 4-5 years most likely. I have recently taken on a new Ph.D. student who will likely be my last. I continued to feel great excitement about science but really about this Ph.D. project. Just allow me a moment to digress on my recent work.

I do cancer research and I am interested in epigenetic programming in somatic cells. Specifically, when a single Cytosine residue get methylated (at CpG sites), it tends to silence the expression of an underlying gene. Such C residue modifications are stable and inherited in daughter cells, largely due to a highly active DNA methylase that operates coordinately with DNA synthesis. Gene silencing in cancer is important because scientists have observed that tumors often have genes that are silenced inappropriately and conversely, other genes have CpG methylation reprogrammed. How and why epigenetic cues are altered with aging is not well understood. A fair amount is known about the DNA methylase that places a methyl-group on cytosine, but far less is known about how the methylations can be revised (removed or reset). In collaboration with a very bright and innovative group in Naples, Italy and independently in my own group, we have published a number of papers showing that when genomic DNA is damaged (both strands are broken) then repaired by a process of homologous recombination, the epigenome can be reprogrammed. Reprogramming takes place over a short (<500bp) segment downstream or 3' of the break. We designed clever GFP (green fluorescent protein) reporters to make this important discovery. The GFP reporter gene has a unique cut site that introduces a single double-strand DNA break in the coding sequence of the GFP cassette. The cell dutifully repairs this break by recombination and the cells turn bright green as GFP is produced in tractable amounts. Sorting the cells based on GFP expression reveals that some cells express high levels of GFP and other cells express very low levels. These expression differences are due to epigenetic reprogramming that attends DNA repair. We got a very interesting result recently. We used "live imaging" confocal microscopy to view the cells as they turn green and divide. We saw a single cell turn bright green (making GFP product) and after cell division, one daughter cell was exceptionally bright and the other was very dim. It is nice to see visual proof ('film at 10!'") showing that our model is correct, although we still have more experiments to fully validate the findings. While a picture is worth a 1000 words (so say the Chinese) a video must be worth at least 10^6 words. Modern technology continues to be a wonder source of inspiration.

Reflecting on the People Influencing My Career Pathway.

I respect myself. This is not arrogance. I am, without any doubt, an overachiever. I really don't deserve anything that I achieved based on some sort of superior intellect. I did it by sheer hard work and perseverance and passion for my objectives and this is the key. My Post-doc advisor had a great expression that aptly applies to the ivory tower of niche knowledge we build in academia. "We learn more and more about less and less" he would say. To put it another way, I know a lot about very little.

Influential people can be positive or negative. In this mini-biography, I choose to focus mostly on the positive influences but recognize that I also learn from negative role models. I appreciate the negative people who stood in my way, since they forced me to find creative pathways and solutions. I learn something from each person I meet. I also learn from students and some may be surprised by this, but everyone can be some sort of influence, be it minor or major. A separate section listing family, mentor, student, friend and colleague influences is woven into the fabric of my story. Writing a separate section would be difficult without getting maudlin and sounding like an acknowledgement in a Ph.D. thesis preface. For this reason, I will not re-list the influential people in my life. I am a sum of the individual parts that impacted my thinking and decision making over the years.

Along the lines of David Letterman, here is my top ten list of behavioral guidelines, regarding what I have learned from my life long journey of education.

1. Place a high value on collegiality and civil behavior. Treat everyone with respect. This might sound like an obvious platitude, but I place high value on this. I tried to bully people and it does not work. If you are in a leadership or mentor's role, your challenge is to get people to work hard and do their job without creating acrimony. This is harder than you might think.

2. Disclose your activities and implicate everyone above you in the chain of command. This is a self-preservation tactic but make sure you have hard evidence (letters, emails, etc.). It can backfire when moving from one university to another, so beware.

3. Don't be risk averse. Life is about taking chances. If you never step out of your safe zone, you are not alive.

4. Related to risk taking, given the option, I tend to ask for forgiveness, rather than asking for permission. Be careful with this one as well... it can bite back.

5. Show enthusiasm for what you do. People may disagree with what you say but with sufficient passion, people will listen.

6. Try hard to communicate at the other person's level, on all matters. As a scientist, I often discuss complex matters with business-oriented audiences and investors. Being able to crystalize complexity into a concise and basic framework is not that hard, with practice. Be responsive to audience response, body language and group behaviors. Adjust accordingly and make eye contact to engage and be inclusive.

7. Be careful with whom you associate in an academic environment, both as colleagues and collaborators. Since some

dogs have fleas, you need to be alert. Usually, your colleagues choose you (when you apply for a tenured job for example). Starting a company, selecting collaborators, creating a new academic department are rare occasions. Who you select to join the lab (Ph.D. students, post-docs and technicians) also qualifies.

8. Research is definitely a niche market that favors those who are creative. New model systems, novel technology with collaborators well outside your field will help. I put together an NIH collaborative Project with an acarologist (a specialty field of entomology that focuses on ticks and mites). We had lunch together and it lead to a tick gene cloning and expression project (that was very well funded). My theory is that "niche" science tends to breed rather tightly focused bands of new information. When two diverse niche projects can find common ground, it defines an overlap of knowledge that is highly creative and novel. In my case, I really enjoy talking to colleagues in diverse fields for this very reason. At OSU, I was a 'graduate school representative' on Ph.D. theses (oral defense) in areas like psychology, pharmacology, English literature, and history (there were others as well). This was great fun and a wonderful chance to learn.

9. Take advantage of sabbatical opportunities whenever possible. DO NOT do a 'stay at home' sabbatical, since this does not serve any purpose other than encourage vacation minded faculty to shirk responsibility (my opinion only). Leave your home base and go to a different lab to re-tool. Learn a new model system, gain new technology insight and work hard to implement the true purpose of the sabbatical. It should all be focused on making you more competitive in research.

10. Finally, avoid negativity in all matters. It is draining and toxic. Stay positive and don't screw people over. It will come back because the world is round! On a related note, use criticism of your work as a source of strength and do not let it erode you (see next section). Non illegitimi carborundum (don't let the bastards wear you down, in Latin; however the actual phrase should read: Noli nothis permittere te terere). You get the picture either way.

It's a Wrap.

In science, you must develop a thick skin against failure. Papers that get rejected, grants not renewed, faltered career plans, lack of jobs, tenure denial, and oppressive, demanding administrators will affect your equanimity. This is why senior faculty become jaded dead wood hulks in my opinion. One of my favorite artists is a folk singer named John Prine. I know and sing many of his songs and am a huge fan. His lyrics really hit home, but there is one song (and line in the chorus) that I love. Through a mutual friend in Scotland (a radio personality who interviews country singers) I obtained a signed card from Mr. Prine that is my personal treasure. I asked Mr. Prine to write the line out in his own hand and sign and date it (see underscored lines below). The song is called "That's the way".

"That's the way that the world goes round

You're up one day and then you're down

It's a half an inch of water and you think you're gonna drown.

That's the way that the world goes round".

In particular, this speaks to so much of what I do as an academic scientist and to the idea that one must be resilient. Failure is in the eye of the beholder and actual definition of failure is either giving up or not reaching beyond your grasp. Be bold and committed with passion. Use negative outcomes (failed grants or rejected papers) to build your resolve and adopt the attitude the "I'm going to do this and nobody can stop my momentum…. I will prove them wrong". J. Craig Ventner (founded TIGR, his own institute and one of the first to sequence the human genome) told this great story. Some years back, he submitted an NIH grant proposing a new sequencing technology that was soundly and repeatedly rejected by the NIH Study Section. He kept the names of the panel members and the harsh negative reviews as a reminder to build his resolve. Many years later, after sequencing the human genome and establishing his private institute (JCVI), he cross-referenced the names of individuals who consistently torpedoed his proposals. Guess what? These same individuals had adopted his technology and were using the same process they had previously rejected. He was not angry, on the contrary, he embraced this seemingly ironic outcome as validation of his ideas. His 'half an inch of water' was the panel of close minded NIH reviewers who could not appreciate novelty that comes with revolutionary science. Clearly, he did NOT drown.

Completed July 15, 2013

Figure 1. My Dad, taken at signal corps
headquarters in Brisbane, Australia, 1942.

Figure 2. Lt. Mark T. Muller, Sr. Brisbane, Australia, 1942.
My Dad loved photography and documented himself and
others using a Leica Model C (circa 1938) which he gave to
me years later (and still works). He took many Kodachrome
35mm slides with this camera all of which are in excellent
shape with minimal fading. I have in my possession several
hundred archival quality Kodachromes from WWII and Korea.
This picture was emailed to me in 2011 from Paul Watson,
an Australian historian and archivist, who found some of my
Father's images in storage. I was not aware that this picture
existed, so it was quite shock to see my Dad at age 27.

Figure 3. Columbia 300 aircraft (N-73TX) where I landed at the highest airport elevation (circa 10,000 feet) in the continental U.S. Many general aviation airplanes have a 'service ceiling' (maximum altitude of 12,000 feet) and cannot take off from this airport (Lake County in Leadville Colorado). On a warm summer day the air density correlates to an altitude of 13000 feet or more. Mount Elbert can be seen in the background (the highest "fourteener" in the Rocky Mountains. The name on the side of the airplane is "Lancair" which is the original manufacturer. Lancair was acquired by the Columbia Aircraft Company and later on by Cessna (currently making the Corvalis, but largely the same airplane).

Figure 4. Flying my Columbia 300 aircraft
to Colorado from Orlando, Florida.

Annette is an excellent co-pilot but is not a pilot. She
has over 1000 hours of flight time logged as my co-
pilot. I have done this flight non-stop, but my bladder
is the weak link, so normally I stop in Arkansas,
Texas or Oklahoma for fuel and bathroom.

Figure 5. Ken and Kimiko Tsutsui with me on "Art Island" during one of my visits to their lab in Okayama Prefecture. The island of Naoshima is known for its many contemporary art museums. For example, the Chiuchu Art Museum ("in the earth") that houses a number of site-specific installations by James Turrell, Walter De Maria, and Cluade Monet. It sits of the highest points of the island, and various exhibits and facets of the museum's architecture take advantage of its commanding view. This is truly an amazing place to visit. I was invited to give a talk at Tsutsui's lab in 2010 and during my stay, we visited Naoshima Island in the inland sea of Japan.

Figure 6. At the Royal Palace in Caserta, Italy with Enrico Avvedimento (far right), my wife (Annette). I did a sabbatical in Enrico's lab at the University of Napoli in Naples (2011). I learned to speak Italian during my stay but now it's pretty rusty. I became involved with a biotechnology company (as a board member) based in Napoli (Prius, like the car). I am still active on the board.

Figure 7. Central Colorado to Daytona Beach Non-stop flight in the Columbia 300. This was a long flight in the airplane, about 7.5 hr. (Better than a 32 hr. drive however).

Figure 8. Upgraded the Instrument Panel with Electronic Flight Instrument Systems (EFIS). The two rectangular instruments in the center view are the new EFIS. While not cheap to install, it is an amazing resource giving real time uplink weather, traffic (other aircraft), terrain, full airport database with all approach plates, coupled to autopilot and GPS.

CHAPTER 12.

Seed Science: A Personal Study of Growth and Development

Tom Seed

Prologue

The dim yellow light from a late afternoon fall sun fills my office, and as I close the computer file on a recent work assignment, I think to myself…. "This is really good stuff these researchers are doing on this contract: the program managers at the NIH will be pleased by this annual progress report".

As I gaze out at the window, I realize how late in the day and how much of it has been spent on reviewing this 'science', and I muse about how nice it would be to have been kayaking or fishing down on the Chesapeake. But…..thinking about science and the exercise of it all, has a special appeal, a special reward, in its' own right [1]. Yes?

I was asked by the editors of this volume to write a short piece about my interest in science and how I found my professional way through, what turned out to be, a life-long adventure.

OK…..no problem…here goes….

My name is Thomas Michael Seed: Tom for short (Figure 1). I am a retired biomedical researcher who has spent some 45 plus years 'in the science game' and still continue to enjoy the exercise of it all (Figure 2). In the following pages I have briefly described 'my story' – the 'how', 'why', and 'to what end'. Hopefully what I say here will be of some use to young students trying to sort out the pros and cons of trying to pursue a career in the sciences.

Regardless of whether or not the reader takes note of my comments here, there are a number of what I consider to be, in general, 'universal truths' of science; and more specifically, in terms of fostering a career in the sciences. The first of these 'truths' is that world of science represents a big and brave new world, filled with opportunities, and founded in a rudimentary hope that, with proper investment and pursuit, better, healthier, more prosperous world societies will emerge. The second, is that science, and the so called 'scientific breakthroughs', are governed by a process akin to a fundamental physical law of 'mass action', namely that the more people get involved in science, the greater the chances are for real and significant progress to be made. The third and final 'truth', is that science and science progress is cumulative by nature and has 'memory'; current 'breakthroughs' are not only dependent on the creative juices of given individuals, but also by a collective foundation of basic information of others, both past and present. The same is true of the educational and mentoring processes involved in developing career scientists and academicians: Analogous to the title and theme of Hilary Rodham-Clinton's book [2], "…..It takes a village to raise a child", it's no secret that it takes 'a community of mentors to develop a career scientist'.

Coming of Age

An introductory comment- I would be negligent if I didn't first mention that, at least from my perspective, that there were few, if any, hints from my childhood that I'd end up having a career in 'science'. There might be lesson here relative to trying to use early childhood experiences as 'predictors' of future career tracks. I suppose that there are indeed a good number of career scientists who knew what they wanted to be very early in their childhood; however, I suspect that the vast majority of working scientists did not grow up thinking about the 'glories' of science or how they'd like to spent professional lives doing scientific work. In speaking to various colleagues over the years, I've come to appreciate the fact that, for most scientists, these career path decisions often came fairly late in development and were often driven by a limited number of life events and/or personal interactions with a select number of mentors.

Vital and general childhood experiences- I was born the third son of John and Connie Seed at the Paterson General Hospital, in Paterson, New Jersey, December 8th 1945. I was good sized (9 lb. 11 oz. - 26 inches long), very strong newborn, as evidenced by the story that I was turning over in my hospital crib on the first few days of life. In any case, after the prescribed period in the hospital, my parents (Figure 3) brought me home to join my two older brothers (eight year old, John Richard 'the elder' brother and 4 year old, William Henry 'the 'middle' brother) (Figure 4) to a pleasant little suburban dwelling in an equally small but charming 'bedroom community' of New York City, namely the village of Glen Rock *[www.glenrocknj.net]*. It was here that I was raised [3] and schooled, initially at Richard Byrd Elementary *[www.glenrock-nj.org]* and, subsequently, at Glen Rock Junior and Senior High Schools *[www.highschool.glenrock.nj.org]*. The best way to describe Glen Rock is to liken it to the old, TV sitcoms about daily family life in the idealized 'modern' suburban community' (but with the obligatory wart or two): if you conger up the image of 'Father Knows Best' or 'Leave it to Beaver' you will get the picture of how I view my childhood. My dad worked as mid-level account manager for New York Bell Telephone and Telegraph Company with offices in the Empire State Building [4]. His chosen profession of accounting was not at all surprising as there had been a long line of accountants, bookkeepers and clerks on the Seed-side of the family. [5]. My mother's situation and place was what I would consider pretty typical of the day- content to stay at home (at least this was the claim) in order to properly raise her three boys, and in general, to run the household. I can't speak for my older brothers, but my mom was the 'hands-on' parent in providing the daily nurturing, guidance and direction, whereas my dad was more 'hands-off' who provided a distant, long-term type of nurturing. My family, my home, and the town in which I was reared all provided for a wonderful childhood with many pleasant memories.

So what does this have to do with my ending up in the 'sciences'? ......probably nothing, nothing at all. But there are elements of my rearing that might be relevant to my eventual career track in science. My parents instilled in their children a sense of free-will and independence in terms of play and problem solving activities. They also insisted that we were physically active, and whenever possible outdoors: the concept of 'taking in fresh air' and 'outdoor exercise' was, in the Seed household, intricately linked to one's general health and well-being.

I greatly enjoyed planning and constructing a wide variety of games and toys that allowed me to imagine myself as a racer of fast cars and boats, as a cowboy, a soldier, fireman, etc.; but never- never as a laboratory 'scientist' who wouldn't see the light of day for days on end. Engineering of new games and toys was a big deal for me: I still remember the fun filled days working new designs for little toy power boats and trains and getting my friends to participate in the testing phase of these toys. Interestingly though, I always enjoyed the solitude of the planning stage, but clearly required my buddies to help in the testing of performance. I was going for a thrilling, rugged outdoor job- like a pro football player, until I recognized that I was way too small, way too slow, and clearly without the essential physical and athletic skills [6].

From an academic standpoint, my high school years were uneventful and clearly unremarkable. If truth be told, I was an unin-
spired, very average student who enjoyed athletics and the social
aspects of school more than the academics.

From the academic point of view, my 'coming of age' came
in three, temporally distinct, phases: first, an academic *awakening*; second an *orientation,* and third, fixing *direction.* The
awakening phase came shortly after graduating from high
school, during the very early phases of my freshman year in
college. This awakening was far from being a gentle process
but more akin to being violently roused from a bad dream.
Through some rather heavy-duty self-examination of my status
and future prospects as a professional, I came to realize just
how essential it was to start to knuckle-down as a student and to
make sure I got through college. The orientation phase came
somewhat later in college in which I decided on the biological
sciences as the major direction for my college work. The third
phase of 'fixing direction' came considerably later during my
graduate studies. While the first phase was driven by me and
fostered by my parents, the second and third phases were the
consequent of some very cogent teachers and mentors.

Developing an Interest in the Sciences

My interest in science certainly did not develop *de novo* but grew
with time, especially during my college years: I was, as some people
would say, a 'classic late-bloomer'. When this interest first developed…. when the 'seed' of interest started to germinate… is hard
to say. I suppose that if you define specific types of a child's play as a
form of rudimentary science, then perhaps such things as working
with my chemistry set during elementary and middle school years,
or perhaps working with outboard motors and racing boats in order to improve overall performance, might be used as evidence of
an earlier time frame for my developing a 'science interest'. As I

have stated earlier and will discuss later, my interest had not been piqued until I was well into my college years: this I attribute to a combination of having a couple of good and interesting college courses and a cadre of supporting advisors and mentors.

Choices of Undergraduate and Graduate School

Choice of college and programs of study. Choices for college and for graduate schools were made for entirely different reasons and under entirely different circumstances. As alluded to above, I was neither the best, nor the most serious of high school students; as a consequence, my choices of topflight schools were limited and clearly not strictly governed by the nature or quality of their science programs. My objective as a high school senior was quite simple and modest: to try to get into the best possible school that would accept me. In this regard, I was extremely grateful in having the University of Connecticut [*www.uconn.edu*] express an interest in having me as a student. Their 'interest' was not because of my 'sterling academic credentials' but rather as an athlete. It so happened I had the good fortunate to play on a very good high school football team (Figure 5) and as a consequence, I was recruited by the university's football program. In good faith and with good intention, I started out playing but dropped out early in my freshman year fearing that I would not be able to keep up academically. Due to the rigors of my course load and a couple of rather ridiculous team requirements (e.g., mandated 'study halls' that fostered anything but appropriate 'study'), I recognized the urgency of my need to settle down quickly and to work hard on my studies. I had a lot of academic ground to make up for and recognized that football was just a game and simply a passing fancy but success or failure in college was essential and would remain with me for life.

I'm happy to report that my 'college plan' did in fact work out: I managed to make it through my initial year of college quite successfully taking a standard requisite load of science and humanity courses. Academic success certainly made up for the fact that my social life took a serious hit at the end of my freshman year, I had significantly improved my study skills and had gained some confidence and strength in my academic abilities. I was ready to move on.

By the end of my sophomore year and having been introduced to a number of very interesting, basic courses in biology and chemistry, I had settled pretty much on pursuing a degree in the 'biological sciences', with a major in bacteriology and a minor in chemistry. I'd like to claim that these choices were made on my own, but in reality they were the result of my having some very thoughtful input from several very gifted teachers. The two individuals that I remember best were Drs. Stanley Wedberg, a full professor and long-time chair of bacteriology, and Lawrence Amundsen who, likewise, was a full professor and (I believe) chair of chemistry. Needless to say I was impressed by these two gentlemen: their teaching acumen and their clear willingness to teach these rather large, introductory courses. I honestly do not think that I would have developed my initial interest in these subjects within having the benefit of sitting through these classes and listening to these professors talk about their subjects.

In addition to the positive influence of these college professors, my family, especially my older brother- Rich 'the elder'- played a key role as well. Rich had gone through all of these academic steps and decision-making processes much earlier and was, by this time, a well-established academic/research biologist on the faculty of Tulane University in New Orleans (Figure 6). I was always intrigued by his comments on his academic profession and by the 'whys and wherefores' of the science he was attempting to pursue. Although I never stated this as such, I'm sure that I saw his joy in what he was doing professionally and thought that it had to be pretty darn good and that I should at least considered this type of professional work as a viable career option.

For the most part and independent on really how I fared in specific courses, I tended to find my science courses to be interesting and quite enjoyable [7], but generally found the humanities to be a bit more tiresome and of little interest [8]. Despite being on a course track to obtain a baccalaureate of science degree, the university mandated, and quite appropriately so, that the science students take a number of courses that fell outside their major subject areas. Well one of those requirements was to take a course or two in the 'Fine Arts': a theater course seemed to be the least objectionable, but little did I realize at the time of course registration, that the class was held clear across campus, at the ungodly time of 8 am, and largely entailed the memorization of minutiae concerning Broadway plays, both past and present. Needless to say, I wasn't positively engaged by any means and ended up with my lowest grade while in college. (I find it a bit ironic now that I really enjoy the theater now and reading reviews about given performances here in the Washington area and in New York). With exception to the above experience with the 'fine arts', I was encouraged by one of my academic advisors (whose name is lost to me) to take some philosophy courses, in order to complement my science classes. The thought here was that philosophical study would 'sharpen the mind' and the creative thought processes associated with science itself. Although a bit skeptical, I went ahead and enrolled in a couple of these courses that ranged in content from 'modern philosophical thought' to 'basic logic'. Low and behold, I loved these courses- the readings, the classes, and associated discussions on various 'life-related thought processes'. Even today, when on rare occasions I'm asked by a parent or youngster about types of courses that might be useful to have for science majors, I point out the utility of including some philosophy courses in their academic programs. This advice was certainly good for me some fifty ago and has served me well since: so, I'm passing the advice forward.

I am, in fact, a believer in the concept that an individual's capacity to reason and to work through problems in a systematic and logical fashion is, in part, a learned trait, and one that can be improved upon. In its most simple form, philosophy is the course study for making such improvements in one's reasoning capacity... clearly a trait essential for the scientific process. (Relative to the discussion on this 'trait', I've listed as a footnote [9] a rather comical episode that occurred within the Seed household that is now part of family lore).

By my junior year I was pretty much ensconced in my bacteriology and chemistry courses and well on my way to collect my science baccalaureate. This was good of course, and I was pleased by this, but it forced me to think about my 'next step' following graduation. Should I continue on to graduate school? Or should I try to get a regular job in a health-aligned industry? How about government service? With the Vietnam War escalating, 'Uncle Sam' was more than willing to 'employ' me...no question about it. To say I was a bit uncertain as to my future is an understatement. However, my 'situation' was by no means 'unique': the vast majority of my college friends were equally unsure as to their fate following graduation [10].

As summer of my junior year rolled around, I found myself looking for part time work that preferably permitted me to gain some 'hands-on' research experience. My brother talked to me about it and suggested that I spend the summer down in his laboratory at Tulane University in New Orleans [www.tulane.edu]. I jumped at the opportunity to do this; I could think of nothing better than having a chance to gain some hands-on laboratory experience, doing a little bit of independent study, and working with my brother. Wow, what could be better than this! [11]. In any case, I was assigned to help out Ed Risby, a senior graduate student in the biology department, and a super nice guy (who subsequently spent his professional career at Tennessee State University in Nashville as the Chairman of Biological Sciences and ended up as Dean of the Graduate Studies and Research) on a problem involving the 'isolation and characterization of a critical enzyme (phosphohexose isomerase) of glucose metabolism within several of the *Trypanosoma* species- rather nasty blood parasites that cause significant health problems (African sleeping sickness) in both domestic animals and in man [12]. Although my stay at Tulane was rather short, it was really a wonderful experience for me and incredibly 'eye-opening' in terms of what might be possible for me career-wise.

Choice of graduate school and associated programs of study. Towards the end of summer between my junior and senior years of college, I had a lengthy discussion with my brother Rich about graduate schools and programs that might be of interest to me. Various schools and programs were discussed and ran a full gamut of training and research opportunities in the general area of 'microbial pathogenesis. As it turned out, Ohio State University *[www.osu.edu]* and its Department of Microbiology turned out to be a winner in my eyes: it had a well-recognized and respected Department of Microbiology, a great graduate training program, and was complemented by some outstanding research programs and research faculty. One of those faculty members was one Professor Julius P Kreier, a veterinarian turned microbiologist, who was running a very productive, vibrant research program on the infectious hemolytic anemias (e.g., malaria, babesia, etc.) of animals and man (Figure 7). Cutting edge technologies were being applied *in vivo* using appropriate animal models of human diseases in order to dissect and characterize the various pathogenic processes of different infectious microbial agents under study. The papers coming out of Dr. Kreier's program at the time were intriguing to me, supporting the thought that this was the type of experimental work that I wanted to pursue. The rest was 'history' as they say.

I initially wrote to Dr. Kreier expressing my interest in his research and in the possibility of working in his laboratory under his direction while pursuing a graduate degree at OSU. Dr. Kreier responded with an encouraging letter and suggested that I not only apply for admission to his department's graduate program at OSU but also to make a short exploratory visit out to Columbus in order to see the campus, tour the department and its facilities, and to meet some of the faculty and graduate students. Although it wasn't stated as such, I believe that Dr. Kreier wanted to have some face-on discussions with me in order to ensure 'due diligence' on reviewing my background and qualifications as a potential graduate student in his department and in his laboratory. In any case the visit went well, and I came away with 'good vibes' about doing my graduate studies at OSU under Dr. Kreier's direction.

Although the plans for my starting graduate studies at OSU were fairly well planned and seemed to be straight forward, the actual carrying out of those plans was another matter. Relocating from the East coast to the Midwest was one issue and finding suitable living quarters was another. Most importantly, however, making sure that my young and lovely, quite pregnant wife, Vicki was 'on-board' and comfortable with all of this. To make a long story short, we managed just fine. We took the move, as it was, a life experience and actually found a bit of pleasure and adventure in all of it. Vicki and I still (some ~45 years later) laugh about the trip and the initial, first couple of days in Columbus (e.g., stopping along a busy highway in the Zanesville area of Ohio with a trailer loaded to the gills, with all our worldly possessions, and climbing over a sizable fence in the highway median in order to shop for some local crockery for our new abode).

We arrived in Columbus very late on a lovely, warm June evening, safe and sound but exhausted from the moving adventure (the route down the Jersey turnpike to the Pennsylvania turnpike over to Wheeling West Virginia through to Zainesville and on to Columbus). We had previously rented a very small but tidy apartment on Stinchcomb Drive (University Apartments) just east of the university and a little way from the football stadium (I mentioned the latter because on nice Saturdays in the fall you could open up the windows and listen to crowd noise coming from the games and bands playing at halftime). Although it was quite late, we managed to find the resident manager of the apartment complex and got the keys to the unit. We mustered up some remaining energy and moved some essentials (a bed, mattress and sheets, and some bathroom items) that night and left the rest of the unpacking for the next day. The following days were consumed with getting the trailer unpacked and returned to the rental agency, our belongings moved into the apartment, sorted and arranged. Besides the 'scouting out' the area for shopping, routes to the university and its various facilities, etc., I presented myself to the Kreier lab and was kindly led around to be introduced to students, staff and select faculty, as well as being assigned desk with some shared lab space. Shortly thereafter I found the Office of the Registrar to register for my courses.

All went well these first few days in Columbus, until I received a bill from the university that included standard line items for tuition, out-of-state residence fees, and associated course fees. Up to that point, I was working under the impression these 'educational expenses' would be covered...by whom, I didn't know and hadn't previously given it much thought. It was like being 'blind-sided' by a 2x4. I didn't know what hit me: up to the time of received this bill, I had only been concerned about 'living expenses' and not school costs, per se. I realized that it simply would not be possible to cover these costs, quarter-after-quarter while in graduate school. I subsequently presented my problem to Dr. Kreier….my teacher….my mentor….who I naively thought would tell me it was OK and provide me with some 'soothing' words of comfort….like 'Tom, don't worry about it, I'll take care of it' (isn't that what a graduate advisor supposed to do?), but instead he simply said 'education is an expensive proposition' and summarily dismissed me, giving me my initial 'JP Kreier life- lesson' (soon to be followed by many more over the next 4 plus years under his tutelage). He wanted me to think about these costs and not to treat them superficially: one's education is indeed expensive and needs to be treated with respect and hard work. Just for record, these 'educational fees' were eventually waived, thanks to help from the department and from Dr. Kreier.

My combined MS/PhD track program was highly structured, fairly intense and comprehensive in terms of packing in essential courses and research into a relatively short, 4 year period (Dissertation titles for the two degrees are as listed [13,14]). Degree requirements set by the both the departmental and by university largely dictated the courses that I took; including a full set of microbiology and immunology courses, coupled with a healthy complement of biochemistry, biophysics and physiology/cell biology. I did, however, have the opportunity to take an elective course or two: one course was a rather unique, semi-independent study course on the electrophysiology of cells and tissues given by Dr. Phil Hollander, an amazing individual who effectively taught his subject largely by way of direct experimental demonstration [15]; a second elective course of interest was 'Scientific photography', that ultimately proved to be very useful to me later on when I was responsible for the running of electron microscopy facilities at several different biomedical research organizations that employed me (Blood Research Laboratory, National American Red

Cross, Bethesda, MD and the Division of Biology and Medicine, Argonne National Laboratory, Argonne, IL).

My education and training at OSU was supported almost entirely by university fellowships and by external research grants: initially, I was supported by a Department of Defense (DOD) research grant that Dr. Kreier had at the time; while toward the end, I was supported by a university fellowship that I had received. Because of a departmental requirement that all graduate students 'teach' for a limited number of academic quarters, I fulfilled this 'requirement' midway by serving as 'teaching assistant' for a short period in a number of large, introductory microbiology courses. Although I really didn't enjoy this 'duty', it did serve one very useful purpose, namely, that teaching to large groups of undergraduates wasn't exactly my 'cup of tea'. It needs to be pointed out that the 'financial support' granted to me was absolutely essential in my being able to work my way through this educational process. Although I was at the time (and still to this day) very grateful for this educational support, it was by no means of sufficient level to provide for the cost of living. Vicki and I not only received additional financial 'life-lines' from our families, but Vicki also resumed teaching within the Columbus city schools in order to make ends meet.

Most of the students like me were expected to work flat out all week, all hours, for 12 months a year. It sounds pretty horrific but, in all honesty, it wasn't that bad, especially since it wasn't all class room work (which I found tedious), but the extended periods of independent research that were mixed into the program gave me some relief. I also recognized that there was a set time-limit to this work schedule, and that eventually, at 'the end of the day', there would be considerable rewards (degrees) granted (Figure 8) Without doubt, it was much harder on my wife, Vicki, than on me; this in terms of being somewhat isolated (from east-coast family), and not having my company but still having the '24/7' responsibility of caring for our infant son, Patrick.

Choice of post-doctoral training programs. This phase of one's academic and professional training is not by any means mandatory, but clearly useful and generally highly recommended in order to enhance one's expertise in a given academic area, while gaining invaluable, 'hands-on' experience in one's chosen line of work. The vast majority of PhDs take this route, and I was no exception. I did my post-doctoral work at the Institute of Pathology at Case Western Reserve University (Cleveland Ohio) [www.case. edu] under the direction of Dr. Masamichi Aikawa (Figure 9). Dr. Aikawa was an internationally recognized anatomical pathologist with a research interest in the ultrastructure of malaria- causative agent(s) and associated disease processes. Because I had done a fair bit of microscopy while working on my PhD project, I had developed an interest in the 'art' and science of both light and electron microscopy, thanks largely to one Dr. Bob Pfister who was not only chair of the OSU's Microbiology Department but also an ultrastructuralist. In any case, Dr. Aikawa enjoyed a reputation of being at the forefront of these technologies when applied to the study of tropical diseases of significance. When Dr. Aikawa offered me the postdoctoral position in his lab, I jumped at the chance to join the group; and what a group it was with such notables as Chuck Sterling who was another postdoc in the lab (currently a professor emeritus at the University of Arizona) and some very accomplished technicians such as John Rabbege. The position, although limited in time, gave me sufficient opportunity to hone my technical skills as well as develop a good working knowledge of the tools and processes of biologically-based microscopy, especially electron microscopy. The latter served me well as it provided me with a number of employment opportunities that I wouldn't have had otherwise.

People and Organizations Influencing the Career Pathway

It's only now when I sit down in a quiet place and think about how I got to where I am today, do I realize and appreciate just how fortunate I am and how many individuals there were along the way providing proper direction, guidance, and encouragement. My first thought is generally related to an answer to a question I raised earlier concerning the requisite support group for a developing scientist. "Yes"…it does indeed take a "village" (a community of advisors and mentors) to 'raise'- to develop- a potential career scientist. …and my second thought is usually related to the fact that "NO", the science neophyte cannot do this independent of a community of supporters. I speak for myself here but do not doubt for a minute that this is also true for everybody else as well.

So what about this 'community of supporters'? Who are they and how did they help?

I consider 'family' as the being absolutely critical in getting the process going: if you consider the analogy of building a 'snowman', one must initially roll a small compact ball of snow over fresh snow in order to build up successive outer layers of snow so that spheres of snow of increasing diameters can be built for use in constructing the snowman itself. By analogy, the same is true of the need of having a solid core of supporters, in my case my 'family', upon which additional, potential layers of support could be collected. Those additional layers grew with time and basically tracked with my schooling. Clearly my immediate family provided the 'core'…the nucleus…for subsequent growth, but my environment was composed of caring and thoughtful teachers, family friends, and neighbors, all of whom provided that extra special, 'beyond the call of duty-sort of help' that children often need in growing up. The list of such individuals (teachers, neighbors, family friends, etc.) is long, but an example or two might be warranted. A 1st grade teacher, Ms. Arlene Fry, worked with me after class for the better part of school year in order to help me overcome a 'minor, but bothersome', speech impediment that I had early on. A 6th grade teacher, Mr. Charles Knipple, encouraged me in my interest in the design and mechanics of model trains. Still later in junior and senior high school, a cadre of teachers and coaches took the time to work with me on a variety of problems. It's only in retrospect that I came to realize the importance of what these folks were trying to do to help me out: at the time, I was neither the most willing, nor the most grateful recipient of their effort. To give you an example, I wasn't very fond of doing homework once school was out; certainly not when I had 'better things to do' with sports and social activities. One of my math teachers - one Karl Krause-who also happened to be an assistant football coach, picked up early during the school year on this bad habit of mine and insisted that I report to him immediately after school for an 'extra special' math class that involved my doing homework then and there. I can still hear him gently bellow "...just do it Seed! ...no questions asked! ...just do it!" Unfortunately, the 'pain' didn't end there, but was carried over to football practice as I was required to run additional wind sprints at practice due to my being 'late' each and every day. I laugh about it now, but I wasn't laughing at the time.

Glen Rock High School (GRHS). My high school days were filled with comparable interactions with other teachers and coaches. I can't say enough about men like Al Deaett, John Cheska, and Ed Bing who were teachers first and coaches second. All of these men took the time to teach essential life skills, especially the virtues of hard work, practice and education. With great pleasure I can remember being taken by these gentlemen, along with fellow classmates, during my junior and senior years in high school to visit and tour various colleges they had personal ties to and that we might be interested in attending (e.g. Amherst, UMass,

Springfield, etc.). This activity was on 'their time' and at 'their expense'.......talk about dedicated teachers and mentors.

University of Connecticut (UCONN). College was another matter entirely. I view this period in my life quite philosophically: no doubt I had a strong sense of 'urgency' concerning my academics and the need to 'keep on the straight and narrow'. My simple, 'have a good time' approach to life that was so dear to me in high school, was not going to serve me well as primary goal of college. By analogy, and in the absence of any real understanding of Darwin and evolutionary biology, the strong selective pressure of my new university environment was profound and clearly at work in terms of my need to adapt in order to survive. Interestingly, a combination of both internal and external forces played a role: first, I was both willing and able to make some changes in my habits, and second, I was positively influenced by a few very good professors who had the capacity to effectively engage students and to bring the subject matter alive and relevant. Again I refer to Dr. Wedberg in microbiology and Dr. Amundsen in organic chemistry.

Tulane University. [www.tulane.edu] Once the 'crisis' for academic survival had subsided, my interest in the biological sciences as a possible career choice was piqued by my summer of '67 New Orleans adventure. Yes, New Orleans proved to be an incredible experience, but I'm referring more to my independent study at Tulane University than to the city itself. Under direction and guidance of a number of very talented people in my brother Rich's lab, i.e., Ed Risby, a graduate student, and a senior technician, Al Gam [16]), I was given, for the very first time, the wonderful opportunity to see first-hand, the 'stuff' that biological science is all about; albeit only a glimpse of not only the very good, creative and technical aspects of biological science, but also by the not so good aspects of the work, as exemplified by disheartening experiments that had 'gone south' (failed for one reason or another).

In addition to learning the 'ins and outs' of daily laboratory life, I was shown that 'science' had another side; a lighter side that could be 'downright fun'. Somehow, somewhere along the way, I had this preconceived notion of what scientists should be, namely 'prim and proper' and always in deep, meaningful thought trying to solve major problems of the world. Quite to the contrary and much to my surprise and amusement, I distinctly remember the 'party- like' atmosphere surrounding field trips to local swamps in order to gather biospecimens of interest (*Trypanosoma sp* susceptible frogs for circadian- related studies). I can't very well go into details here but let me say that these 'field trips' were second to none in terms of their being a 'good old boy' southern type of party.

This time I spent at Tulane undoubtedly constituted the actual 'initiating' event in my eventual track to a career in research: for this time, this opportunity, I'll be forever thankful. Because of the very positive nature of this training experience, I've tried throughout my career to 'pay this debt forward', specifically in terms giving other young students the opportunity to work in my lab and to try to help them gain experience in what a professional life in research is like and what benefits it might offer as a career choice.

Ohio State University (OSU) [www.osu.edu]. Grad school at OSU fostered my general education of the biological sciences, while solidifying my research interests in various parasitic blood diseases of both man and animals. Clearly and without doubt, Dr. Julius P Kreier, my graduate program advisor and mentor, was my guiding light during my stay at OSU and beyond (Figure 7).

As strong an influence Dr. Kreier was during this period of my life, he certainly was not the only person/persons that had a role to play in deciding where I was going. Factors external to OSU and my family were to play a role- a key role: the US Selective Service and the local draft board in Hackensack NJ played THE principal role in the unfolding melodrama by making a series of decisions that put my life into a tailspin. As I mentioned earlier, when I graduated from college in the spring of '68, the Vietnam war was in full swing, educational deferments were disallowed, and there was great uncertainty about the future for literally all recent male college graduates: I was no exception, but I decided to push forward with my plans to attend graduate school at OSU and to 'let the chips fall where they may'. Within a few short months following our arrival in Columbus and my starting my class work and working in the lab, I was notified by my draft board that my deferment had been cancelled and that I was to report to a local army base (Wilberforce, OH) for my initial processing into military service. Although not exactly 'overjoyed' by the prospects of military service, especially when that service almost invariably entailed a fun-filled, breath-taking adventure of a military 'holiday' spent in Vietnam, I went for this initial processing as I was directed...no questions asked. In any case, several weeks went by, and I was making appropriate family arrangements for my young wife Vicki and our newborn son Patrick to move back East to be with our families thinking that I would soon get a notice directing me to report for active duty. But what I received instead was a notice from the draft board saying that they had canceled a previous draft decision and had granted a new deferment. If you've ever played the board game 'Monopoly', this notice served as my 'pass go' card; it effectively served as my 'passport' to an advanced education. To this day, I still don't know why the draft board reversed itself, but I was later told that the geographic area that I was being drafted from was a 'high volume' area and that the draft quota had exceeded their manpower quota. Thinking about it now, I wondered if I would have gone back to school following this military service: I have many good friends who fought in Vietnam, and some came home unchanged while others came home severely affected. It's hard to say what I would have done or how I would have fared; but without doubt, it's a bit unnerving to think about the possibilities and how apparently so minor events can influence the course of one's life [17].

At OSU, working in Dr. Kreier's lab was a life experience in its own right- a little United Nations, composed of representative students and fellows from around the world. It was a busy place, filled with folks of vastly different cultural and ethnic backgrounds, trying their best to get an education while working to develop their scientific skills. I should mention that a number of professors like Drs. Hollander, Pfister, and Chorpenning provide more than a modicum of essential mentoring during my OSU days, while my fellow grad students and lab partners provided an almost daily education. These individuals include: Dick Prior, Carol Rotile, John Mansfield, Ian Swann, Ram Mohan, Sabah al-Abbassy, John Gnau, Ted Green, Mohamed Mahmoud, Lou Wehrle, etc.. Still to this day, I view Dr. Julius P Kreier a kind of 'el maestro' - a conductor of the orchestra, trying get all of his musicians (students) up to speed, while at the same time trying to produce an ear-pleasing piece of fine music (some good science). Clearly, this was a tough job, knowing the cast of characters and the wide range of skill levels of various lab members; but at the 'end-of-the day', students eventually graduated, learned some biological science, and acquired essential skill sets along the way.

Case-Western Reserve University (CWRU) [www.case.edu] situated near the southern shore of Lake Erie in Cleveland, Ohio is where I did my post-graduate training in biological ultrastructure at the Institute of Pathology, under the direction of Dr. Masamichi Aikawa (Figure 9). My stay at CWRU was quite short, but all so sweet: I was supported by an NIH postdoctoral training fellowship that provided me with some leeway in the laboratory to pursue my own interests, but without having the responsibilities of securing my own independent financial support for the work. Further, Dr. Aikawa was a gracious mentor in terms of the technical training in the general field of biological ultrastructure, with specific mentoring in the advanced ultrastructural studies of malaria parasites. I still marvel at his willingness to provide his postdocs with full access to laboratory equipment and supplies, as well as providing extra technical support when needed. I was fortunate to have the opportunity to work alongside of a number of very talented postdocs like Dr. Chuck Sterling, currently a professor emeritus at the University of Arizona, and staff like Mr. John Rabbege, a real life adventurer who supported his 'habit' of rather far-flung wilderness exploration by running the lab and overseeing much of Dr. Aikawa's personal research projects. Hard work ruled the day in Aikawa's lab, but it was not without some extracurricular events thrown in. I remember the golfing outings on nice fall and spring days, with some serious, but good natured competition that Dr. Aikawa always managed to bring out in me and my fellow lab mates while enjoying a beer or two following the round. We would kid Masamichi about his very novel and loose interpretation of the informal rule casual golf players will often employ after making a bad initial "T" shot, namely that of 'taking a Mulligan'. Masamichi would claim multiple "mulligans" on a single hole until he got off a T shot that he liked. Since we all were having good time, and he was the BOSS, we always let it slide. On the '19th hole, we would inevitably argue about each other's score and who played the 'best'. I would generally chime in and claim superior play, but in all honesty, I was clearly and consistently the worst player within in the group.

Vicki and I and our two children, Patrick and Meredith Anne, born at Children's hospital shortly after moving to Cleveland in summer of '72, really enjoyed our stay in Cleveland, but my very limited income, coupled with some medical expenses, prompted me to look for a permanent, better paying research position, and I indeed found a suitable position back East in Washington DC area.

National American Red Cross (NARC). [www.redcrossblood.org] Toward the end of my appointment at CWRU and in the process of actively looking for a new position, I came across an advertisement placed by Blood Research Laboratory (BRL) of the National American Red Cross, located in Bethesda, Maryland and adjacent to the NIH, for an PhD level research scientist with training in biological ultrastructure and a research focus in hematology and hematopathology. "Boy" I thought, this job looks very promising and with a few letters and a couple of phone calls, I was off to my first real job interview. Little did I think that some 22 years later, I would be traveling back again to Bethesda for another job interview, for a more senior position at the National Navel Medical Center and its Armed Forces Radiobiology Research Institute. As it turned out, the ARC/BRL position was offered to me and proved to be exceedingly useful to me as a young investigator [18].

This initial position at ARC's Blood Research facility turned out to be quite challenging, but at the same time represented an incredible learning experience for me as a young professional. I reported directly to the director of laboratory, Dr. Graham Jamieson, who was not only an internationally recognized hematologist specializing in the processes and mechanisms of blood homeostasis, but quite a talented administrator as well. Graham was relatively young, scientifically aggressive but always fair when it came to his handling of personnel matters. What he expected out of me was a functioning EM facility that at the time was lacking at the BRL, for basic ultrastructural analyses of blood elements and various infectious agents (viral agents) of concern. In order to support this assignment, I was given ample though empty lab space and a modest, but workable budget for the purchase of essential equipment (EM scopes) and supplies, and for hiring technical help. Although I had a few missteps along the way, the overall objective to develop a fully functional facility was achieved within the first year or so of my employment. In retrospect, I marvel at the confidence Dr. Jamieson and his organization had in me to actually deliver on these expectations.

This short period of employment was quite productive for me scientifically. New associations and new collaborative relationships were formed: ARC staff members such as Len Friedman (bioengineering), Stephan Mironescu (cryobiology), Jim Prahl (immunochemistry), Roger Dodd (Immunology/virology), and Marty Jett (cell biology) all contributed significantly to my growth as a young scientist. Many of my older collaborative relationships continued as well and contributed to a productive early start of my professional career.

As in other places of employment, it wasn't all work, all the time: collaborative bonding with colleagues sometimes entailed social events, and more often than not, these proved to be a lot of fun. I have numerous stories that might be told, but probably shouldn't, so I'll leave the story telling for another day. I will say this however that I've come to love the Chesapeake Bay and all the outdoor sport that it offers. I was introduced by a good neighbor and ARC/BRL colleague, Roger Dodd, currently, director of ARC's Holland's blood research laboratory, to the very fine art of simultaneous 'surf fishing' and beer drinking. Actual catching of fish was optional.

Argonne National Laboratory (ANL). [www.anl.gov]. The way I found my way to Argonne National Laboratory, a University of Chicago operated, DOE-funded research facility that lies a short distance southwest of the city of Chicago, provides testimony to the utility of having a strong supporting network of mentors and colleagues. While I was still at CWRU in Cleveland and starting to look around for permanent, full time position, I was contacted by a Dr. Thomas E Fritz, a veterinary pathologist, who was working at ANL in the Division of Biological and Medical Research (Figure 10). It so happened that Dr. Fritz (Tom) was an old grad school pal of my PhD mentor, Julius Kreier, and was in need of acquiring some 'new blood' (postdocs) for his research group and was calling to see whether I might be interested in working at ANL. After some pleasantries, I thanked him for his call, and for his interest in having me join his group, but I was quite 'up-front' with him and told him I wasn't really looking for another postdoc but was most interested in obtaining a more permanent research staff position. He understood what I was saying, and we ended the conversation by agreeing to 'keep in touch'. Well, for the next couple of years, while I was at the ARC's Blood Research facility busily doing my 'thing', the ANL folks (Tom and his boss, Dr. Bill Norris) did keep in touch and eventually offered me a full time position. Although I was quite content with my position at the ARC's Blood Lab, and my family was nicely settled and thriving in Bethesda, the job being offered was hard to dismiss: first, working at ANL would be significant step up in terms of its reputation, the physical plant itself, its operator (University of Chicago) and its' incredibly rich talent pool and bountiful resources; second, the job being offered had a dual function, namely as staff researcher within Norris's radiation toxicity group but also as head of a large, extremely well equipped, fully staffed electron microscopy facility; and third, there was the issue of money, namely a nice boost in salary was to be just north of a 50% raise of what I was making at the ARC. To make a long story short, I accepted the offer, and we were off to Chicago and to a new life in Illinois. Little did I suspect at the time that my stay at ANL would last for 20 plus years.

As you can imagine over the course of 20 years a lot of things can happen. To paraphrase a title from an old Clint Eastwood movie, ANL provided the "the good, the bad, and the ugly". Because of political winds coming from Washington and associated pressures on DOE funding, at any moment in time, specific programs were deemed either 'hot' or 'cold': 'survival times' of investigators could be tracked easily, with some precision, based on program assignment. More often than not, those poor souls that happen to working in low interest areas, regardless of the quality of their research, generally lost funding and were phased out - generally not in a slow, gentle fashion, but often quite abruptly and sometimes quite savagely. As a consequence, researchers often found themselves being 'pushed and pulled', almost 'willy-nilly', according to Washington's 'research initiative de jour' [19]: I was certainly no exception to this process. It was only during the last couple of years at ANL did I feel that this 'process' significantly impacted my work. Programmatic shifts that served to downplay radiation biology in general, but more specifically, limited use of large animal models for given radiobiological experiments, provided a good deal of discomfort and dissatisfaction It was as they say "....the 'perfect storm' and I was caught up in it.

Despite this and as I said earlier, a good part developing and sustaining a career as a research scientist involves learning how to adapt to changing work environments and funding climates. I changed, I adapted, and I certainly survived: as it turned out, it was all for the better. Although I view my period at ANL with some ambivalence, there's no question that it served me well, especially during this 'maturing' stage of my career. I not only took advantage of the opportunity to learn a lot of new biology, especially in terms of radiation biology and associated cellular and molecular biology, but also to hone fledgling administrative skills; e.g., in term of running labs, dealing with personnel (technicians, students, postdocs, visiting fellows, etc.) and working on an assortment of committees (divisional, lab- wide, and outside, national and international committees)(Figures 11-15). With all of this I learned and gained experience that would help me latter down-the-road in other employment venues. All of this was not done solely by myself, in a vacuum, but rather through the considerable help from many people that I worked with on a daily basis; from my superiors, from my colleagues and from the staff [20]. I am especially indebted to Tom Fritz (Figure 12), who through 'thick' and thin', provided not only critical administrative support and guidance, but also essential education in veterinary pathology and in large animal biology. Dave Tolle is another good friend, patient as the day is long, who provided me with a 'hands-on' education in clinical hematology.

As I've mentioned previously, a collection of *individuals* played very a significant role in my scientific development, but now I'd like to suggest that the *institutions* have played an equally influential role in my career. Without question, ANL provided mentoring in the more *practical* aspects of my science 'education' and my maturing as a research scientist. During my time at ANL, I was fortunate enough to have DOE, 'el patron', 'bless' my work, allowing me a fair amount of freedom to pursue my own research interests but always within a general, but well defined framework that entailed 'descriptions and mechanisms of radiation injury'. As a result, I was able to sustain adequate funding over the course of my ANL tenure, allowing me to maintain not only a modest research operation (laboratory of radiation hematology), but more importantly, to keep a long-term focus on the biologic problems that really interested me (i.e., 'characterizing the nature of chronic radiation-induced leukemogenesis'). My year-to-year funding was generally more than sufficient to support a full time technician or two (Ms. Lillian Kaspar was my senior laboratory technician for most of my time at ANL) an occasional postdoc, and several students (Figure 11). I should also mention that the funding of the EM facility came from internal divisional funds and was independent of funding of my 'radiation hematology' lab (Figure 15).

With such combined support, I was able to accomplish a fair amount of science (i.e., as assessed by common metrics of science productivity; e.g., numbers of journal articles, reviews, invited presentations, etc.) [21]. But beyond all of this (and relevant to the theme of this eBook), this support by my 'ANL/

DOE-mentor' provided me a grand opportunity to 'pay-forward' to numerous young students, postdocs, and visiting fellows, some of the research and training opportunities that were generously afforded to me by my past 'mentors'. It's hard to say actually the extent to which these students benefited from the ANL lab experience, but judging by their subsequent education and current professional positions, I would like to think the experience was a positive one. A couple of examples might be instructive in this regard: Frederick (Rick) Domann spent the better part of a year in my lab immediately after graduating from the University of Wisconsin examining the nature of canine osteoclasts and their response to ionizing irradiation. Rick is currently a full professor at the University of Iowa and is a currently studying various aspects of molecular radiobiology. Derek Lundberg started in my lab during his high school years and continued throughout college (University of Chicago) on all his inter-semester breaks on a very difficult canine cytogenetics problem. Derek received his PhD in neurobiology from the University of Rochester (New York) and is now a full professor (senior lecturer) at University of Tasmania School of Medicine, in Australia. The several postdocs that I had in the lab all continued their professional science careers and all were highly successful: Donna Buchholz as an immunologist/cell biologist and senior research manager at Abbott Labs, Highland Park, IL, a major pharmaceutical organization; Kris Goltry as a tissue stem cell specialist and group leader at Aastrom Biosciences, Ann Arbor, MI, a relatively small biotech company; and George Niiro as an academician, professor and chair of anatomy at Midwestern University, an Osteopathic medical school in Downers Grove, IL

Despite the often 'trials and tribulations' of my ANL tenure, I always found comfort knowing that my family was comfortable and content living in the quaint village of Glen Ellyn; a wonderfully little community that lay directly west of the Chicago that provided all of the amenities one could hope for in a small town. Despite moving around quite often, both our adult children, Patrick and Anne, still consider Glen Ellyn 'home' (Figure 16).

Armed Forces Radiation Research Institute (AFRRI) [www.usuhs. mil/afrri]. It soon became time to move on as my work was winding down at ANL. I was contacted by an old friend and colleague, John Ainsworth, who had previously worked at ANL and had moved early on to the West coast to a position at the Donner Lab at UC Berkeley and subsequently, moved to the East coast to assume the position of scientific director of the Armed Forces Radiobiology Research Institute, or AFRRI for short (Figure 17). During the same period that DOE was reprogramming its funding in radiation biology to make room for new initiatives, the 'research de jour', so was DOD. In a moment of 'enlightenment' by the deep thinking, senior managers in the DOD hierarchy, a concept arose that 'radiation biomedical studies were no longer necessary and hence, no longer needed to be funded. Everything that needed to be learned about the biomedical effects of radiation exposure had already been well established. This type of critical thinking (military intelligence at its' best) reminds me of the story about the serious discussion by government bureaucrats toward the end of the 19th century about closing the US patent office, since '....that all of the important gadgets, devices, and ideas have already been thought of and had been already patented!' Oh boy I thought, ...another ill-informed change in support of basic, fundamentally important science by yet another government agency, which historically had supported radiation biology and medicine related studies. As John and I were discussing these issues, and the rather dire situation that confronted AFRRI, John indicated that he could really use an ally in mounting a defense- a scientifically based defense- against the onslaughts of these budgetary and lab closure threats. The end result of our conversations was a 'non-binding' job offer, but a promise that I'd have a more than sufficient professional challenge on my hands and I'd have a lot of fun in taking it on.

So once again, after talking to Vicki, we were off to Bethesda and a new enterprise at the National Naval Medical facility. As it turned out, it didn't take very long to convince DOD to continue funding AFRRI and its R&D work on 'radiation injuries and associated countermeasures'. CAPT Eric Kearsley, and later COL Bob Eng, were instrumental in engaging top military brass and convincing them how foolhardy it would be to close down AFRRI, since it was the sole research facility within the DOD during this type of work. At the same time, E. John Ainsworth worked the same issues and pressed the same arguments, with the top civilian research managers within DOD. After countless briefings, site visits, and workshops, the battle turned, and the 'war was soon won'. These three gentlemen taught me the essence of planning, executing, and winning such a critical funding battle with a major government agency. It was quite an education to say the least. Still to this day, I marvel at their collective skill and tenacity in waging this this battle- a battle many thought was a lost cause. Besides my serving as an aide and as 'subject matter expert', my main job was to keep the 'home fires' burning; namely to continue to run a major component of the institute's basic science program (medical countermeasures) while the funding situation stabilized. Despite the pressures associated with the rather dramatic reduction in personnel and related resources, the institute's overall productivity needed to be maintained: thanks to a core staff of very competent, dedicated, and hard-working scientists this mission was accomplished and accomplished quite successfully. Although numerous staff were involved in AFRRI's science program during these dark, shaky days, a select few need special mention, including: Drs. Bill Blakely, Dave Ledney, Sree Kumar, Mark Whitnall, Vasan Sriniivasan, Alex Miller, Tom Elliott, Mike Landuaer, and Vijay

Singh, along with Ms. Cindi Inal, and Mr. Jason Deen and Mr. Ray Toles.

Now in retrospect, I consider myself extremely fortunate and to an extent very lucky in having the opportunity to work at AFRRI. Although AFRRI was, without doubt, a smaller, less prominent research organization than was ANL, still I was afforded at AFRRI a much larger role in managing a sizable bioscience program- one that clearly had some relevance in terms of national security and defense. As an 'institutional mentor', AFRRI gave me brand new perspectives on my science and on the drivers of that science. Although in my mind, this science (countermeasures of acute radiation injury) was fairly basic radiobiology, we sold it to our customers (various funding arms within DOD) as something quite applied....almost ready to go and something essential for the 'war fighter' (protection of) (Figure 18). However, the reality was (and still is) our science products (countermeasure agents/devices) were generally quite primitive, and generally in need of extensive development prior to being 'fielded'. The military itself dictated, either directly or indirectly, what medical products should be placed in the 'war fighters' pack, but it ceded the oversight and regulatory authority over drug, biologics, and device development to US Food and Drug administration (FDA). The bottom line was (and still is) that much of the AFRRI's science was impacted by the FDA and the agency's guidelines on product development. It was certainly a different world to me: 'free-wheeling' science was a 'no, no' and replaced by a rather bureaucratic process of rules and regulations that had to be adhered to.

Other gov't labs and agency's often were involved in AFRRI's science; so much so, that the acronyms often dominated one's conversation with colleagues, to the extent that it would often sound like we were talking about the 'alphabet soup' the cafeteria was serving that day rather than talking about our science. (e.g., Hey

Reflections & Connections - Personal Journeys Through the Life Sciences

Joe, I missed you at the meeting, reps from CDC, NIH, HHS, DTRA, DARPA, and OHER were all there and ...so forth and so on. ...say what?) .

Another interesting element of the mentoring process at AFRRI related to the very, 'military nature' of the organization: military-speak, -gestures, and overall -demeanors ruled the work-place. About a half the of the institute's staff (~200) were military so saluting was very big, and ubiquitously practiced: even the civilian staff would forget themselves from time to time and give some surprised visitor a hardy 'salute'. I was no different, and sometimes even extended this habit to home (a habit, I might add, that family members didn't find amusing).

Let me tell you one more story about being mentored by AFRRI: I was tasked with the duty of fostering military science-radiobiological science- between our NATO allies. For the most part, the duty assignment was not as daunting as one might expect as the groundwork for such a collaborative network had been previously laid by others over a number of years, and the number of NATO countries that actually had viable radiobiological research programs were minimal, although several ally countries (e.g., UK, Germany, France, and Netherlands) had rather robust programs. Actual collaborative efforts were kept to a minimum, and the working group's product was largely informational by nature. I still can remember sitting in a meeting in Paris and listening to our WG's secretary take a role call and noting the high fraction of delegates from countries with essential no ongoing, military-relevant, radiobiological research programs (Figure 19). What these delegates sought was information...plain and simple....they didn't want to left without a 'seat at the table'. By the way, Paris is nice and the French, even its military, sure know how to host a nice garden party.

Catholic University of America (CUA). [www.cua.edu] It's the end of September 2003 and I'm about to set off on a new employment adventure. It's a long story, but I'll try to keep it short: John Ainsworth, my long-term colleague, a very good friend, and my AFRRI boss (now deceased), decided to retire to the sunny climes of California a wee bit earlier than anticipated. The only reason I was at AFRRI in the first place was to work with John on some challenging science and administrative issues that confronted the institute and had no intention to stay at AFRRI and remain within the DOD system for the duration. Also the search for John's replacement and the involvement of the Uniform Services of the Health Sciences University (USUHS) negatively impacted me, and I got 'antsy' and anxious to 'move on'.

I had previously established a good working relationship with one Ted Levowitz (deceased), a professor of physics at CUA, and certainly one of the more interesting characters I've come to know over my years in science. Although Ted and I shared a common interest in the biophysics of radiation, both the ionizing- and non-ionizing types of radiation, Ted's real claim to fame and fortune, came through his early seminal work on the physics of vitreous glass and the capture and retention of radioisotopes and heavy metals. Through Ted's research efforts (and patents), both he and his university, CUA, made a ton of money, sufficient to support, among other things, a major research institute, Vitreous State Laboratory, on the CUA campus. At some point in our interactions, Ted suggested that I move over to CUA and join him in the VSL, a beautiful, relatively new facility with a lot of lab and office space and what appeared to be good supporting resources. It was a generous and tempting offer, and one that I eventually turned out accepting. All I had to do was to bring sufficient research funds in order to support my new lab. No problem I thought as I had a sizable, transportable DOD contract; but what I didn't appreciate at the time, and clearly underestimated, was the handicap that I'd be working in relative isolation (as least initially) at a university with no real history in the field of radiobiology.

Nevertheless, I viewed this opportunity as 'one grand experiment', filled with endless possibilities and challenges. In the end however, my 'grand experiment' was only partially successful: I did manage to setup a functional, fairly productive radiobiology/hematology lab at VSL but failed in terms of 'growing the lab' as I had anticipated, due largely to an insufficient infrastructure at CUA. Further, I was unable to convince CUA's senior management of the utility of supporting this area of science and lending some financial support. The 'isolation' factor also contributed to my thinking that my CUA experiment would be short lived, and indeed I was 'spot on'.

I did learn an important lesson from this CUA experiment; namely that, I needed to 'keep my day job' (as a researcher), not to try to make a living being a 'salesman'. Seriously though, I also took away from my stay at CUA some very pleasant memories of working with some wonderful people. Of all the places that I've worked, CUA had wonderfully pleasant, good spirited staff and students: if I'd have only been a Biblical academician, CUA would have been my 'little bit of heaven'.

Radiation Effects Research Foundation (RERF) [http://www.rerf.jp]. RERF is a 'cooperative Japan-US research organization' (Figure 20) that is jointly managed by agencies of the two countries. On the US side, the foundation is managed by the Nuclear & Radiation Studies Board of the National Academies of Science (NAS) *[http://dels.nas.edu/nrsb]* and is funded by DOE. The foundation (and its predecessor, the Atomic Bomb Casualty Commission/ABCC) has a very long history going back to end of World War II and the bombing of Hiroshima and Nagasaki. RERF's research centerpiece is an human epidemiology study (currently the longest standing, continuous epidemiology study of record) that seeks to characterize the long-term health effects and associated health risks associated with the acute radiation exposure of the residents and their children as a consequence of the atomic bombing of these two cities [22]. Through decades of work by the Institute and its dedicated staff, standards of radiation-associated health risk assessments have been developed, promulgated, and subsequently employed by world's community of nuclear regulatory agencies and advisory committees.

RERF was in the midst of reorganization of its senior management and in need of an Associate Scientific Director. The recruitment of candidates was handled by the NRSB-National Academy of Science, specifically by Dr. Evan Douple, a program manager. After some standard vetting, I was offered the position; a position that I didn't take very long deciding to accept [23]. I was well aware of what this position had to offer: I had traveled to Japan any number of times previously to attend meetings and interact with Japanese colleagues, and I had always found Japan to be a very pleasant, accommodating place.

This position was unlike any of my previous positions and I was more than a bit nervous and uncertain on a number of fronts. I was fairly confident in my ability to handle the science associated with RERF's two basic science programs in radiobiology for which I was to be responsible, but I was indeed concerned about a possible 'cultural divide', specifically in terms of possible major differences between how science is managed and administered in Japan versus that in the US. As it turned out, the latter was a legitimate concern and one that turned out to be a bit problematic. I should also add that my concerns extended beyond the workplace and into my family life. In contrast to me, Vicki had never been to Japan and this was a total unknown for her: I was concerned about how she would do in this new, foreign environment. Well much of the latter concern was diminished by some serious mentoring by a collection of very kindly folks, both stateside, as well as in in Japan. NRSB-NAS had a long history of prepping US staff prior to placing them overseas. We were tutored on cultural mores, appropriate behavior, and general points of etiquette. Although Vicki and I initially laughed at the prospects of going through these 'cultural training sessions', they proved to be extremely useful. Despite the cultural training, I seemed to have a penchant to 'screw-up' and do something, or say something that was a cultural 'no, no'. Unfortunately, I have way too many examples to share, but this one might suffice- the 'New Year card' blunder. It was my first year at RERF, the New Year (2006) was drawing near, and I knew it was customary for New Year holiday cards to be sent out, wishing friends, colleagues, and staff alike 'best wishes' for good health and prosperity for the coming year. ...pretty standard stuff, right? At least I thought so, but what I wasn't aware of is the convention that such holiday cards were NOT to be sent to individuals who had lost a family member during the year. Well, I was oblivious to the latter 'rule', and blissfully sent out my holiday cards to the staff, at least one of whom, had indeed had suffered the loss of a family member. ....oops! In any case, my secretary, Fujiwara- san (Mahoko Fujiwara), finally caught this slip in etiquette and offered to take care of these mailings for me for all future holidays: I bowed, thanked her in my very best Japanese, and accepted her kind offer [24].

When I first arrived in Hiroshima in early in December 2005 (Vicki was to join me latter in March) and reported for work at RERF [25], I was greeted by Charles Waldren, a seasoned administrator and a well-respected radiobiologist. Charles was just completing a four year stint at the lab initially as the 'associate director of science' and subsequently as the 'chief of science' and was due to move back to the states at the end of year. For two to three weeks, Charles worked with me and brought me up to speed on a wide range of administrative, scientific and personnel issues related to RERF and its work environment. This information was not only intended to help me, but to pass along to the new incoming 'Vice Chairman', Roy Shore, an epidemiologist from NYU. I'm not sure how well I would have fared without Charles's brief, but intense tutoring, but I'm sure I wouldn't have done nearly as well.

During the off-hours in this initial period, Charles and his wife Diane, were very kind and generous in showing me around the city and instructing me in the essentials about such mundane items as grocery shopping, washing and dry cleaning services, types and locations of restaurants (Figure 21). These folks weren't the only individuals who lent a hand during my first couple months at RERF: Doug Solvie, the Associate Director of Finance, made sure that my lodging was adequate, that I had access to local banking services, and, most importantly, that I had proper authorization(s) from the government to work in Japan for an extended period.

Senior management was extraordinarily helpful as well, not only initially, but throughout my stay in Japan: Dr. Toshiteru Okubo, the chairman of RERF, provided a full view of science management at the institute; Mr. Takanobu Teramoto, a permanent director, gave me a glimpse of some of the facility's legalities and governing principals. By contrast, Roy Shore was as much a neophyte as I, but because our common situation as 'new kids on the block', we teamed up and actively shared our thoughts on the problems encountered daily (Figure 22).

Because I was tasked with overseeing the institute's basic science departments and wanting to be an active and 'hands-on' administrator, I needed to interface with departmental staff but differences in language and lab culture presented potential obstacles. I was very fortunate to have the help and the friendship of Dr. Kei Nakachi, head of radiation biology and gifted researcher, who made sure that I had full access to staff and facilities and a working knowledge of, and standing invitations to, all ongoing lab activities, including social events such as, the Bonenkai, the traditional, fun-filled, 'end of year' parties (Figures 23-25). Without doubt RERF was an able 'teacher' and 'mentor', and I did in fact learn from the organization. As in any successful student-teacher relationship, both parties need to be fully engaged and willing to cooperate in the learning process. Considering RERF and its staff collectively as 'the teacher' and I as 'the student', I felt that we were both engaged and cooperative. I was intrigued by the organization and curious about its basic functions, especially the processes that governed research on a daily basis: I wanted to know more. For example, in the US, the vast majority of researchers live and die by their ability to secure research grants and contracts. Salaries for staff, plus often sizable overhead costs levied by institutions, are all nicely folded into these grants/contracts. The lion's share of the burden falls on the *individual* researcher in the US. I learned just how different the Japanese play this game: institutions like RERF are principally responsible for covering the costs of their staff; money for research in any given area goes to the institution and not to the researcher per se. The institution has the right and the obligation to divide those funds as it deems appropriate. The bottom line is that the fully employed Japanese researcher generally doesn't have to worry about bringing in enough grant/contract money to pay his or her salary or that of laboratory staff 'year-in and year-out'. This is all covered by the organization. As a consequence, the urgency of applying for and securing grants and contracts is not an issue; 'time' is generally not a factor, especially in terms of accomplishing 'time-dated' given research goals of a given grant/contract. Whether 'at the end of the day', these differences substantially influence the quality and quantity of government funded science in Japan vs. the US remain to be seen. Nevertheless, it was indeed an education.

What I brought home from Japan, and my work at RERF, extended well beyond lab science: it was a personal enrichment. There was the initial round of 'courtesy visits of introduction' that Roy Shore and I were obliged to make to various 'persons of note' within the Hiroshima and Nagasaki communities: these visits included those to the city mayors, university presidents, and to the heads of medical organizations that were directly involved in medical concerns of the atomic bomb survivors, the so-called 'hibakusha'. I still vividly remember one of those visits to the head of a medical organization in Nagasaki and the story he related to us about the bombing of the city on that fateful day, August 9 1945. Apparently his family lived directly in the city itself, but as good fortunate would have it, he had been sent out of town by his parents to visit relatives. Being old enough to be aware of what had happened on that day and fearing for his family's safety, he managed to make his way back into the heart of the city, through an indescribable path of personal misery and destruction, only to be unable to find his love ones or even his home. In addition, there were more visits to places of national importance such as shrines, temples, castles- 'shiro' or 'jo', Peace Parks, etc. than I can count, many of which were listed as 'national treasures' (Figure 26) . The Japanese absolutely love these 'treasures', and they assume an almost reverent posture toward them. Maintaining their cultural heritage, especially in terms of basic arts and crafts (e.g., pottery, painting, weaving etc.), plays an important role in the country and the best and most talented of the artisans who practiced these crafts are treated like 'rock stars'- hence the bestowed, honorific title 'a living national treasure'. On rare occasions, we were treated to meeting a number of these talented folks and seeing them apply their artistic skills. Despite the unquestionable pleasure of having the chance to take in these varied cultural treasures at every possible opportunity, the list of 'must sees' seemed never ending.

By the end of calendar year, we were ready to depart and ready to resume our lives in the states. It was a great time, but it was time to *go home* (Figure 27).

Tech Micro Services (toms). I had always intended that my position at RERF would be my last full time, permanent job: after leaving my post at RERF, I planned on enjoying more of the great outdoor life on the Chesapeake Bay. To date, I have managed to do just this, but complemented retirement with a limited amount of consulting work in both the private and public sectors. The consulting work is done under the name of Tech Micro Services Co. Although the consulting work is often interesting, quite often I find it a bit frustrating in that the 'consultants' role- my role- is simply to offer the best possible advice to the client, who may or may not accept that advice. As clients go, government agencies are generally better in this way than are pharmaceutical companies, especially the small biotech firms. In addition to the consulting work, I'm still engaged in some committee work, e.g., the working group of the International Commission of Radiological Protection/ICRP, which is tasked with developing guidance in the relevance of the radiobiology of 'tissue stem cells' in assessing health risks associated with radiation exposure. Further, I try to help a colleague or two (e.g., Vijay Singh at AFRRI and currently the Acting Chair of Radiobiology at USUHS) who on occasion will seek my counsel on various radiobiological issues and directions.

Reflections, Lessons Learned

This short, descriptive study of 'seed science' started off by asking a paraphrased, rhetorical question raised by Hilary Clinton (among others) as to whether or not it takes a 'village' (community of mentors) to raise a 'child' (a scientist/researcher). With the benefit of time and age, I would say the answer is clearly and unambiguously 'yes'. I fully realize that a life experience of a single individual represents an 'n' of one and would, under most experimental settings, tend to invalidate any given answer. Nevertheless, I think one would appreciate the fact that it's hard to dismiss the realities of one's own life experiences. For me personally, I do not believe that I would have enjoyed the benefits of working as a researcher/scientist for most of my adult life, without being 'raised' and guided by a collection of people and organizations that I have collectively called 'mentors'.

Was I fortunate to have such a collection of 'mentors' helping me and instructing me at every difficult turn in my career? You bet I was…..no question about it! But I was also blessed by the time and place of my development: the 1960s were a turbulent period in our nation's history, troubled not only by war in Southeast Asia, but also by the threat of the Soviet Union and its rising military power and its scientific prowess. Science and education were not only on the 'radar screen', but on the 'front-burner' of the nation's cookery. I hope that I'm not over stating the situation, but it seems to me that there was a general buy- in by the nation as to the utility of science and science education, research and development. I would dearly love to see that same national 'mind set' today; but unfortunately I can't honestly say that I see it.

I believe that a broad 'paint brush' of change is needed within our society in order to foster a proper science mind-set of our citizenry. Those necessary changes certainly include improved education of our children, starting from elementary school and extending all the way through college, not only in terms of the hard elements of science but also about the utility of that science in providing the vehicle for an improved life. The second major element of change would involve a rethinking of how we, as citizens, chose to allocate basic resources and assets. In financial parlance, federal dollars going toward science has proven to be a very good investment for society, often with high yields of return contrasting to the relatively poor yields provided by the rather massive expenditures afforded to national defense and security. Financial support of science at the national level is mainly the domain of our federal government; the fed has the power to create a 'growth environment' for national science; but it also has the ability to suppress such growth. Unfortunately we are at a time in our nation's history where science and education are all too often considered as 'luxury, nice-to-have items', but not as absolutely essential items. Of course, I beg to differ with this attitude, as I consider an investment in science and education to be absolutely essential to our nation's health and well-being, not only for the present, but also deep into the future. I apologize about this 'soap box' rhetoric here, but my thought is that a bump-up in fed dollars and the jobs and industry it would create would make the mentor's job to talented, science-oriented youngsters a lot easier if science and science education were deemed 'national priorities' and that there would be sufficient demand for their talents and skills within the market place.

Time will tell as to whether or not the future brings a 'science-enlightened' US citizenry and an equally enlightened and responsible US government. However, I have no doubt that globally, science and science education will grow and prosper. Hopefully not at the expense of the US and its citizens.

References and Notes

[1] Note- "Thinking about science" is not an uncommon phrase used often in speaking about the (es) of science. The reader is referred to an interesting account of Max Delbruck's scientific life that's similarly entitled "Thinking about Science: Max Delbruck and the origins of molecular biology", by Ernst P Fischer and Carol Lipson, WW Norton & Co, New York, 1988.

[2] Note- The best-selling book by Hilary Rodham Clinton entitled "It takes a village: And other lessons children teach us", published by Simon and Schuster, 1996

[3] Note – My summer months were spent away from Glen Rock at our family's summer cottage at Lake Hopatcong, located in the northwestern part of the state.

[4] Note – A comment about my thoughts on my Dad's commuting to work each day in order to provide for the 'family'.

Growing up, I always thought that his daily commute (i.e., one that entailed an initial travel leg by train from the Glen Rock station to the Hudson River ferry and then by subway to midtown Manhattan was a 'piece of cake', pleasant and fairly enjoyable, but I later learned that this commute wasn't all that easy or enjoyable. I learned this by accompanying him on his commute into the city for about a week during a mid-semester break during my junior year in college: I had a research term paper that was due at the end of the semester and I needed to access one of city's medical school libraries. I found the daily trip to be a tough one, not at all pleasant, and vowed at that point that I wouldn't be trapped into any job that required such daily travel. By and large, I kept this vow and deliberately limited the distance/time between work and home.

[5] Note - By contrast to the males my father's side of the family who were largely accountants and book keepers, the males on my mother's side, i.e., her father, grandfather, etc. were all jewelry craftsmen.

[6] Note- Although, I did in fact grow out of my youthful dreams of being a cowboy, fireman, etc. I still, to this day, dream about what it would be like to be a professional athlete. I fear that I'm a 'lost cause', in terms of my love of sports.

[7] Note- BA track science courses taken at UCONN- 1964-1968: Intro. Microbiology, Fund. Microbiology, Adv. Gen. Microbiology, Dairy/food Microbiology, Public Health Microbiology, Bacteriology Seminar, Independent Research Project (taken at Tulane University), Exp. Virology, Gen. Zoology, Gen. Parasitology, Histology, Human Biology/Human Biology Lab, Introductory .Botany, Intr. Chemistry, Organic Chemistry 1&2, Quant. Anal. Chemistry, Biochemistry, Conservation & Natural Resources, Physical Geology, Physics,

[8] Note- BA track, non-science courses- English 1&2, Intr. Philosophy, Fund. Logic, Intro. History/philosophy of Religion, Economics, Psychology 1&2, Sociology /Anthropology, Topics of

Mod. Math, Calculus, Intr. Theatre, Spanish 1&2 (Spanish 2 was taken at Rutgers University *[www.rutgers.edu])*, Intro. Business Law.

[9] Note- Relative to the discussion on the 'trait' of 'thinking about a given problem', I offer this amusing little family story, that is now part the Seed family lore. First, I need to state that all of the males in the greater Seed family have sometimes been accused of 'cogitating' too much, especially when females in the family want chores done in a reasonable period. My father really 'excelled' at in this processing- cogitating about given problem (chore) he was supposed to solve while sitting in the backyard (and out of range of my mother) in a comfortable lawn chair and enjoying a good smoke while doing his 'problem solving'. Did I say that this is a Y-linked genetic/familial trait? I say this due to the clear evidence provided by family lore: the story goes that my dad and his David (Rich 'the elder's son and my nephew) were sitting out in back of the house and enjoying an extraordinary nice spring weather. All was quiet and here comes mom who says: "I've been looking for you two boys: what are you doing?" Have you started on those chores yet?" Well, my dad feigned being deaf and didn't respond, but my young and precocious nephew quickly answered by saying "......Gramma, we've been very, very busy ..... thinking about what we are going to think about!' I guess the moral of this story is that one can at some point overdue the 'thinking part of problem solving' and actually get on with the process actually doing the work.

[10] Note- My college roommate and life-long friend, Pat Harrell had joined the ROTC early during his freshman year in school, largely to help defray the cost of college. Because of his ROTC involvement, he was one of those few individuals who knew exactly what the future held in store: following graduation he was commissioned as a 2nd lieutenant in the marines and did a tour of duty in Vietnam. Fortunately, he managed to get through this war-time posting OK

and came home unscathed. Pat, and his wife Carla, live in the greater Hartford area of Connecticut, CT, not far from their adult children.

[11] Note- Vicki, my wife of some 45 years and mother of my now adult children- Patrick and Anne, was born and raised in the same town as I, Glen Rock, NJ. Our families were well acquainted, with family members still retaining close ties even today. Although we went through high school together and were friends, we hadn't seriously dated during our early college years, but latter married during our senior years.

[12] Note- The publication that resulted from my independent study at Tulane during the summer of 1968: Risby EL, Seed TM, Seed JR. *Trypanosoma gambiense, T. rhodesiense, T. brucei, T. equiperdum,* and *T. lewisi:* purification and properties of phosphohexose isomerase. Exp Parasitol. 25(1): 101-106, 1969.

[13] MSc Thesis, 1969, entitled "Autoimmune reactions in chickens with *Plasmodium gallinaceum*: The isolation and characterization of a lipid from trypsinized erythrocytes which reacts with serum from acutely infected chickens, Academic Faculty of Microbial and Cellular Biology, The Ohio State University, Columbus Ohio, 1969.

[14] PhD Thesis, 1972, entitled "Erythrocyte membranes alterations and associated plasma changes induced by *Plasmodium gallinaceum* infection, Department of Microbiology, The Ohio State University, Columbus, Ohio, 1972.

[15] Note- Toward the end of my PhD program at OSU, I was looking for potential post-doctoral fellowships and the opportunity arose for me to continue my work with Dr. Hollander on measures of 'ion potentials/ion fluxes' in normal and malaria-infected erythrocytes. I did seriously consider this opportunity, but my interest in working with Dr. Masamichi Aikawa at CWRU in Cleveland on the ultrastructure of malaria and malaria infected tissues was even greater.

[16] Note- The world of science can sometimes seem daunting by its scope and history, but also, at times, quite small and 'home-spun' by nature. Mr. Albert (Al) Gam, a senior technician in the lab, casually mentioned to me while we were chatting about 'science', that the great and famous Albert Einstein was a relative of his and would often show up at family gatherings. Although, Professor/Dr. Albert Einstein is unquestionably one of the greatest scientists that ever lived, Al (Gam) said he would just call him 'Uncle Albert'.

[17] Note- Relative to my 'educational deferment' for military service, I still, even today, feel a bit uncomfortable in the fact that I had received an 'educational deferment' while many of my high school and college friends were being drafted into military service. I take some comfort, however, from the fact that I latter 'served' for close to a decade at 'the pleasure of the DoD' as one of its' loyal public servants. I'm not sure that this 'tour of duty' while at AFRRI was equivalent to being drafted, but due to the work environment (e.g., overseas field assignments) that the DoD often placed me in, I certainly felt that it was (equivalent service) and that at the 'end of the day' I had served my country.

[18] Note- Just as a side comment, the ARC job was one of the few professional positions, if not the only position, that I've managed to obtain without the help of some professional connections and the benefit of a professional 'network'. What are the 'odds'? Not good, I grant you and I certainly wouldn't recommend this job seeking approach (applying to solely to job advertisements) as a primary, employment-securing method.

[19] Comment- I learned a good lesion about the life in the sciences early on from my 'ANL mentor' (and it isn't by any means all pure and smart and 'ivory tower-ish' in nature) watching a series of outstanding labs and investigators within the DOE system, who at the time were doing 'cutting edge' research and later recognized to be seminal work on the molecular and genetic bases

of carcinogenesis, be phased out, mainly because Washington program managers couldn't understanding the 'relevance' of the work. About a decade later, and after the Nobel's were being awarded to investigators on work (oncogene work) directly related to earlier DOE efforts, I learned with a bit of amusement the same Washington cadre of DOE managers were asking the question 'why the DOE labs' were not doing these types of groundbreaking studies. .....go figure!

[20] Comment- A partial listing of <u>ANL-associated</u> mentors, colleagues, associates: Dr. Tim O'Conner (virologist, biology division director); Dr. Doug Grahn (radiation biologist/geneticist, biology division director); Dr. Tom Fritz (veterinary pathologist & supervisor, pathology mentor); Dr. Bill Norris (radiation biologist/radiochemist, radiation toxicity group leader); Mr. David Tolle (clinical lab supervisor, clinical hematology instructor); Ms. Lillian Kaspar (group technician, experimental hematology); Ms. Sue Cullen-Myers (group technician, clinical hematology; Dr. Cal Poole (staff veterinarian, animal facilities); Mr. Ted Chub (chief technician, EM facilities); Dr. Ted Tamisian (retired senior scientist and EM expert); Dr. E. John Ainsworth (radiation biologist, neutron/gamma toxicity studies group leader); Dr. RJ Michael Fry (radiation biologist/pathologist, carcinogenesis group leader); Dr. Bruce Carnes (biostatistician, biostatistics- toxicology group); Dr. Malcom MacCoss (biochemist, molecular biophysics staff member); Dr. Dave Grdina (radiation biologist); Dr. Gayle Woloschak; Dr. Steve Danluyk; Dr. Fred Stevens (biophysicist/immunochemist). A partial listing of <u>external colleagues and collaborators</u>: Dr. Kate Carr (Queens Univ., Belfast, N. Ireland; Dr. Ken Anderson (Rush St-Luke's Medical Ctr., Chicago, IL); Dr. Om Johari (SEM Inc.); Dr. Dan Oldsfield (DePaul Univ., Chicago, IL); Dr. M LeBeau (Univ. Chicago, IL); Dr. Inoue (Tokuyama Univ. , Tokuyama, JP); Dr. Tamenoi (Chiba Univ., Chiba, JP); Dr. Ted Fliedner (Univ. Ulm, Ulm, GR); Dr. Julius Kreier (OSU, Columbus,

OH); Dr. J Rich Seed (Texas A&M Univ., College Station, TX); Dr. W. Nothdorft (Univ. of Ulm, Ulm, GR); Dr. Marv E Frazier (PNL, Richland, WA); Dr. Bill Knospe (Rush St-Luke's Medical Ctr., Chicago, IL); Dr. Z. Somosy (Frederick-Curie Institute, Budapest, HU); Dr. Dick Schiedner (PNL, Richland, WA).

[21] Seed T. Publications as cited in 'Pub Med' . www.ncbi.nlm. nih.gov/**pubmed.**

[22] Reference. Effects of Atomic Radiation. A Half-century of Studies from Hiroshima and Nagasaki by William J Schull, John Wiley and Sons, New York, USA, 1995, .

[23] Note- A comment on my thinking of the job opportunity in Japan- For many years, I watched with a bit of envy visiting researchers from various US and foreign institutions on sabbaticals working in my lab, or right 'down the hall', and not being afforded a comparable professional opportunity. In principal, some of my previous employers (e.g., institutions such as ANL) would allow their professional staff to take sabbaticals, but generally no one would (or could) take advantage of this option: for in practice, if a researcher left his lab for an extended period, that lab space, its equipment, and supporting funds would soon disappear. The bottom line was sabbaticals were rarely taken, and if they were, it was at great personal risk to one's professional health (actually a polar opposite of what the sabbatical option was intended for)

[24] Ms. Mahoko Fujiwara (Fujiwara-san) was not only my very talented, hard-working secretary, but also my principal 'in country' etiquette coach. As previously established, I was not always 'the best of students'. Fortunately, I was blessed by having a number of kindly souls who tried to help me out and to guide me on various cultural issues.

[25] Note- As part of the original plan, I was to go initially to Japan in December by myself, get settled, and then return to states in February and to pick up Vicki and make the trip back to Hiroshima with her.

[26] Note- Hibakusha personal stories can be readily obtained via the web; e.g., *[http://www.voanews.com/content/a-13-2005-08-05-voa38-67539217/285768.html*

Figures and legends

Figure 1. A scientific presentation by TM Seed (or as I commonly think of it as ' ...doing a science-based song & dance)

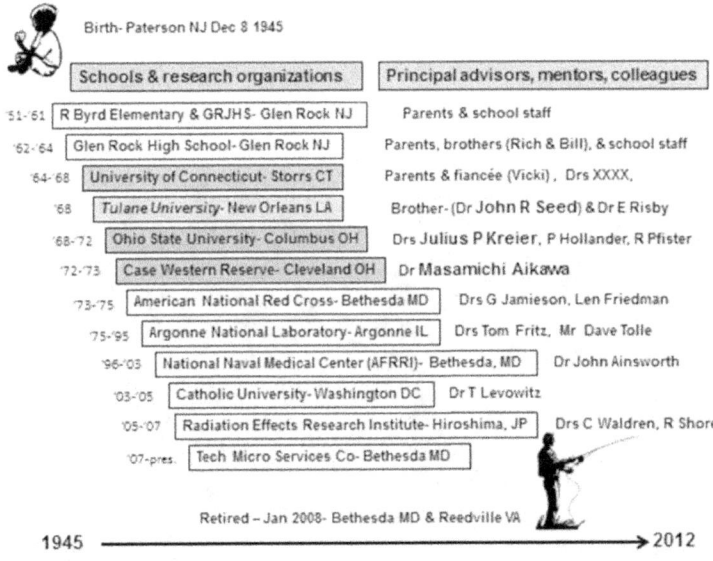

Birth- Paterson NJ Dec 8 1945

Schools & research organizations		Principal advisors, mentors, colleagues
'51-'61	R Byrd Elementary & GRJHS- Glen Rock NJ	Parents & school staff
'62-'64	Glen Rock High School- Glen Rock NJ	Parents, brothers (Rich & Bill), & school staff
'64-'68	University of Connecticut- Storrs CT	Parents & fiancée (Vicki) , Drs XXXX,
'68	Tulane University- New Orleans LA	Brother- (Dr John R Seed) & Dr E Risby
'68-'72	Ohio State University- Columbus OH	Drs Julius P Kreier, P Hollander, R Pfister
'72-'73	Case Western Reserve- Cleveland OH	Dr Masamichi Aikawa
'73-'75	American National Red Cross- Bethesda MD	Drs G Jamieson, Len Friedman
'75-'95	Argonne National Laboratory- Argonne IL	Drs Tom Fritz, Mr Dave Tolle
'96-'03	National Naval Medical Center (AFRRI)- Bethesda, MD	Dr John Ainsworth
'03-'05	Catholic University- Washington DC	Dr T Levowitz
'05-'07	Radiation Effects Research Institute- Hiroshima, JP	Drs C Waldren, R Shore
'07-pres.	Tech Micro Services Co- Bethesda MD	

Retired – Jan 2008- Bethesda MD & Reedville VA

1945 ——————————————————————→ 2012

Figure 2. An education and employment-based time-line

Figure 3. Photograph (circa early 1950s') of my parents, Connie and John, along with my brother Bill enjoying a nice day in the park.

Figure 4. Early photo (circa 1950) of my two older brothers, Bill (on the left) and Rich (in the middle) and I at home and all dressed in our 'Sunday best'. Both brothers were, what I loosely consider now to be my 'mentors in-residence': Bill ably taught me some basic survival skills any young man needed to know growing up; while Rich, the elder brother, actually provided quite useful mentoring and guidance for my fledgling interest in 'science' while still in college.

Figure 5. Glen Rock High School's football team of 1963. The team went 7-1-1 for the season and was awarded a 'Group III High School Championship' title. The coaches shown (both to far left- Al Deaett and John Cheska- and right, Larry Smith) played significant 'mentoring roles' during these formative high school years. I'm seated in the first row on the far right side with jersey 41. The young man seated on the opposite end of this row with the number 91 is Mike Zdziarski, a life-long friend (recently passed away), as is his wife Barbara: Barbara and Mike are our daughter's (Anne) Godparents.

Figure 6. My 'mentoring' big brother Rich (Dr. John Richard Seed) and his wife Judy (my sister-in-law) circa e 1960's.

Figure 7. Photos of Julius P Kreier (left), along with his wife Ruth (deceased) on holiday trip to Playa del Carmen, Mexico. For me, Julius epitomizes the often used phrase of 'being an effective mentor', as he taught in the most meaningful way, not only the fundamentals of biology and good research practices, but also basic life skills. From my perspective, he is without peers in terms of being a gifted academic, teacher, and gentlemen. I consider myself extremely fortunate to have had the opportunity to study under his direction and guidance.

Figure 8. Graduation day at OSU, 1972. Vicki is 'multi-tasking' by trying on my 'cap and colors', holding on to Patrick, and 'carrying' our yet-to-born' 'Anne.

Portrait of Masamichi Aikawa,
MD, PhD
(September 1931–April 2004)

Figure 9 Masamichi Aikawa, MD, PhD (deceased). A portrait of
my mentor at Case Western Reserve University. Masamichi,
an anatomical pathologist by training, enjoyed an international
reputation in the field of electron microscopy as it related
to malaria. I will always be grateful for his instruction and
guidance concerning the 'fine art' of biological ultrastructure.

Figure 10. Division of Biological and Medical Research, ANL, Argonne IL (Top photo). The bottom photo shows the white Argonne deer that roam free on the site.

Figure 11. Radiation hematology lab at ANL with Kris Goltry (postdoctoral fellow) and Bobbi Jorkos (undergraduate student)

Figure 12. ANL associates, Dr. Tom Fritz and Dr. Itsuro
Taminoi are shown standing with me in front of research
poster at an international radiation research meeting in
Edinburgh Scotland UK. Itsuro, a researcher and educator
from the Chiba University, had previously worked in my lab
a 'visiting research fellow', while Tom Fritz, a veterinary
pathologist by training, was both a colleague and supervisor.

Figure 13. International colleagues visiting ANL, circa early 1990's: Drs. Itsuro Inoue and Yoko Hirabyashi from Yokahama University, Yokahama Japan. Currently, Drs. Inoue and Hirabyashi are at the National Institutes of Health Sciences in Tokyo, Japan

Figure 14. A collection of radiobiologists attending a research meeting in Delft, the Netherlands, at TNO, a Dutch government Research facility. This photo brings to mind the often hear statement about the difficulty of trying to get scientists together and to agree- i.e., "....like trying to herd a bunch of cats". In this collection of folks a number of very well-known radiobiologists are shown: Dirk van Beekum (front row and center, Johan Broerse, (left of vanBeekum), Ted Fliedner (back row, behind vanBeekum). I'm at the far left, while colleagues Tom Fritz is near the center, and John Ainsworth is standing in the first row, third in from the right.

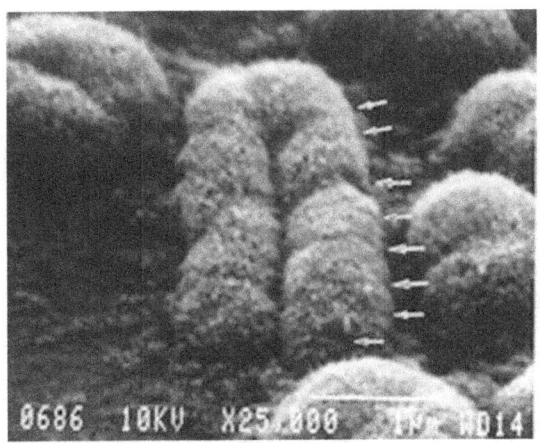

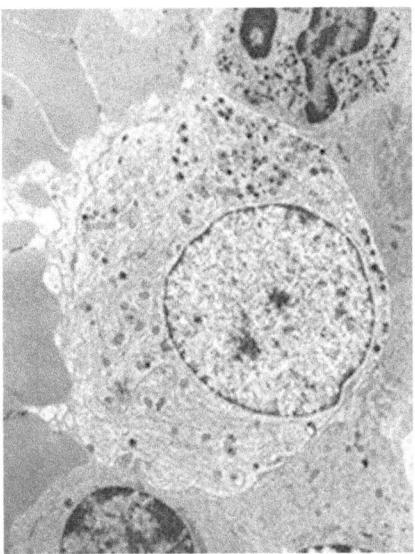

Figure 15. The art and science of biological ultrastructure as viewed by electron microscopy. The panel on the left shows by transmission microscopy an immature type of white cell found in the bone marrow, while the right hand panel shows by scanning microscopy a high resolution image of a canine chromosome (#1) following 'banding'

Figure 16. The 'Seed family' at home in Glen Ellen, Illinois for a holiday and all wearing favorite college sweat-shirts. Patrick has his own family now (his wife Kim and two wonderful daughters) and currently lives in Chapel Hill, NC. He is a faculty member at Duke University, Durham, NC. Anne (Meredith Anne) is a preschool/afterschool teacher with a local child care organization in the Bethesda area.

Figure 17. E John Ainsworth and wife, Carolyn: John (deceased) was a long-time friend and colleague dating back to my ANL days in Chicago. Carolyn currently resides in the San Francisco bay area, close to her lovely family. John was the scientific director at AFRRI and was responsible for bringing me to the DOD organization.

Radiation Casualty Management: Protection from Ionizing Radiation

The drug 5-Androstenediol boosts the immune system and increases resistance to infection and will be fielded in coordination with FDA and private industry.

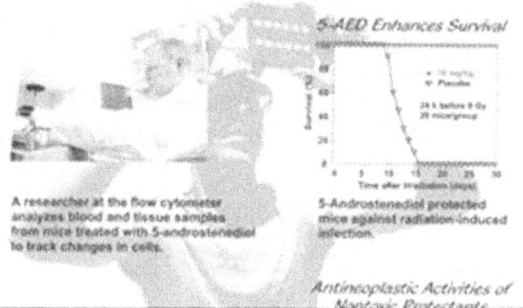

Figure 18. *A* poster 'advertising' the positive attributes of one of our medical countermeasures developed at AFRRI

Figure 19. NATO Radiation Research Working Group,
Paris, France, circa 2000.Radiation scientists from
participating nations included: UK, France, Germany,
Netherlands, Canada, and the US (as well as one non-
NATO representative from Turkey). Col Bob Eng, AFRRI's
military director, is standing in the first row to the left of me,
while Maj Gerry Vavrina, a US military aide and secretary
to the group is the back row, second from the right.

Figure 20. Cover page of a RERF brochure: Radiation Effects Research Foundation: A brief description. RERF facilities are located on a small hill and park called Hijiyama Koen and overlooks the city of Hiroshima.

Figure 21. A last farewell traditional Japanese dinner at a local Hiroshima establishment for Dr. Charles Waldren and his wife, Diane prior to their departure back to the states, December 2005. Charles is on the right, Diane is center, followed by Dr. Okubo and Doug Solvie.

Figure 22. Dinning out with RERF directors, circa 2007 (left to right) Roy Shore (Vice Chairman/Chief of Research); Toshiteru Okubo (Chairman); Burton Bennett (Former Director); Takanobu Teramoto (Permanent Director); TM Seed (Associate Chief of Research).

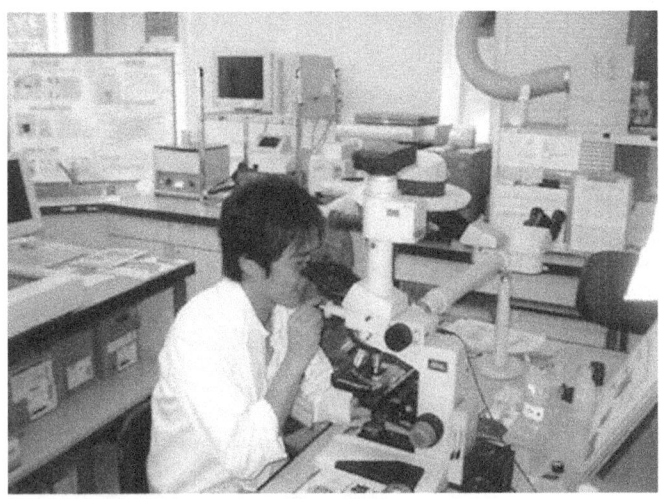

Figure 23. Radiobiology lab technician examining tissue specimens. (The photo taken at time of an 'open house' at RERF, circa 2007Figure 24. End of year party- 'Bonenkai' – Radiology's staff having some fun doing a traditional (hardly traditional) Japanese 'song and dance' routine.

Figure 24. End of the year party- 'Bonenkai' – Radiology's staff having some fun doing a traditional (hardly traditional) 'song and dance' routine.

Figure 25. RERF Hiking Club members celebrating reaching the summit of the day's climb. (left to right) Hayakawa-san and Wada-san are holding the club flag; Dr. Okubo stands to the far right.

Figure 26. Temple on Mijima Island (a national treasure)
in the Seto Sea close to city of Hiroshima

Figure 27. Retired and paddling off into the sunset.

CHAPTER 13.

My Career

Dan Tomas Spira

Early Years

I was born as a result of an accident. No, it's not a joke. January 1932 was a very cold month in Central Europe and my mother, who did not trust the young village physician - her husband - my father - with the delivery of the baby, decided to travel to the city where she grew up. On the icy road the car slipped, overturned, but my pregnant mother and my father got out unharmed. Luckily the car was like the Ford T, one of those sturdy square boxes. Villagers turned the car over, added oil, and my parents safely arrived at the house of my grandparents. A couple of days later, I was born (Fig.1).

My early childhood was uneventful in the new house of the village physician, where three languages were spoken simultaneously. Thus I grew up with German, Slovak and Hungarian all mixed up. I began my school years in the local public school, but already in 1940 at the end of the second grade my parents were told that I, as a Jewish child, could not continue in the public school. Fig.2. Thus my third grade was in the Jewish school in the town of Nitra, where my grandparents lived. That also didn't last for long. The new Slovak State, constructed by Hitler on the ruins of the only true democracy in continental Europe, Czechoslovakia, started to issue anti-Jewish laws. Gold, jewelry, fur coats, sport equipment, radios many other items, including pets, had been confiscated. The way to school was not safe, children and even their teachers from the public schools were throwing stones or at least insults at us on our way to the Jewish school.

In October 1941 my father was taken to a work camp for Jews intended to receive a large number of inmates, before the decision to deport all Slovak Jews to Poland was made. Early in 1942, when I was ten years old, and had just started fourth grade, my mother, sister and I were also brought with many other Jews, to the camp where my father acted as the camp physician. After deporting more than 5.000 people, the deportations to the death camps stopped and we became inmates of "Work Camp For Jews" in Novaky, where we spent the next two and a half years. As the camp was a family camp, all adults worked ten hours a day and we children, up to the age of fourteen were kept busy by inmate teachers. It was not a real school, with no regular classrooms or rigid curricula, but a lot of thought and effort has been invested in the education of the children. We had accelerated physical education, mainly running and climbing, to keep us as fit as possible for the unknown future. The intellectual side was not neglected. There were lectures, math, reading and writing exercises, even under the so difficult circumstances. All this was possible because the authorities understood that people will work harder if they know that their children are being taken care of (Fig. 3). And indeed they did work hard. The camp produced uniforms for the army and the police, furniture, central heating facilities, office supplies and many other products which were delivered to the state, We were unpaid slave labor. People worked as hard as they could in the hope that the economic considerations of the government will save the inmates from deportation to the dreaded death camps in Poland. Indeed, after a year or so, the conditions in the Camp were quite good, even the food rations were nearly sufficient and "daily life" became quite acceptable.

My father, Dr. Jacob Spira was the camp physician, responsible for the health of the inmates, but as I now understand, mainly responsible for preventing any infectious disease within the camp that could endanger the surrounding population. This was 1942, only 23 years after the first world war with its epidemics that swept through Europe and the world. There were 2-3 more physicians in the camp, an indeed they succeeded in preventing any major outbreaks of disease. In his testimony after the war, about the conditions in the camp my father mentions tuberculosis, under-nourishment and a few other problems that did not seriously influence life in the camp and certainly did not spread disease in the surroundings.

A flash back from those times: My father always brought "home" from the infirmary a bottle with pills that must have been very important or secret. I (the future microbiologist) still remember the label; "CIBAZOL". It must have been an early *Sulfathiaz*ol by the company CIBA, later known as CIBA-Geigy. It was, I presume, smuggled into the camp by the Jewish underground, and no doubt very effective by preventing all kinds of sores and infections. In an interview he gave to "Yad Vashem", the Holocaust Martyrs Remembrance Authority in Israel. He emphasized that no epidemic erupted in the three year of the camps existence, and that the number of newborn babies was equal to the number of people dying of natural causes.

I want to highlight one incident that came to my attention years after father's death in Israel, in 1982. After the fall of the communist regime many documents from the Holocaust era were transferred to Israel. Among them, I found a verbatim transcript of an official meeting at one of the fascist Slovak government ministries. My father is on the list of invited people and talks about the health status of the inmates in the Work Camp for Jews. My father, a meek, soft spoken person, who seldom rose his voice, gave there a fiery speech, emphasizing the situation of women in the camp. He DEMANDS either to lower the work load or to add 200 more calories in food to women working as much as men. I do not know exactly what the forum was, there are acronyms of the place and participants, but who can decipher them today. How my father an inmate in a camp, a Jew, addressing an official governmental body of the fascist Anti-semitic Slovak State had the audacity and the courage to talk so bluntly is to me a mystery to this day. Of course I have no idea whether any of his demands were ever fulfilled.

We were liberated from the camp suddenly, when on August 29,1944, part of the Slovak army and the anti-fascist underground, also very active in the camp, declared "The Slovak National Uprising" against the fascist government and the Germans. The men of the camp joined the fighting forces and the families moved to the "free" areas of the country. It took two months for the German army and the SS to occupy the whole area of Slovakia and bloodily suppress the uprising. The fighters, the anti-fascists and the Jews had only one option that was to withdraw to the mountains and mountain hamlets, already under heavy snow. The assumption of the partisans, was, that the German army will not dare to pursue them into those areas. Indeed the Germans, now close to the approaching Russian "east front", were interested mainly in the communication lines (main roads and railroads) and did not bother to fight the partizan units in the mountains. Never the less, the Germans twice came to the hamlet where we were. In one case a SS unit caught us in the forest where we tried to hide when we understood that the Germans are coming. The Germans took us to one of the few houses in the small hamlet in the mountains. Two of the soldiers even slept in the same room as our family. We had the impression that they knew that we were Jewish, telling us that "all whom we caught in the morning were shot". The next day they left, not without saluting and shaking hands with all of us. Thus, I dare to believe that I am the only Jewish kid that had been saluted by two SS men and their commanding officer.

After this, we left the hamlet which had become unsafe and found, with two other families a "bunker" in the woods. It was a hole in the ground with a roof of branches, some dirt and a lots of snow. It had a tiny iron stove, to keep us from freezing to death. How we survived for nearly six winter months living in a hole dug into the forest ground, at a mountain altitude of 3000-4000 feet, is something I still cannot understand. My father, the physician, treated the wounded partizans and thus, from time to time got some potatoes, beans and even a piece of sausage. Apparently, it was enough to keep our family and even other people hiding with us, alive for this whole time. The Russian army arrived in April 1945. We descended from the mountains and returned weak, infested with lice and with torn clothes back to the village where I grew up. Good people, recognizing the doctor and his family, gave us clothes, food and shelter, and the three year nightmare was over.

Looking back on my early years I realize how unnatural my childhood was. No recollections of elementary school and very little of the high school years. More than school or family affairs, those years were dominated by political upheavals, tragic circumstances and a constant fight for mere survival. Those were years of mortal danger, fright, hunger and cold; with very little "childhood" if any at all.

The village children with whom I grew up and started school became hostile when the Slovak State adopted fascist and anti-Semitic ideology. The children, as their parents, accepted the state line, due to the false feeling of freedom and independence, which Hitler's propaganda connected to "cleansing" the society from Jews.

The contacts I made with children and teachers in the Jewish School were also shattered, when after a year and a half there, in early 1942 (fourth grade) the majority of the class and all the teachers disappeared, being deported to the death camps, not one (to my knowledge) ever to come back.

Even those kids that I met and lived with in the labor camp, were scattered by the German occupation. Some of the survivors of the brutal murder; following the uprising left the country immediately after the war ended. When the communist regime took over, some become became communists and remained there, but only very few.

The fact that our family, father, mother, sister and myself remained together and jointly survived all the difficult times of the second World War is, in my view, the major factor that my inner world remained comparatively stable, and I was able to develop and live a normal life. My sister Vera, two and a half years my senior suffered much more than I both physically and emotionally, being older and being a teenager and a young woman. For her, this unnatural world of mortal danger, but the added danger of sexual abuse in a completely lawless and chaotic world was bad.

In the camp she had to work as the adults did, although she was barely thirteen. After the camp years, in the mountains she feared not only the Germans, but the Russian partizans as well. Our parents succeeded to installing some calm thinking and thoughtful planning in these extreme situations. I would like to emphasize that our parents, as many other Holocaust survivors never spoke about the period of 1939 - 1945 with us.

Post-War Period

During the summer of 1945, I as many other children, had to catch up with lost school years. September 1, 1945, I went back to school. This good beginning again didn't last very long. In March 1948 the communists gained full control of the reestablished

Czechoslovak Republic. According to Soviet interest, they supported the young state of Israel not only supplying arms and ammunition, but also by letting the Jews of the country immigrate to Israel. Thus, in February 1949, in the middle of my tenth school year, I came to a new country, new surroundings and a new language practically unknown to me.

At the age of 17 I was a member of a group of youngsters, all Holocaust survivors, accepted by a Kibbutz for one year, where we had to work for half a day and study Hebrew and the local culture for the rest of the day. At this age my whole horizon changed. After immigrating to Israel I had to erase my past in order to build a new normal life. That needed adaptation and a lot of it: new language, nature climate, the communal way of life in a kibbutz, all this added to the normal physiological and emotional stress of the years of young adulthood. I remained in the Kibbutz with the group, working as a carpenter for two more years, and then left it to finish high school, in order to get to university studies.

It took nearly two years to get through all the matriculation exams in Hebrew. During this time I also worked as an intern in a laboratory producing BCG vaccine for the young state of Israel which was heavily riddled with tuberculosis. Of course I could not work in the sterile production area but was busy bringing supplies and doing the necessary chores. This experience brought me to the decision to abandon my dream of following in my father's footsteps into medicine, and I choose instead microbiology as my future. In 1954 I enrolled as a student at the "Hebrew University of Jerusalem.

University

Finally, at the age of 22, I was registered as a regular student of The Hebrew University, Division of Microbiology. One of 15 000 students attending the only University, at that time in the country. A regular student like all the others, not "the Jew", not the "newcomer" to Israel from the Holocaust, just "STUDENT".

To explain the parts that will follow, I will say a few words about the University. The system at that time was completely different from the American higher education system. Israel had no College at the B.A. level, thus the students actually enrolled for a Masters degree, which in the sciences meant a serious research project after, the compulsory curriculum of four years. The Medical School Program lasted 7 years which ended with an MD title.

My first encounter with the university was not very friendly. The school year started late due to a strike by the staff because they did not receive their salaries on time. The institution had yet to reorganize itself after the 1949 war of independence. All teaching in the divided city of Jerusalem was in makeshift quarters, rented spaces, and in prefab wooden structures imported from Sweden. The original buildings of the campus on Mount Scopus were in the area controlled by the Kingdom of Jordan, and were not accessible to Israelis.

I had to find lodging in Jerusalem. In 1954, Jerusalem was a small city, completely unknown to me. Luckily I met a group of older students, one of them a member of the group in the kibbutz. With their help I found a room on the edge of the ultra-orthodox area of the city, not too far from the temporary quarters where sciences were taught. They also introduced me to the student "mensa" the very cheap but quite awful eating facility offered by the university. The year 1954 was still part of the "tsena" austerity regime of the young state, when food, clothing and many other goods were rationed and available only with coupons that all residents received monthly from the authorities. There were also some restaurants in the town, but who had the money to eat there?

My very first university class was in a former warehouse all painted black. The place was overfilled, with students of medicine, microbiology, biology and agriculture, who jointly received the introduction to biology. All the chairs were occupied and many had to sit on the stairs. The coats, wet from the morning rain were drying on simple nails in the walls. Well, this was far from my image of the elegant universities in Europe, which I knew only from stories.

The lectures were a revelation for me. They were not like the technical high school level, but attempted to transmit to the students the real problems of basic biology, chemistry, physics etc., as were known at that time. Here too, at least at the beginning I felt an outsider. Not yet fluent in the language as one who did not grow up in the country and did not experience a full course of the Israeli education in Hebrew.

Following the lectures was difficult not only due to the language but also because it was intellectually so challenging. I felt as, I guess many university students do, that I had to catch up and do more and better than was necessary in the former years of school. The great revelation was the library, although not all books had been transferred from the original campus. It had an impressive collection of texts in all fields of science. I spent long hours in the library reading whatever I understood and could put my hands on, slowly understanding what science has to offer.

Reading mainly English!? I don't understand how I could read English without ever properly learning the language. Already when I was 5-6 years old my family tried to learn English with the hope of emigrating to the U.S, where my father's whole family had lived since 1920. My grandparents and six siblings of my father lived in Cleveland, Ohio, but the circumstances prevented us from joining them. I do remember a set of vinyl records of some Berlitz English courses which we listened to, but have no recollection of ever trying to talk English in my childhood. For my high school matriculation in Israel I needed English as a foreign language, but it was literature, Shakespeare's Macbeth and some selected poems and essays which we had to analyze. There was very little spoken English. I guess I learned more from "The Readers Digest" than from Shakespeare and more from the news on the radio than with contact with English speakers. I certainly had ionary with every page I read, and often had to guess the meaning of words from the context. Still reading English was easier for me than reading Hebrew due the simple fact that the later is written in a completely different alphabet than the one I learned as a child in my early years. Actually, I learned English last of the five languages I am fluent in but to this day prefer it in daily reading and writing. Some say that if senile dementia should arrive, the language last learned evaporates first. One gain from writing these lines in English is the knowledge that apparently "senile dementia" has not yet developed.

The first year was by and large uneventful. I needed some help with calculus, became interested in chemistry, hated my lodging, cold in the winter and hot in the summer and liked evenings of classical music (from records) in the student union. Exams passed to my satisfaction, far from the best in the class, but certainly not at the very bottom. During the summer I found a job in the National Arboretum, helping with experiments on water consumption by trees in open space or behind wind breakers of rows of Cypress trees.

The second year was much more interesting, and more significant than the first. The lectures covered cellular aspects of metabolism, biochemistry, physiology, genetics, including the first mentioning of DNA. This was quite new knowledge that DNA was the true and main molecule of all that that transmitted all in Mendelian genetics. The second year was also rich in lab exercises and experiments, which I loved, thus more fun than the first year.

I also got to know the other students better and slowly became one of them, I became "Israeli". During this year a most significant decision and change in my life happened: I GOT MARRIED.

Cipora

When I was studying for the High School matriculation and worked in the BCG vaccine production lab, I met Cipora, a medical lab technician, working in the neighboring lab. Nearly sixty years passed, but I still can see her as I first met her, in an oversized lab coat, the most beautiful girl I in the world (in my eyes). It was love at first sight. I believe that my work "gained" from this event, as I started using the chemical analytical balance in her laboratory whether necessary or not. We were married two years later, a small, intimate wedding as was customary in those days. We had a similar background, both from Slovakia, although she came from a quite religious family, well known in the country due to the big antique book store they owned in the capital. Cipora's father was deported to Auschwitz early in 1942 and her older sister was the youngest partizan in Slovakia (only 16 years old), captured and executed by the Germans after the Slovak National Uprising, in December 1944. We lived happily, raised three children in a tightly knit family, saw them through school and University. In 1987, soon after the two sons got married, and the first two grandsons were born Cipora woke one morning to a massive cerebral bleeding, never to gain consciousness. Thirty years of happiness ended abruptly. After this nothing was as before (Fig.4).

After we got married we desperately needed a larger income than Cipora's salary alone. Dr. Avivah Zuckerman of the Department of Parasitology advertised on a billboard that she was looking for help in her lab. It was before I had taken any courses in parasitology and did not know Dr. Zuckerman. After a thorough interview I got the job and thus the line of my future career was sealed.

Avivah, (I never called her by this, her first name) returned from months of work abroad, collecting material from different host - malaria combinations, to prove that infected animals loose more blood than one could predict from the number of malaria parasites in blood cells. This was a cardinal question in Malariology: what causes such, a at times life threatening anemia in patients, suffering "only" from a light bout of the disease? My job at this time was to examine and count infected blood cells in smears from a variety of birds and monkeys infected with appropriate malaria species. This simple student job was the beginning of a life time fascination and interest leading to my professional career.

One possible explanation of the accelerated blood loss in malaria was the enlarged spleen common to all cases of human or animal malaria. To test this: my M.Sc. thesis two years later was: "Blood loss and replacement in splenectomised malarious rats". The results did not prove that a big spleen - the graveyard of red blood cells - is the direct cause of excessive blood loss.

I turned to test an immunological explanation of the phenomenon. Are normal RBC attacked by antimalarial antibody in sick rats? The tool at my disposal at that time was the "Coomb's test", used to diagnose Rh incompatibility in new born babies where RBC are attacked by incompatible maternal antibody. All went well, the sick rats were positive, the control rats negative; until one day a control rat suddenly became positive. The lab log showed that this rat had an accident a few days earlier: it's tail was damaged, apparently cut by the heavy cage lid. Blood stains were seen on the cage bedding. This rat taught me an important lesson. Never accept results without re-testing and re-checking them, not even when a plausible explanation is at hand. I found that anemic animals turn "Coomb's" positive, and discovered that positivity was not due to antibody on the red blood cells, but to transferrin, the protein that supplies young RBC with iron. Thus my thesis did not solve the question why malarious animals and people develop such life threatening anemia, but I saw in the complexity of this disease an opportunity for further research. This sentence, now fifty years later while malaria is still killing millions of people every year with no solution in sight, may sound ridiculous. In the fifties some of my teachers asked me why? Why a young person wants to go into a field that does not exist? There is no more malaria!!! Even the WHO was convinced and announced an exact plan when this disease is expected to disappear from the whole world.

Doctoral Studies

After receiving my Masters degree the next question arose, a result of the new wave of immunology entering into the biological thinking and new techniques emerging. For my Ph.D. thesis I chose to investigate, whether the erythrocyte destruction is not connected to antigens identical in the parasite and on the host red blood cell. Blood cell antigens have been known for a long time, a diversity of bacterial antigens has been described, but there was no mention in the literature (at least not in the literature available to me) about antigenic complexity of Protozoan parasites or their relation to normal RBC antigens.

Analyzing malaria antigens directly led to the next questions: vaccination, difference in strains, is malaria in Africa the same as in Asia? These are the early sixties when America became involved in Vietnam; Mc Gregor in the Gambia showed that antibody can protect against the disease. A logical step but certainly new and unprecedented at our university: Prof. Zuckerman wrote a grant application to the U.S. Navy and we received a big grant to study and compare antigens of malaria species and of African and Asian strains of human malaria parasites.

I traveled first to the London School of Tropical Medicine and Hygiene, to increase my experience in purifying malaria antigens. I learned there how to adapt my technique for working with micro amounts of rat blood to the larger quantities of blood obtainable from man. Of course the first "dry run" was with my own blood and then blood from infected monkeys discarded from chemotherapy experiments at the London School.

London was an enormous experience for me. The first encounter with the "Big Wide World". London the center of culture, science, of the "British Empire" (well, not any more but still in the aftermath of its glory). My U.S. Navy Grant 'per diem' enabled me to see great theater, Zefirelli's first modern production of the opera Turandot and the great Joan Sutherland in Lucia d'Lamermoor at Covent Garden. I participated at a meeting of "The Royal Society of Tropical Medicine" where I met the leading names in European parasitology. What more could this youngster ask for? Where could he have dreamt of all this? Only one cloud, a big dark cloud hung over all this excitement: more than three months without my wife and children, including the newborn daughter. In London I became aware of how strongly and thoroughly I was bound to my family. The psychologists would certainly diagnose this as PTSD, after my turbulent Holocaust experience in childhood. Well, if that was all that remained from those days, I should be happy.

I did not participate in the second leg of the project, the collecting of *P. falciparum* from discarded infected human placentae in the Gambia, I just could not leave home for another protracted period, thus Yosef Hamburger, the new graduate student, in the department took my place and returned with excellent samples of African human malaria. My next trip, about a year later was to Bangkok, Thailand. This was already 1966 and the U.S. army established an impressive research center, the SEATO Laboratories directed by Robert S. Desowitz, the renown parasitologist and sea shell collector. Working conditions there were excellent, a car and driver from the lab took me to malaria clinics in villages or to the regional Red Cross Hospital in the small town of Sri Racha where I got remnants of blood samples taken from patients for routine laboratory tests, many of them containing malaria parasites. I concentrated and purified the parasites the next day in the lab and froze the samples. After three months of hard work in the heat of the pre- and monsoon season, with two thermos containers full of samples of Asian human malaria in dry ice, I was on my way to yet another experience: my first visit to the US.

Prof. Zuckerman and I stopped briefly in Hong Kong (she, as a US citizen was not allowed at that time, to purchase anything made in China, I spent all the little money I had saved in Thailand on Chinese art). From there we traveled for about a week to Japan, again an overwhelming experience. In less than one year, London the "capital" of Europe, exotic beautiful Bangkok still barely touched by western style development, Hong Kong and then Kyoto, Nara and Tokyo. The whole ancient and recent imperial grandeur of Japan was simply overwhelming. My, Hong Kong purchased new camera, worked overtime to conserve the impressions for eternity.

America here I come! First to Washington DC, a Malaria Conference at the Walter Reed Army Institute of Medical Research, where I was invited to report on the "Antigenic Analysis of Malaria Parasites". Suddenly I met all the great names of malaria research in flesh. Those who wrote the text books, set the foundations of modern malaria research, all listened to Dan Spira's lecture (is he really an adult among adults or still the kid he once was? how come that all these people are listening with such interest?--I have this feeling sometimes even today, although I have already passed my eightieth birthday). One of the participants at that meeting was Paul H. Silverman from the University of Illinois at Urbana, who had recently received the first big civilian NIH grant on malaria research. He was recruiting people for the new line of research in his department. Paul offered me a post-doctoral position in Urbana

.I spent another month at the Davis Cal Primate Center, collecting monkey parasites from animals discarded after chemotherapy studies. I returned home to finish the write-up of my dissertation on: Antigenic Analysis of Malaria parasites, As we were ready to go to Urbana, we got caught up by the Six Day war in Israel. A couple of weeks after the war ended we moved to Urbana for two exciting years.

Post-Doctoral Period

We rented a house in Urbana (our first after small, rented apartments). The children were enrolled in school - one son in the 3rd grade and the other begun his 1st grade there. We worried how they would manage with a new language and a different alphabet, but by Christmas we were told that both were completely integrated. (Some of our American friends claim that one of my sons still has a distinctly "Midwestern" accent).

I was lucky that shortly after our arrival in Urbana, the University of Illinois was celebrating its' centenary. That was an opportunity to meet important and interesting invited speakers including the great Francis Crick, "father" of DNA. I started to work on cellular responses to malaria, which kept me busy even after returning to Israel. I was fascinated by the idea that malaria was believed to behave at least at that time similarly to cancer in that cell mediated immunity was not sufficient to deal with the disease but demanded such an effort that other immune responses were suppressed. As with many interesting ideas, this one also went, where all mistakes go.

Our intellectual activity at Urbana was within a group that was centered around the author and anthropologist Oscar Lewis with his theories about poverty as a social phenomenon. It fitted well with the activity on and off campus. It was 1968 a year of campus upheavals and anti-Vietnam war demonstrations, the flower children, Chicago, Woodstock and all the rest. We were in Urbana when Martin Luther King Jr. and Bobby Kennedy were assassinated. Another political event which had significance for us was the Prague Spring and subsequent invasion of Czechoslovakia by the Soviet-Block forces. Thus 1967-1969 were a concentrated school of local and global politics and social changes. It was also the time when "affirmative action" was implemented. Thus we spent a very interesting period at Urbana. Lots of activity around us, many new ideas in the air and new friendships, successful integration of our children in English and in the American way of life.

We returned to an Israel quite different from the one we had left two years earlier. Some borders disappeared, Jerusalem became a large, binational city and hopes were high for a settlement with our neighbors. Research emphasis at work also shifted as Israeli soldiers served in the uninhabited parts of the Jordan valley where *Leishmaniasis* was prevalent. There was an urgent need to develop drugs, vaccination or any other means of protection against the disease. The other change in the department was the arrival of Charles Greenblatt from the NIH, who emigrated to Israel with his family. We started collaborating on the urgent local problems of *Leishmaniasis* in Israel. Our tight friendship continues to this day.

One of the first things we had to do was to standardize "vaccination" against *Leishmania*. For many years Prof. Saul Adler, the great man of Leishmaniasis in the Middle East, used to "vaccinate" by inflicting an ulcer on the arm by injecting parasites kept in permanent culture. This immunized the patient from further infections on unwanted areas of the body. We found out that this ancient technique was done without any controls and precautions. Considering, that we saw cases of multiple ulcers on young soldiers, who had not been vaccinated, unaware of the protection by long middle eastern garb we had to use "Leishmanisation" was all we had - infection with virulent parasites, as used by Adler and many others around the world. The soldiers were infected before serving in the affected areas, protecting them from multiple *lesions*. The parasites we now use are grown in a serum free sterile culture and monitored for possible contamination before use.

Work on malaria continued. Yossi Hamburger performed a series of fractionations of rodent malaria parasites and succeeded to induce significant immune reactions with certain fractions. He finished his Ph.D. and left for a post doc with Julius Kreier at OSU. Kobi Golenser also finished his thesis and did his post doc with Prof Meuwissen as part of a joined Holland - Israel project, at the Catholic University of Nijmegen Holland.

University Tenure

I continued with Leishmania research and with my Ph.D. student Emanuela Handman worked on in-vitro cultivation of *Leishmania* amastigotes in mouse macrophages. A beautiful collection of pictures by scanning EM still awaits publication.

I returned to malaria research in 1974 when Jim Jensen brought us *in vitro* malaria cultures, recently developed at the Rockefeller University. With Kobi Golenser who returned from his post doc, we started the first malaria cultures in Israel, when it was still very new to the rest of the world. The first project was based on a fairly large population of Jews from Kurdistan, known for their high rate of G6PD deficiency. With Jacqueline Miller we tested the growth of parasites in normal and deficient blood cells and came to the conclusion that parasite survival in the deficient RBC is seriously impaired, although the deficient red blood cells did support the development of the parasite. Malarial cultures are still kept in the department, nearly 40 years later.

A few years later I spent half a year in Jensen's laboratory at Michigan State University and was about to begin teaching one semester in Venezuela. On the way there, a telegram informed us, that due to student revolt, the university would be closed. So I missed an opportunity to visit South America and be exposed to the local culture and problems there. Instead, we went for a few months to Tubingen, Germany. There I established malaria cultures and worked with a couple of industrious young people, on iron chelators as potential anti-malaria drugs.

For us going to Germany was like a jump into a dark cold bath. Both Cipora and I were concerned of visiting Germany and meeting Germans. Although both of us were fluent in the language, our childhood experiences were quite traumatic. My experience there, in the lab with young people was excellent, but for the whole time we found it difficult to communicate with the older generation. We just could not relate to people who were older and may have been active during the years of the Third Reich.

We returned home, and I continued to work on G6PD deficient red blood cells and on iron chelators as potential anti-malaria drugs. Part of this was in collaboration with the Mahidol University in Bangkok, a renewed opportunity to visit Thailand. Bangkok wasn't the quiet, beautiful oriental city I remembered from my first visit there. Now a mixture of an oriental city and of a very busy westernized city with much less charm than in the sixties existed. At about this time, I came to the conclusion that malaria vaccination and even chemotherapy should be left to the "big players". As an Israeli I had limited accessibility to many malaria infected countries - without having the opportunity for field work or collaboration, malaria research was difficult.

HIV, which popped up in the 80ies, was the new hot field in research and with it *Cryptosporidium,* a ubiquitous parasite, but dangerous to people with a diminished immune response. I decided to switch my focus (Leishmania was taken care of by Chuck Greenblat), and turned my attention to *Cryptosporidium* and *Giardia*. *Giardia was* a problem for children in Israel and all over the world. Luckily, a group of physicians at the department of pediatrics at the Hadassah Medical School Hospital started an effort to understand certain aspects of giardiasis with Richard Deckelbaum, the senior member of the group who later left Hadassah, to head the pediatric nutrition unit at Columbia University Medical School in New York. We sent a grant proposal to the NIH dealing with the natural history of *Giardia* and *Cryptosporidium* in newborn babies. We proposed to compare the first infections and the clinical picture in Beduin children intimately exposed to contact with cattle and other domestic animals, and children from the same area, living in a modern urban environment.

The Personal Aspect

This is the place for two side steps in writing my memoirs, until now more or less in a chronological order.

In Israel, we lived happily, saw the children to grow up and leave the house. Shortly after building the joint project and receiving generous funding from NIH, however my inner world was shattered. Thirty years of happiness ended abruptly, when my wife suddenly died. After this nothing was as before. I was alone, with big plans and hopes for the last years of my career. An unfathomable trauma nearly paralyzed me. I tried to act, function and live as before, but to no avail. I thought that after a certain time of grief, the loss will remain but life will continue as before. Nothing was as "before" but life did continue. I functioned in on "automatic drive" mode, performed my duties at the university, even a couple of months after Cipora's death I organized a meeting of the "Mediterranean Parasitologists". I thought that to leave Jerusalem for some time would help and went for a short sabbatical to New York. But that changed only very little. As my friend Chuck Greenblat used to say: 'the starch was gone'. I was 55 years old, more than 10 years before retirement and knew that things are not going as they should, but I did not know what to do. I went through a period of rage, "how come the world goes on as if nothing happened"? Why is the sun shining and the earth turning"? I guess these and other thoughts are common to all who suffer great loss. I also went through a period of suicidal thinking, not really suicidal but certainly expecting to die soon. The fact that I got through all this, sane and more or less unharmed is only thanks to my tight family, my children and their spouses, my friends and colleagues and possibly my innate and proven lust for life.

(I am writing these lines on the day, when Cipora's eightieth birthday should have been).

International Collaboration Projects.

During my career I was lucky to participate in several large international collaboration projects, two with lasting impact. The first one, Holland- Israel collaboration was initiated in the early sixties, during some scientific meeting by Avivah Zuckerman and Christoph Jerusalem of the Catholic University in Nijmegen Holland. There were joint meetings, workshops and exchanges between the participants. Poor Christoph, when he visited us and was asked for his name, he replied and was always told, "OK, sir you are in Jerusalem but what is your name" The Dutch government was interested in international collaborations to help developing countries. This project included at its start two laboratories in Nijmegen, that of Ch. Jerusalem in Anatomy and the lab of Joep Meowissen in Microbiology. Avivah's lab in Jerusalem (where I was the senior graduate student, Yosef Hamburger the junior and Jacob Golenser at the beginning of his graduate work. The international collaboration included the group of Prof. Oshunkoya of Ibadan, Nigeria and later also that of Thivi Ponnudurai from Colombo, Sri Lanka (Fig. 6).

It was a true 'technology transfer', a term popular during the late fifties and sixties. On top of jointly planned experiments on malaria immunology, we helped to introduce new techniques and I even gave a course of immunology in Ibadan, Collaboration with Sri Lanka was more complicated. That country, under the leadership of Mrs. Bandaranaike was a member of the "block of independent nations", together with the Nehru family of India, Abdul Nasser of Egypt and Tito of Yugoslavia. They were not exactly friendly to Israel and the "west" at large. The problem of restricted collaboration was solved when ethnic fighting erupted in Sri Lanka and our friend Ponnudurai, a Tamil and Christian, had to flee the country with his family and settle in Holland. For me this multi-national, multi-ethnic, multi-religious collaboration and friendship was a cardinal life experience. To see and feel that in this thorn world after the second world war, full of real and propaganda induced hatred among people and nations, simple quick and uncomplicated friendships can sprout just by having a cup of coffee or even better, jointly injecting some mice with a tested drug, shaped my unwavering belief in mankind.

The second multi-national project was far bigger and politically much more complicated. President Sadat of Egypt, after the "October War" in 1973, disastrous both to Egypt and to Israel, decided to end the tension between the two countries. Before it became publicly known, the US Agency for International Development prepared itself for such an event. Indeed before Sadat came to Israel there were meetings about scientific peaceful collaboration between the two countries. Dr. Sanford Kuvin of Palm Beach Fl. the founder and supporter of the "Kuvin Centre for the Study of Infectious and Tropical Disease" in our department, worked very hard to convince the Egyptian authorities and US AID was the first joint project on: "Arthropod Borne Diseases". And so it was. One big incentive for the Egyptian authorities to join forces with Israel was an epidemic of Rift Valley Fever that decimated their cattle and sheep herds. As Dr. Kuvin used to say "mosquitoes and diseases do not recognize borders". The Israeli veterinary authorities, fearing that the epidemic could cross the borders, were ready with effective measures to prevent this. Although, before the collaboration agreements were ready and funded, the epidemic ended, and the virus temporarily disappeared. Other problems still had to be solved. The first joint planning meeting of the working groups was in June 1982 in Sweden, neutral ground for all involved. (Fig.7). The meeting went well, but no close contact or personal friendships between the Egyptians and the Israelis developed, but serious discussions and planning of labor division were achieved with the help of US and European colleagues. How ironic that during the last of the joint sessions I got a phone call that war broke out between Israel and Lebanon. Personally a cause for great stress, as we had both sons in the army. For five days we could not get a flight back home nor a word as to where and how the boys were. Luckily, there were no injuries nor did the war prematurely ended the project. Our boys returned safely and the collaboration went on as planned.

The AID supported project continued for more than ten years. There were meetings in Egypt and in Israel, more than 100 publications on Rift valley virus, *Leishmania*, Malaria and tick borne fever with joint Egyptian-Israeli authorship appeared. There were workshops on specific problems here and in Egypt, but the main lesson was that even "arch-enemies" on the political level can scientifically cooperate and be good friends on the personal level. After us, projects in agriculture and oceanography followed. I had only one problem during one of my visits in Cairo. My friend Sherif El Said the PI of the Egyptian group took me with him on his daily rounds of the best of Cairo's night clubs and bars. Unfortunately I could not claim that I am a Muslim and therefore not drink alcohol (which he certainly was, but it did not prevent him from drinking), and so I am not sure that my lecture the next morning was coherent and instructive. Nevertheless I received from the president of En Shams University in Cairo an impressive pewter plate inscribed with thanks for my collaboration with and help to the University (Fig. 8). This experience, so unique and unexpected, the long friendly discussions on private matters as well as on politics and mainly science, again strengthened my belief that people can be friends whatever the politics or backgrounds are. Some of the Egyptians even were at the wedding of our son in Jerusalem and many wrote me condoling letters, when Cipora passed away.

The third project I was involved in was the simplest and therefore less exciting than the two others. The project jointly written by Deckelbaum, Prof. Nagan of the Ben Gurion University in Beer Sheva and myself, was presented to the NIH by Richard Deckelbaum of Columbia University. The project was funded and a collaboration between my laboratory together with Hadassahs' department of pediatrics in Jerusalem, the Soroka Hospital in Beer Sheva and Columbia Medical School in New York began. The scientific collaboration was excellent, new insights were achieved, the chance of infection of babies and young children by *Giardia* and *Cryptosporidium* thoroughly studied and the clinical picture in the two populations compared. The homogeneity of the participants made it less interesting for me. We all were products of the same scientific way of thinking and dealing with problems. It was the "American school" where most of the participants were trained. We understood each other, knew what to expect from each other, thus, although it was a successful multi-center, multi-national project, what seemed so exciting in the other collaborative projects was lacking here. As I already wrote, this project started soon after Cipora's death. It may be that my mental state caused me to have less good feelings about the project rather than the project had defects.

Later Years and Retirement

The last years before retiring were mainly devoted to administration: I became head of the department and later of the Institute of Microbiology at the Hebrew University Hadassah Medical School. The Institute had more than 35 senior academic staff, teaching both at the Medical School and Faculty of Science. My research zeal hampered by administrative duties and personal problems, shrunk an so did my lab. I was still teaching and had 1-2 graduate students working with me, there were also post docs from Germany, the US, Australia and others but activity was reduced .

The last year before retiring I spent in Australia at the Queensland Institute of Medical Research in Brisbane. I chose to go there in order to learn and practice the techniques of molecular biology, a completely new field and for me, a different laboratory technology. It was a beautiful year in a beautiful country and a city that did everything to prove that it is not a faraway place in a hot, rainy north. I worked there with Jacqueline Upcroft, dealing with *Giardia* DNA. An excellent opportunity to learn new techniques, and thus get into the new exciting field of modern science.

At the congress of the International Society of Protozoology in 1998, after hard lobbying, it was decided that the next International Congress of Protozoology would be held in Jerusalem. We started the preparations as soon as we returned from the congress, but again politics interfered (Fig. 9). As the Chair of the Israeli Society, I was in charge of the preparations. The beginning went well but as we were about to send out the invitations, the so called Intifada, started particularly in Jerusalem. It was clear that no one would dare to come to a city where bomb threats and bombing of busses were a daily issue. We had to find another place and thus the 2001 Jerusalem Congress of the International Society of Protozoology was transferred to Salzburg, Austria. The congress was a success, but not easy for me to lead and chair, as three months earlier I had major surgery and still carried its aftermaths (Fig.10)

Retirement

My retirement date was October 1st, 2000, the first day of school after my 78th birthday. That is the rule at the University although one can continue teaching as a volunteer and work as long as he has sufficient funding. Indeed, I continued both. There were still two graduate students and my lectures were in demand. Three years later another blow of fate came which changed my life completely. It started one day while I was driving, I wondered how come the line marking the edge of the road was not straight. Actually no line I looked at was straight. I was diagnosed as suffering from AMD (age dependent macular degeneration), a disease that inevitably lead to blindness. Indeed my vision deteriorated quickly, the treatments available in 2003 did not stop the deterioration of vision and so I had to stop teaching and limit my activity in the lab. A couple of years later an effective treatment emerged, luckily before I lost all sight, thus I still can safely walk on the street and use this computer for writing, after it has been adapted for a visually impaired person. This change in life slowed me down but did not take away my lust for life and did not spoil moments of happiness and good spirits.

Of course my life did not end by retiring from the university, even after giving up nearly all activity in the department. I was concerned about the danger that my eyesight might deteriorate further. I decided to go and live in a Sheltered Living apartment house in Jerusalem. Thus, whenever help was needed it was available and I did not have to bother my children or friends all the time. Not long after I settled in the new place I got to know one of the resident ladies in the house, and something immediately clicked between us. A new connection, a new love developed. Miriam (Marianne) Karmon, the widow of a geography professor and herself the chief cartographer of the Hebrew University became my partner and companion. Seven years have passed, my family accepted her as their own and thus even in the late years, with a new love and happiness my life became much more enjoyable (.Fig.11).

Reflections

Looking back at my life I can be satisfied. It was interesting, full of surprises, of novelty, of action, problems and difficulties but also of some achievements. It was a life with a family, with love and many, many beautiful moments of pure enjoyment and pleasure. But here I am writing about my professional life, my career and not about my private life. About that I have a few thoughts and comments.

Life in academia is not easy. It is a constant struggle to be the best, to be original, to move on and never to slow down or to fail. This might be true for many fields of life but in academia it has an additional component: one constantly has to obtain funding for one's own and group's activity. This also might sound familiar to many, but the problems that I and my colleagues in Israel face, are more complicated. Yes, we had to compete at the same international funding bodies as everybody else in the world. But just during my own professionally active years, Israel was involved in seven wars. Not long wars like WW II or Vietnam, but still wars. Either you, your students or your children were involved in the war. You were exposed to the stress of the situation, and the stress was huge, but also was the task to protect our culture, your protocols and the whole laboratory, so that the work of years would not be lost. You still had to keep the deadlines of grant submissions and reports, war or no war. I am not complaining, just wondering how we, the whole Israeli scientific community, carrying all this additional burden, did manage to compete for grants, for recognition among peers and still achieve an acceptable intellectual and scientific level. Looking at the younger generations, we apparently didn't fail them in teaching. Our students and post docs are all over the globe, some in the best Universities and institutes. I want to repeat I am not complaining, just wondering. I am proud of this collective achievement. It might even be that the additional stress helped. It added to the natural driving force in each man and woman for the proverbial "better, higher, and faster". I am also thinking of thousands of scientists all around this world, with fights, demonstrations, insecurity and murder around how will we progress and succeed?

So let me just repeat that I wholly and fully enjoyed my life, every minute of it, and I am infinitely grateful to all, and there are many, that accompanied me and still accompany me and were with me for all these interesting eighty years.

Fig.1. Sitting on the Repaired car, with My Sister

Fig.2. Prior to the Holocaust with
Grandparents, Parents and Sister.

Fig.3. Entrance to the Novaky Work Camp
(Cover of a book about the camp)

Fig.4. With Cipora

Fig.5. Receiving Ph.D. Diploma from Dean Rachmilewitz

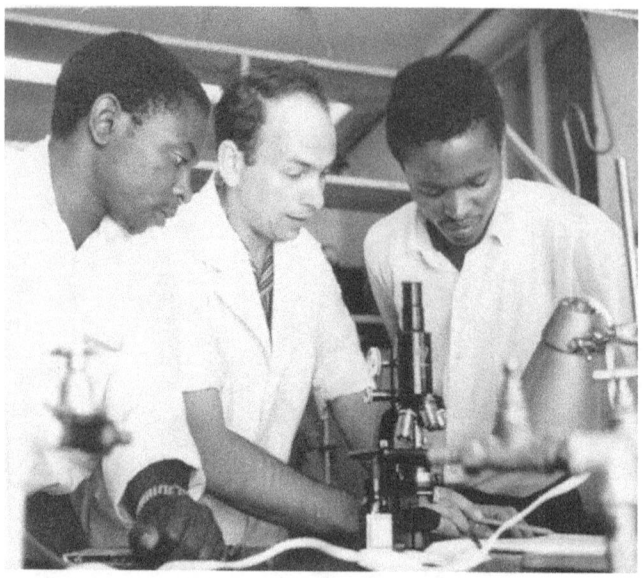

Fig.6. Laboratory Instruction with the Students in Nigeria.

Fig.7. First Meeting with Egyptian
Scientists. Second Row from Left:

A. Spielmann, Sherif El Said, J, Jensen, D. Spira.

Fig.8. Pewter Plate from En-Shams University, Cairo.

Fig.9. International Commission of the Society of Protozoology.

Fig.10. Opening XI congress of the
Society of Protozoology, 2001.

Fig.11. Miriam Karmon, My New partner.

TRIBUTES

Volumes I & II

Our Influencers & Mentors
Dr. Norman Borlaug Tribute
Dr. Murray F. Buell Tribute
Dr. Linda Styer Caldas Tribute
Dr. Otto J. Crocomo - For Those Who Made A Difference In My Scientific Life
Dr. Otto J. Crocomo - Para Aqueles Que Influenciaram Minha Carreira Cientifica
Dr. David A. Evans Tribute
Dr. David E. Fairbrothers
Dr. Percy Cyril Garnham Tribute
Dr. Leonard George Goodman Tribute
Dr. James E. Gunckel & Jean Longworth Gunckel Tribute
Dr. Roger F. Keller Tribute
Dr. Miodrag Ristic Tribute
Dr. Maro R. Sondahl Tribute
Dr. T.S. Subramanian and Meenakshi Subramanian Tribute
Dr. Clara Gertrude Weishaupt

Norman Borlaug Tribute

By MATT CURRY and BETSY BLANEY – Associated Press Writers

Submitted by: William R. Sharp

Agricultural scientist Norman Borlaug, the father of the "green revolution" who won the Nobel Peace Prize for his role in combating world hunger and saving hundreds of millions of lives, died Saturday in Texas, a Texas A&M University spokeswoman said. He was 95.

Borlaug died just before 11 p.m. Saturday at his home in Dallas from cancer complications said school spokeswoman Kathleen Phillips. Phillips said Borlaug's granddaughter told her about his death. Borlaug was a distinguished professor at the university in College Station.

The Nobel committee honored Borlaug in 1970 for his contributions to high-yield crop varieties and bringing other agricultural innovations to the developing world. Many experts credit the green revolution with averting global famine during the second half of the 20th century and saving perhaps 1 billion lives.

Thanks to the green revolution, world food production more than doubled between 1960 and 1990. In Pakistan and India, two of the nations that benefited most from the new crop varieties, grain yields more than quadrupled over the period.

"We would like his life to be a model for making a difference in the lives of others and to bring about efforts to end human misery for all mankind," his children said in a statement. "One of his favorite quotes was, 'Reach for the stars. Although you will never touch them, if you reach hard enough, you will find that you get a little 'star dust' on you in the process.'"

Equal parts scientist and humanitarian, the Iowa-born Borlaug realized improved crop varieties were just part of the answer, and pressed governments for farmer-friendly economic policies and improved infrastructure to make markets accessible. A 2006 book about Borlaug is titled "The Man Who Fed the World."

"He has probably done more and is known by fewer people than anybody that has done that much," said Dr. Ed Runge, retired head of Texas A&M University's Department of Soil and Crop Sciences and a close friend who persuaded Borlaug teach at the school. "He made the world a better place _ a much better place. He had people helping him, but he was the driving force."

Borlaug began the work that led to his Nobel in Mexico at the end of World War II. There he used innovative breeding techniques to produce disease-resistant varieties of wheat that produced much more grain than traditional strains.

He and others later took those varieties and similarly improved strains of rice and corn to Asia, the Middle East, South America and Africa.

"More than any other single person of his age, he has helped to provide bread for a hungry world," Nobel Peace Prize committee chairman Aase Lionaes said in presenting the award to Borlaug. "We have made this choice in the hope that providing bread will also give the world peace."

During the 1950s and 1960s, public health improvements fueled a population boom in underdeveloped nations, leading to concerns that agricultural systems could not keep up with growing food demand. Borlaug's work often is credited with expanding agriculture at just the moment such an increase in production was most needed.

"We got this thing going quite rapidly," Borlaug told The Associated Press in a 2000 interview. "It came as a surprise that something from a Third World country like Mexico could have such an impact."

His successes in the 1960s came just as books like "The Population Bomb" were warning readers that mass starvation was inevitable.

"Three or four decades ago, when we were trying to move technology into India, Pakistan and China, they said nothing could be done to save these people, that the population had to die off," he said in 2004.

Borlaug often said wheat was only a vehicle for his real interest, which was to improve people's lives.

"We must recognize the fact that adequate food is only the first requisite for life," he said in his Nobel acceptance speech. "For a decent and humane life we must also provide an opportunity for good education, remunerative employment, comfortable housing, good clothing and effective and compassionate medical care."

In Mexico, Borlaug was known both for his skill in breeding plants and for his eagerness to labor in the fields himself, rather than to let assistants do all the hard work.

He remained active well into his 90s, campaigning for the use of biotechnology to fight hunger and working on a project to fight poverty and starvation in Africa by teaching new drought-resistant farming methods.

"We still have a large number of miserable, hungry people and this contributes to world instability," Borlaug said in May 2006 at an Asian Development Bank forum in the Philippines. "Human misery is explosive, and you better not forget that."

Norman Ernest Borlaug was born March 25, 1914, on a farm near Cresco, Iowa, and educated through the eighth grade in a one-room schoolhouse.

"I was born out of the soil of Howard County," he said. "It was that black soil of the Great Depression that led me to a career in agriculture."

Murray F. Buell Tribute

Submitted by: William R. Sharp

Murray F. Buell died July 2, 1975 while on a field trip in the New Jersey Pine Barrens. He had recently assumed the leadership of a natural resource study and at the time of his death was actively engaged in activities that delighted him throughout his lifetime: working with students, studying vegetation, and advancing the cause of conservation.

Murray Buell's influence on ecology was deep, constant and long sustained. This quiet, patient scholar was reared in a liberal New England family, and studied at the Loomis School, Cornell University, and the University of Minnesota. After studying under W. S. Cooper, he started his professional career at North Carolina State University in 1935. There he began his notable work on the paleoecology of bogs, plant succession, and tension zones between vegetation types. In 1947, he moved to Rutgers University where he eventually became Professor of Botany and Director of the William L. Hutcheson Forest. He devoted great effort in setting aside this forest and in making it into a major ecological study area and one of the best studied woods in North America. Well before it was fashionable, he initiated important studies linking ecology to land-use management. Two decades ago he and his students studied the impact of people on park ecosystems, investigated the ecology of power line right of ways and the use of fire on forest and hydrologic management. He also made intensive studies of the structure and dynamics of vegetation in and around New Jersey, and now the State is among the best known ecological regions in North America. Among his last works is the book *Vegetation of New Jersey* coauthored with Beryl Robichaud.

After his retirement from Rutgers in 1971 he served as a visiting professor of ecology at Yale, the University of Minnesota,

Georgia, Arizona, California Davis, California Santa Barbara, Montana and Colorado State.

Perhaps Murray Buell's greatest impact on ecology was achieved as a teacher. A gentle and thoughtful man, he was considerate of his students, yet demanding of excellence. His influence on undergraduates resulted in a steady stream of students flowing to graduate schools, while the ecology program he initiated at Rutgers attracted scores of students from throughout North America. In Murray Buell they found a stimulating teacher concerned not only about the study of ecology, but about them as individuals. His life touched many hundreds of North American ecologists through the Rutgers Ecology Seminar that he initiated and sponsored. In the many summers he taught at the University of Minnesota's Lake Itasca Biological Station, he recognized exceptionally promising young students. Often the fortunate person was hired as an assistant, transported across the country in his car, fed chicken dinners and given a thorough introduction to life as a field ecologist. A summer at Itasca was the beginning for at least a half dozen current full professors of Ecology. Murray Buell's relationship with his students did not end with the award of a diploma. He actively followed their careers, acted as a sounding board for ideas and decisions, and provided wise counsel when asked.

No recounting of Murray Buell's career could be complete without mention of his wife, Dr. Helen Foot Buell. Murray and Helen Buell worked as a team and between them maintained a lively and inquisitive interest in all things around them. Many of us were fortunate to pass through their sphere of interest.

Murray Buell labored long and hard for the Ecological Society of America. He served as Associate Editor of Ecology and Ecological Monographs, Secretary, Vice President and as President in 1961-62. At the time of his death he was Chairman of the ESA Awards Committee. Less obvious, but nonetheless important, was his contribution to the drafting of the new constitution and by-laws of the ESA and his contribution to the early development of the Institute of Ecology. For his multifaceted contributions to ecology, Murray F. Buell was named Eminent Ecologist by the Society in 1971.

The loss of this scholar is great, but there is some satisfaction in knowing that Murray Buell died while fully active and in pursuit of the things he loved.

F. Herbert Bormann
Paul G. Pearson

List of Paleoecology Graduate Students
Murray F. Buell
provided by Allen M. Solomon
June 25, 2003
• John Cantlon
• Peter Comanor
• Ralph Good
• Kathy Harmon
• <u>Calvin Heusser</u>
• William A. Niering
• Bill Reiners
• Allen M. Solomon

Partial Bibliography of Palynology and Paleoecology Papers of Murray F. Buell provided by Allen M. Solomon June 25, 2003
• Buell, M.F. 1939. Peat formation in the Carolina Bays. Bull. Torrey Bot. Club 66:483-487.
• Buell, M.F. 1945. Late Pleistocene forest of southeastern North Carolina. Torreya 45:117-118.
• Buell, M.F. 1946. Jerome Bog, a peat-filled "Carolina Bay." Bull. Torrey Bot. Club 73:24-33.

- Buell, M.F. 1946. Size-frequency study of fossil pine pollen compared with herbarium-preserved pollen. Am. J. Botany 33:510-516.
- Buell, M.F. 1946. A size-frequency study of *Pinus banksiana* pollen. J. Elisha Mitchell Scientific Soc. 62:221-228.
- Buell, M.F. 1947. Mass dissemination of pine pollen. J. Elisha Mitchell Scientific Soc. 63:163-167.
- Buell, M.F. 1970. Time of origin of New Jersey Pine Barrens bogs. Bull. Torrey Bot. Club 97:105-108.
- William S. Cooper and Helen Foot. 1932. Reconstruction of a late-Pleistocene biotic community in Minneapolis, Minnesota. Ecology 13:63-72. ESA Bull. 56(4): 26, 1975. Information retrieved on October 19, 2013 from http://www.palynology.org/murray-f-buell

A Tribute to Dr. Linda Styer Caldas

Authored By Sally Miller, Chip Styer and Sandy Styer

Linda Styer Caldas was born December 10, 1945 in Appleton, Minnesota, USA, the eldest of five children of Dr. Donald James and Carol Opal Hancock Styer. Linda's father was a career army dentist, and as a typical military family, the Styers moved often. Linda lived in several different states and in Germany prior to beginning undergraduate studies at the University of Colorado at Boulder. When her family returned from Germany and settled in the Washington, D.C. area, Linda transferred to George Washington University, where she studied biological sciences. Linda was a brilliant student and highly enthusiastic about science. She graduated from GWU in 1967 and entered the Department of Botany at The Ohio State University as a graduate student the same year in the laboratory of Dr. Carroll Swanson. She studied elm tree physiology, and completed her M. S. research in 1969 with her thesis entitled "Diurnal changes in radius of trunk and water potential of leaves of *Ulmus americana* L.".

Linda then entered the burgeoning field of plant tissue culture, specifically wild carrot tissue culture, for her PhD research. Dr. Rod Sharp, then a lecturer in the Department of Microbiology at OSU, met Linda in 1969 when they shared basement labs in the Botany and Zoology building. Rod remembers Linda as a star graduate student, extremely bright, energetic and passionate about her graduate program. "Linda was more like a faculty member than a graduate student. I remember Linda's participation in the weekly departmental seminar program when she delivered a remarkable presentation about plant embryology followed by a huge round of applause and a robust question and answer session. She was a gifted science writer and was always editing manuscripts for fellow students." While a graduate student, Linda met Ruy de Araujo Caldas, another bright and capable OSU graduate student in the same lab. Linda and Ruy married in 1971, and then moved to Brazil to begin their academic careers. Linda defended her dissertation entitled "Effects of Various Hormones on the Production of Embryoids of Wild Carrot (*Daucus carota*)" before a committee composed of both OSU and University of Sao Paulo - ESALQ faculty following her move to Piracicaba.

With Rod Sharp, Dr. Otto Crocomo (who at the time was the research coordinator of the Plant Biochemistry Sector of the Center for Nuclear Energy in Agriculture), and Ruy, Linda pioneered plant tissue culture research in Brazil. They launched collaborative research programs developing cell cultures for important cultivars of tropical crops with special interest in citrus, cocoa, coffee, beans, palm, and sugarcane. These programs were initiated in collaboration with geneticists and plant breeders at the University of Sao Paulo ESALQ Campus and the Institute of Agronomy in Campinas, Sao Paulo. Rod marveled that "Linda developed fluency in Portuguese within six weeks of her arrival and was participating in research meetings at CENA and USP-ESALQ– an amazing feat!"

In 1972, Linda and Ruy moved to the University of Brasilia (UnB), where Linda served on the faculty of the Botany Department until her retirement in 1998. She became a naturalized Brazilian citizen, and with Ruy had three children: Pedro, Cristina and Juliana.

Linda was a highly productive researcher in plant physiology and continued her pioneering work in plant tissue culture at UnB. She published more than 40 peer-reviewed journal papers, seven book chapters and five books, and served on several editorial boards. She chaired the UnB Botany Department from 1991 – 1993, and served as vice-director of the Biosciences Institute of UnB from 1994-1998. She was also a member of the Brazilian

Agency for Education Development and Research (CNPq). She remained very active in academia after her retirement, including organizing the VII Brazilian Congress of Plant Physiology in 1999. She was awarded the title of Professor Emeritas of UnB in 2006. She spent two years as a professor at Catholic University of Brasilia after her retirement helping the university develop its biological sciences curriculum. Linda also collaborated with Bioplanta Technology Ltd., one of the first biotechnology companies in Brazil.

Linda was a dedicated teacher. Remarkably, after only a few years in Brazil, she wrote a biology textbook in Portuguese *"Principios Biológicos- uma Introdução"* to fill a much-needed niche in undergraduate education. This was the first textbook of its kind in Portuguese in Brazil. She also advised many M.Sc. and Ph.D. graduate students in plant science, particularly in botany and ecology. Linda was an early adopter of student-centered learning and committed her considerable energy and enthusiasm to her students.

Within a few years after their arrival in Brasilia, Linda and Ruy purchased a farm in Cristalina, where she fell in love with and became dedicated to the preservation of the Cerrado, the beautiful plains with tropical woodlands and scrub vegetation, an ecoregion found only in Brazil. Linda also recognized the risk of loss of Cerrado plant species by aggressive agricultural development. She created, and served as President of, the Cerrado Seeds Network, a preservation group funded by the National Fund for the Environment/Ministry of Environment in 2001[3]. On August 30, 2007, the Botanical Garden of Brasilia dedicated a garden in her memory to recognize her many contributions to the preservation of the Cerrado. Accounts of the dedication of the space are excerpted as follows:

"The space dedicated to Dra. Caldas is located in the Cerrado Medicinal Garden, next to the Visitor Center, an area of 500 m² of Cerrado maintained with native trees and species catalogued for medical use. Also, another seed she planted, the education efforts about the Cerrado, will be immortalized in the area, which will be visited over time from students and external audiences for environmental education programs. 'She was and always will be a model in the Federal District for research on the Cerrado. This is a way to make her energy and dedication to the perpetuation of the Cerrado known to future generations' said Jeanito Gentilini, director of Botanical Garden.

Dr. Kumiko Mizuta, 67, arrived in the federal capital in 1969 to join the faculty of UnB. In 1972 she met Linda, and together they built a relationship of friendship and dedication to biology and the Cerrado. 'What struck me the most was her dedication to work, her willingness to share knowledge and the captivating way she treated everyone around her' recalled Kumiko. Kumiko was beside Linda until the time that breast cancer, detected at an advanced stage, took the life of the researcher. 'She fought until the last moment and did not stop working. I remember her walking through the Cerrado and teaching even with the pains of the disease' said her friend.

Her children, Pedro, 35, Cristina, 33, and Juliana, 22, chose to follow in Linda's footsteps. 'My mom passed on her dedication and love for education and the environment as examples, always present in our experience', said Juliana Caldas, recently graduated in biological sciences from UnB. Pedro, who takes care of the family farm, opted for a major in agronomy and Cristina also followed the career of biologist and science writer. 'We are flattered by the honor of the garden dedication. For us it is very important to see her remembered the way she would like: in the Cerrado and sharing her knowledge with others', said Juliana.

Cerrado Seeds Network

With young researchers from Acesita Energetica, Linda developed mini-cutting (a cultural practice) for clonal multiplication of eucalyptus, used today industrially in Brazil, which leads the world in this technology. 'With a sliver of the plant, she could produce thousands of clones', Kumiko explained. Linda also published important books such as *Tissue Culture and Genetic Transformation of Plants*, a reference in biotechnological applications throughout Brazil. One of her final achievements was the creation of the Non-Government Organization (NGO) Cerrado Seeds Network.

For Gustavo Souto Maior, UnB professor and president of the Environmental Institute Brasilia, the inauguration of the Cerrado garden space represents the rapprochement between academia and conservation areas in DF. 'There are places to do research that are very rich, but have been forgotten. We want to revive the interaction between academia and the environment, following the examples that Linda has left us. Hundreds of people who visit the garden can read and reflect on a phrase written by Cristina, the daughter of the researcher, which summarizes the history of her mother: Cerrado: respect, know and love'." [1]

"About ten seedlings were planted at the inauguration of the space. Linda's friend Celina de Oliveira Martin believes the Botanical Garden is indeed the ideal place to celebrate Linda's achievements and contributions to biology. 'The wind, the water, the trees and the calm of this place sing her name. Whoever comes here meditating will be able to feel her energy and love of nature's beauty. Hopefully it touches other people and lets them continue the work of preserving the Cerrado she began' said Celina." [2]

In 2005, Linda returned to the US to spend several weeks with her father in Huntsville, Alabama, and with her siblings and their families across the US. With her usual energy, she took on the task of organizing hundreds of Styer family photos and documents, and wrote detailed histories of the family. These histories are a treasure to Linda's children, grandchildren, siblings, nieces and nephews, and to future generations. After Linda's death, her children found that she had written down the things for which she would like to be remembered: "the ability to be original and advocate innovative ideas, unrestricted love dedicated to friends and family, and contribution to biology, education and conservation of the Cerrado".[2]

We will always remember the kindness, creativity, energy, charisma and beautiful soul of our sister and friend, Linda Styer Caldas.

We thank Alba Clivati McIntyre and Claudio Vrisman for assistance with translation.

Notes

[1]http://pib.socioambiental.org/pt/noticias?id=48967

[2]http://www.secom.unb.br/unbagencia/ag0807-76.htm

[3]http://www.iesambi.org.br/parcerias_arquivos/triste_perda_lindacaldas.htm

Linda Styer Caldas Photograph 1

Linda Styer Caldas Photograph 2

For Those Who Made a Difference in My Scientific Life

Authored By Otto J. Crocomo

Excerpt from the Address delivered in the ceremony of granting the Fernando Costa Medal by the Association of Agricultural Engineers of the State of São Paulo on 15 June 2012.

The streetcar went down XV de Novembro Street, snaking through José Pinto de Almeida Street, reached the São Joao Street and entered the sacred territory of the agricultural sciences and parked next to a building bearing the word "Chimica" surmounting the facade of the Hall of Chemistry at the Escola Superior de Agricultura "Luiz de Queiroz"–ESALQ, University of São Paulo, in Piracicaba.

I got out of the streetcar, walked up the staircase, and entered its spacious lounge. As I did many times, my gaze followed once again the top of the wall bearing several bronze medallions showing the faces of illustrious ancestors whose decisions were crucial to further substantiate the already magnificent Agriculture School. The Fernando Costa Medallion had been there since the Pavilion was opened in 1930.

I was, at that time, attending the first year of ESALQ and I looked for Professor Euripedes Malavolta. I started working in his laboratories as an intern under his orientation. It was in that same laboratories that five years later, in June 1958, already an Agronomy Engineer, I was invited by Prof. Malavolta to submit to a contest for the "Livre Docente Degree" although not yet employed by the University of Sao Paulo.

That moment was a very important fact of my entire life. Had it not been for the confidence placed on me by Professor Malavolta I would have never been a recipient of the Fernando Costa Medal. So now, I pay a tribute to his memory (photo 1).

We all need one another. We are not juxtaposed individuals, each one living their own lives oblivious to what happens to their surroundings. As in all areas of the universe of human knowledge, in the world of biological sciences in which I live, an idea can arise in a single mind, but the realization of it requires many other minds to reason and observe with critical and keen eyes the results of the experiments: there needs to be many other hands to manipulate the test tubes and culture flasks, and many other feet to support the bodies that stand by the laboratory counters or walk between the lines, as if they were backstreets, and to separate and identify creeping or slender plants in the agricultural fields with the desirable agronomic characteristics.

As I use that metaphor I remember Demosthenes Santos Correa, my first Professor of Chemistry. With him I entered the realm of chemical reactions and laboratory techniques, which involved long hours, and I recall how he enthusiastically encouraged my participation in debates of Chemistry during the three years of my high school course. I am also remembering now Jose Dall Pozzo Arzolla, my Professor of Organic Chemistry at ESALQ. With him I got in touch with the dynamism of theoretical and practical classes in the laboratory, developing refined procedures for conducting experiments and the statistical interpretations of the results. I am remembering also Admar Cervellini, who greatly supported me in my research using radioisotopes, as Diretor of the Center Nuclear Energy in Agriculture – CENA. Much of that research such as mineral nutrition of plants was conducted at CENA in collaboration with my very good friend Andre Martin Louis Neptune. Also to the memories of these four great men I pay my tribute.

In the 50s and 60s of the last century the biochemistry of plants was emergent in Brazil and there I was entrenched in the biochemical intricacies responsible for the life of the plant cells. I remember one of my brightest student, Luiz Carlos Basso: long hours, day and night, spent making and repeating experiments, including the one that led to the discovery of a new enzymatic activity. My tribute to him.

I would like to pay tribute to William Rod Sharp who, in June 1971, arriving from the United States of America, joined me to introduce in Brazil the techniques of cell and plant tissue culture in agriculture, building since then, a fruitful relationship and scientific exchange between us (photo 2).

At that time, in June 1981, it was created at ESALQ, the Center for Agricultural Biotechnology –CEBTEC, thanks to the unconditional support I received from Aristeu Mendes Peixoto, Joaquim Jose de Camargo Engler and Paulo Fernando Cidade de Araujo. CEBTEC has always been supported by various ESALQ and FEALQ directors, among them Joao Lucio de Azevedo, Antonio Roque Dechen and Justo Moretti Filho who is no longer among us. I pay also tribute to them.

With the collaboration of my teammates and technicians at CEBTEC, mainly Helaine Carrer, Enio Tiago de Oliveira and Luiz Antonio Gallo, it was possible to give life to the real meaning of plant biotechnology: from cell to a viable plant, using the traditional improvement technology if necessary, and finally to commercialization. This was done with various plant species at CEBTEC in a very happy union among biochemistry, molecular biology and cell biology in our R & D projects followed by the transfer of technology to the private sector.

I need to emphasize the *sine qua non* collaboration I received from one of my greatest friends that began when we were undergraduate students at ESALQ, in the 50s, Ary A. Salibe, expert in vírus citrus. This collaboration made it possible for CEBTEC at ESALQ and the Faculty of Agronomy of UNESP, in Botucatu, State of São Paulo, in the 90s, the tripartite project with a private company, using the technique of micrografting, to produce "orange pear" plants resistant to the "tristeza" vírus. These plants are now being cultivated in an agricultural field – a striking example of a genuine biotechnological product. To Ary A. Salibe, who passed away in 2013, my tribute (photo 3).

There is not enough space to mention each one of my undergraduate and graduate students, my collaborators in Brazil and abroad; nevertheless, each one is present in my memory and in my heart.

I'd like to quote Albert Einstein: "A hundred times every day I remind myself that my inner and outer life depend on the labors of other men, living and dead, and that I must exert myself in order to give in the same measure as I have received."

PHOTO 1

OTTO (LEFT) AND MALAVOLTA (RIGHT), NOVEMBER 1990

PHOTO 2

ROD AND OTTO WORKING ON THE BOOK, AND DIVA.

OTTO'S HOME. PIRACICABA, SEPTEMBER 2013

PHOTO 3

ARY (LEFT) AND OTTO (RIGHT). SHOWING SAMPLES
OF THE ORANGE "PEAR" FREE OF THE "TRISTEZA"
VIRUS. IN THE FIELD. VOTORANTIM CITRUS
FARM. ITAPETININGA, SP, BRAZIL, 2003.

Para Aqueles Que Influenciaram Minha Carreira Carreira Cientificia

Authored By Otto J. Crocomo

Baseado no Discurso de agradecimento ao ser agraciado com a "Medalha Fernando Costa" pela Associação dos Engenheiros Agrônomos do Estado de São Paulo em 15 de junho de 2012.

O bonde desceu a Rua XV de Novembro, serpenteou pela Rua José Pinto de Almeida, alcançou a Rua São João, adentrou o território sagrado das ciências agronômicas e estacionou próximo a um prédio que ostenta a palavra "Chimica" encimando a fachada do Pavilhão de Química da Escola Superior de Agricultura "Luiz de Queiroz" da Universidade de São Paulo, em Piracicaba.

Desci do bonde, subi a escadaria, entrei em seu amplo saguão. Meu olhar uma vez mais dirigiu-se para o alto da parede que ostenta vários medalhões de bronze mostrando as faces de ilustres antepassados cujas decisões foram cruciais para alicerçar ainda mais a já, naquela época, magnífica Escola Agrícola. O medalhão de Fernando Costa lá está, desde quando o Pavilhão foi inaugurado em 1930.

Estava eu, naquele momento, cursando o primeuiro ano da ESALQ e naquela manhã de novembro de 1953 procurei pelo Professor Eurípedes Malavolta. Passei a trabalhar nos seus laboratórios como estagiário, sob sua orientação. Foi nesse mesmo Pavilhão que, 5 anos mais tarde, em junho de 1958, já sendo eu engenheiro agrônomo, recebi o convite do Professor Malavolta para prestar concurso para "Livre-Docente", apesar de ainda não pertencer ao quadro de contratados pela Universidade de São Paulo. Não fora a confiança em mim depositada pelo Professor Malavolta a mim não teria sido outorgada a Medalha "Fernando Costa". A ele, portanto, a minha sincera homenagem (foto 1).

Todos nós precisamos uns dos outros. Não somos indivíduos justapostos, cada um vivendo sua própria vida alheio ao que se passa em seu entorno. Como em todas as áreas do universo do conhecimento humano, também no mundo das ciências biológicas em que vivi e ainda vivo, uma ideia pode surgir em uma única mente, mas a concretização e as provas de sua veracidade exigem muitas outras mentes que raciocinem e observem com olhos críticos e clínicos os resultados dos experimentos, muitas outras mãos que manipulem os tubos de ensaio e os frascos de cultura e muitos outros pés que sustentem os corpos que ficam diante das bancadas dos laboratórios ou caminhem pelas entrelinhas, como se ruelas fossem, a separar e identificar plantas esguias ou rasteiras, nos campos agrícolas, com características agronômicas desejáveis.

Ao usar essa metáfora, lembro-me de Demósthenes Santos Correa, meu primeiro Professor de Química. Com ele penetrei no reino das intrincadas reações químicas, com horas e horas de práticas de laboratório, incentivando-me a participar de debates de Química durante os 3 anos de meu Curso Colegial. Lembro-me também de José Dall Pozzo Arzolla, meu Professor de Química Orgânica no Curso de Graduação na ESALQ. Com ele familiarizei-me com o dinamismo das aulas teóricas e práticas, desenvolvendo refinados processos experimentais e aplicando métodos estatísticos na interpretação dos resultados. Estou também lembrando-me de Admar Cervellini que, como Diretor do Centro de Energia na Agricultura – CENA, me proporcionou todas as facilidades para a realização de experimentos utilizando radioisótopos. Muitos desse experimentos, como aqueles sobre nutrição mineral de plantas, foram conduzidos coma colaboração de meu grande amigo André Martin Louis Neptune. A todos esse 4 homens as minhas homenagens.

Nas décadas de 50 e 60 do século passado a bioquímica de plantas era incipiente no Brasil e lá estava eu entranhado nos meandros bioquímicos responsáveis pela vida das plantas. Recordo-me de um dos meus mais brilhantes orientados Luiz Carlos Basso: longas horas, de dia e de noite, passamos fazendo e repetindo experimentos dentre os quais aquele que nos levou à descoberta de uma nova atividade enzimática em células vegetais, nos finais dos anos 60 e início dos anos 70. A ele minhas homenagens.

Alguém a quem não poderia deixar de prestar minha homenagem é William Rod Sharp que, em junho de 1971, vindo dos Estados Unidos da América do Norte, colaborou comigo na introdução no Brasil das técnicas de cultura de células e tecidos de plantas em agricultura. Essa nossa atividade foi importante para que a ESALQ e o CENA contribuíssem sobremaneira para o desenvolvimento da biotecnologia de plantas no Brasil (foto 2).

Nessa época, em julho de 1981, foi criado na ESALQ o Centro de Biotecnologia Agrícola –CEBTEC, graças ao apoio de Aristeu Mendes Peixoto, Joaquim José de Camargo Engler e Paulo .Fernando Cidade de Araujo. Esse Centrp sempre foi apoiado pelos vários Diretores da ESALQ e da Fundação de Estudos Agrários Luiz de Queiroz – FEALQ, entre eles João Lúcio de Azevedo, Antonio Roque Dechen e Justoi Moretti Filho, o qual já não mais se encontra entre nós. Minhas homenagens a cada um deles.

Com a colaboração de Helaine Carrer, Enio Tiago de Oliveira e Luiz Antonio Gallo, meus atuais continuadores no CEBTEC, é que foi possível dar vida ao real significado da biotecnologia de plantas: da célula à planta viável, seguido, se necessário, da metodologia tradicional de melhoramento, e finalmente à comercialização. Assim foi feito com várias espécies de plantas, vivenciando-se no CEBTEC uma feliz união entre a bioquímica, a biologia molecular e a biologia celular nos nossos Projetos de Pesquisa e Desenvolvimento e de transferência tecnologia à iniciativa privada. Aqui eu lhes presto minhas homenagens.

Tenho de enfatizar a colaboração *sine qua non* de Ary Aparecido Salibe na realização de um projeto tripartite entre a USP, a UNESP e empresa privada, para obtenção de plantas de laranja "pera" livre do "vírus da tristeza", utilizando a técnica de cultivo de microenxertos. As plantas estão sendo cultivadas em campo agrícola, e seus produtos comercializados. Exemplo marcante de um produto genuinamente biotecnológico. Salibe faleceu em novembro de 2013. Para ele, a minha homenagem (foto 3).

Não há espaço suficiente para mencionar e homenagear todos os meus alunos de graduação e de pós-graduação, os meus colaboradores do Brasil e do Exterior. Cada um deles está presente em minha memória e em meu coração.

Quero citar Albert Einstein:

Centenas de vezes todos os dias eu me conscientizo de que a minha vida interior e exterior depende do trabalho de outros homens, vivos ou que já não mais estão entre nós, e que eu devo devolver à humanidade na mesma medida o que eu dela recebi".

PHOTO 1

OTTO (À ESQUERDA) E MALAVOLTA (À DIREITA). NOVEMBRO, 1990

PHOTO 2

ROD E OTTO EDITANDO O LIVRO, COM DIVA.

RESIDÊNCIA DE OTTO. PIRACICABA, SETEMBRO 2013

PHOTO 3

ARY (À ESQUERDA) E OTTO (À DIREITA). LARANJA "PERA"
LIVRE DO VIRUS DA "TRISTEZA". NO CAMPO. FAZENDA DE
CITRUS DA VOTORANTIM. ITAPETININGA, SP, BRASIL, 2003.

David A. Evans Tribute

Submitted By William R. Sharp

NEW BRUNSWICK, N.J., June 5 /PRNewswire/ -- David A. Evans Ph.D., Chief Executive Officer of WellGen, Inc., passed away on June 1, 2006, following a brief illness. Dr. Evans, 54, was a well-regarded scientist and business leader in the food, biotechnology and nutrigenomics industries. WellGen, Inc. is a privately owned biotechnology company using nutrigenomics to develop ingredients that reduce the risk and severity of disease for the food, therapeutics, and dietary supplement markets. Richard Laster, Chairman of WellGen, said, "We are all shocked and deeply saddened by this terrible, unexpected occurrence. Dave Evans was a wonderful colleague and friend. He was highly intelligent and filled with energy and new ideas. Over the years, Dave had earned tremendous respect throughout our industry as a serious scientist and businessman, and had been doing an outstanding job guiding WellGen towards a very promising future. We will greatly miss him." Mr. Laster stated that the Board of Directors will meet soon to decide on the longer-term leadership of the Company. In the interim Mr. Arthur Finnel, WellGen's Chief Financial Officer, will assume day-to-day operating responsibilities for the Company. Dr. Evans joined WellGen in 1999 as CEO and President. From 1981-1999, Dr. Evans was with DNA Plant Technology Corporation, where he was Co-Founder and Executive Vice President of Business and Product Development. While at DNAP, he was on the Board of Directors of several joint ventures with DuPont, Union Carbide, and others. He introduced several new products into DNAP's $300 million produce operation, from research through production to market introduction.

Dr. Evans earned his B.S., M.S. and Ph.D. from Ohio State University in Genetics, and completed the PMD at Harvard Business School. He was formerly Assistant Professor, State

University of New York, and Research Manager, Campbell Soup Co. Dr. Evans was an inventor on 12 US patents and published over 100 scientific papers. He was an adjunct faculty member of the Department of Biology at Rutgers University from 1983-1994. In addition to his business activities, Dave pursued many interests. He was an avid birder, completed the 2005 New York City Marathon and played racquetball daily. WellGen, Inc., is based in New Brunswick, NJ, and developing products for food, therapeutics, and dietary supplement markets. WellGen's technical platform is a method of screening the effect of food and related substances on the expression of genes associated with human health conditions. The company has developed proprietary substances that help reduce risk and severity for a variety of diseases.

Source:
Information retrieved October 19, 2013 from http://m. prnewswire.com/news-releases/dr-david-a-evans-chief-executive-of-wellgen-inc-passes-away-after-sudden-illness-55878842.html

David E. Fairbrothers, 1925-2012 Tribute

Authored By David Lee and Dennis Stevenson

David E. Fairbrothers, a long-time Rutgers professor and eminent botanist and systematist, passed away on October 29th, 2012, after a lengthy illness. He had a distinguished academic career in the field of plant molecular systematics, and was a leader in the conservation of plants and natural areas, particularly in his home state of New Jersey. David was born and raised in Absecon during the depression years, part of a family of commercial fishermen and duck hunters, and he grew up close to nature. Soon after graduating from high school in 1943, he joined the Army when he turned eighteen. An excellent marksman as a young man, he became a Sergeant Rifleman and Squad Leader of the L Company, 376th Regiment of the 94th Infantry Division. Landing at Utah Beach on the second day of the Allied invasion, his company fought its way across France and was in the middle of the infamous Battle of the Bulge during the bitterly cold winter of 1944-45. In a frozen pothole during that battle, he suffered severe frostbite of his lower legs and barely escaped having both of them amputated, injuries that affected him the rest of his life. After hospital recuperation, David was stationed in Prague; after the German surrender on May 8th of 1945, he helped supervise train convoys of starving and ill refugees returning to their homes.

David was discharged from the Army in February of 1946 at the age of 20, and he took advantage of the G.I. Bill to enter Syracuse University. He met his future wife Marge while in school, and they married in 1949. He graduated in 1950 and immediately enrolled as a graduate student in Botany at Cornell University, under the direction of Robert T. Clausen. David worked in the area of grass systematics, completing a Master's degree (*A Cytotaxonomic Investigation within the Genus Echinochloa*) in 1952, and his Ph.D. (*Relationships in the Capillaria Group of Panicum*)[1] in 1954. David was recruited for a faculty position at Rutgers University by the eminent plant morphologist and head of the Graduate School, Marion L. Johnson. David began his 34 year career at Rutgers, the autumn semester of that year. He was the department's taxonomist and Director of the Chrysler Herbarium; during his tenure, its collections increased from 37,000 to over 140,000 specimens. His successful career at Rutgers was marked by two traits: (1) an intimate knowledge of plant and habitat diversity in the small but biologically rich state of New Jersey; and (2) an interest in employing new and multiple techniques (perhaps influenced by his earlier use of cytogenetics) in plant systematics. He was in the right place at the right time. Rutgers was the home of the Serological Museum, which had contributed significantly to advances in zoological systematics, particularly among birds, from the leadership of Alan Boyden and his junior colleague Ralph De Falco.[2] With the help of Marion Johnson, David learned the immunological techniques Boyden and De Falco had employed in studying animals, and they applied them to problems in plant systematics.[3] Boyden died in 1962, and Johnson in 1964; Fairbrothers then developed an independent research program in plant molecular systematics that utilized a growing arsenal of techniques, starting with immunology, adding polyacrylamide gel electrophoresis (PAGE) and isoelectric focusing, and secondary compounds (terpenoids and flavonoids) to a range of taxonomic problems, from population variation within species, to hybridization and introgression between closely related species, and to phylogenetic relationships among different families. This laboratory operated until his retirement in 1988, and it was the setting for the training of 29 graduate students. David was an excellent mentor, supportive of students and always available for discussions. His students will treasure the memories of field trips in his station wagon, cruising up the New York Thruway as David ticked off the names of roadside grasses and composites. The lab was also the temporary home of nine faculty members visiting on sabbaticals, and six post-doctoral fellows. The majority of his 122 peer-reviewed articles were in the field of plant molecular systematics. This body of research, along with the laboratory at the University of Texas, formed the backdrop for the revolution in systematics that came with the application of techniques for the analysis of DNA, starting with DNA hybridization (actually quite reminiscent of the serological research) and then sequencing.[4] Although David is considered by most to be a flowering plant taxonomist, his interests were actually more eclectic as evidenced by his co-authorship of the *Ferns of New Jersey* and, at the time of his illness, his interest in and study of the marine algae of the state parks in New Jersey.

Two of David's close friends were Arthur Cronquist (1919-1992) of the New York Botanical Garden and Armen Takhtajan (1910-2009) of the Komarov Institute in St. Petersburg. Whenever Armen was in New York, David visited Armen there and/or he travelled to Rutgers. David frequently attended the Torrey Botanical Club meetings at the New York Botanical Garden accompanied by Rutgers students. He and the students often went for the day, and thus the students could use the herbarium and interact with other botanists. David met Art Cronquist for discussions about Art's system of classification and David's deep knowledge of the New Jersey Flora when Art was revising Gleason and Cronquist's *Manual of the Vascular plants of North-eastern United States and Adjacent Canada*. As part of his commitment to northeastern botany, from 1990-1998, David Fairbrothers served as the Torrey Botanical Club representative on the Botanical Science Committee of the Board of Managers at the New York Botanical Garden. He was also a long-time mentor for the Flora of New

Jersey Project (www.njflora.org) which has close links with both NYB and the Torrey Botanical Society.

David's deep knowledge of New Jersey botany was a mother lode for projects that his students pursued, and some of them involved work on endangered species and habitats. In time, he became more focused on practical issues of endangered species and habitat management, and this coincided with the environmental movement of the 1970s. For decades, the Chrysler Herbarium had grown in the range of collections, particularly of endangered species and habitats. It eventually became the resource that allowed Fairbrothers and the herbarium manager, Mary Hough, to complete (to our knowledge) the first state description of threatened and endangered plant species.[5] This publication was influential in the modification of the Endangered Species Act, first passed by Congress at the end of 1973 and modified in 1975, to include plants, and to stimulate other states to conduct similar surveys. As a south Jersey native, David had great affection for the Pinelands and deep knowledge of its plants and natural history; he worked with others to protect this special area. He helped prepare "A Plan for a Pinelands National Preserve," and presented it to the U.S. Senate (through its Parks and Recreation Sub-Committee) and assisted substantially in the passage of the act authorizing it in 1978. His study of endangered and threated plants in the pinelands led to two publications that were instrumental in the establishment of the comprehensive management plan for the reserve,[6,7] which explicitly mentioned the initial 54 species to be protected. This act established the first Federal Reserve, similar in intent to the Catskills and Adirondack Parks in New York, but partly under the umbrella of the National Park Service and managed by the state of New Jersey. Later, the pinelands were added to the Federal Natural Preserve system, and then to the UNESCO Global Biosphere Reserve system in 1988.

Later in his tenure at Rutgers, he performed more administrative service, inaugurating the establishment of the Department of Biology (with 89 faculty members) as its first chairperson. Although he continued to be involved at Rutgers, through advising and consulting, he retired as a Distinguished Professor in 1988. His activity in endangered plant and pinelands issues at the state and federal levels continued well into his retirement. He and Marge moved south to Toms River, and David lent his support to the protection of natural communities in nearby Island Beach State Park. There, he helped with the development of the Emily de Camp Herbarium at the Forked River Interpretive Center, and helped to document plants and communities at the park. David and Marge frequently visited their son and daughter, spouses, and five grandchildren. They pursued new interests. Familiar with the history of New Jersey, David became interested in the antique glass produced in the state, then in silver overlay antique glass, and they both continued studying and collecting other antiques further afield. He became a sought-after lecturer on these subjects. Because of his declining health, he and Marge moved to Lebanon, NH, in 2010 to be near their son. David died two years later, at the age of 87 and after 63 years of marriage.

In recognition of his accomplishments, David received several awards. In addition to a variety of teaching, research and administrative awards at Rutgers, he was awarded the Rutgers Medallion in 1988. The Chrysler Herbarium and other collections were re-organized as part of the university biodiversity collections, and a symposium and banquet were held in his honor in 2005, to launch the fundraising effort to establish the David E. Fairbrothers Plant Resources Center. The Botanical Society of America presented him with its Merit Award in 1989, in commemoration of his research discoveries and service to the society. For his contributions to conservation in New Jersey, The Garden Club of New Jersey awarded him its Gold Medal in 2008, and the

Pinelands Preservation Alliance placed him in its Pine Barrens Hall of Fame, also in 2008.

He had a long, productive and happy life, and he will be deeply missed by his family, many former students, professional colleagues and personal friends.

David Lee
Department of Biological Sciences
Florida International University
Miami, FL 33155

Dennis Stevenson
New York Botanical Garden
Bronx, NY 10458
Republished from Plant Science Bulletin

References

Fairbrothers, David E. 1953. Relationships in the Capillaria Group of *Panicum* in Arizona and New Mexico. *American Journal of Botany* 40:708-714.

Strasser, Bruno J. 2010. Laboratories, museums and the comparative perspective: Alan A Boyden's quest for objectivity in serological taxonomy, 1924-1962. 2010. *Historical Studies in the Natural Sciences* 40:149-182.

Fairbrothers, David E. and Marion A. Johnson. 1961. The precipitin reaction as an indicator of relationships in some grasses. *Recent Advances in Botany* (University of Toronto Press, Toronto), pp. 116-120.

1. Jensen, Ü. and David E. Fairbrothers, eds. 1983. *Proteins and Nucleic Acids in Plant Systematics*. Springer Verlag, Heidelberg, 408 P.

2. Fairbrothers, D. E. and M. Y. Hough. 1973. Rare or endangered vascular plants of New Jersey. *New Jersey State Museum of Science Notes* 14:1-53.

3. Fairbrothers, D. E. 1979. Endangered, threatened, and rare vascular plants of the Pine Barrens and their biogeography, pp. 395-405. In R. T. T. Forman, ed.: *Pine Barrens: Ecosystem and Landscapes.* Academic Press, New York. (Forman was at Rutgers 1966-1988, and was an important ally in the campaign to save the Everglades; he then moved to the Harvard School of Design as its landscape ecologist).

4. Caiazza, N. and D. E. Fairbrothers. 1980. Threatened and endangered vascular plant species of the New Jersey Pinelands and their habitats. Prepared for the New Jersey Pinelands Commission, New Lisbon, NJ.

Figure 1. David with long-time friend Armend
Takhtajan at Rutgers University in 1968.

Figure 2. David and Marge Fairbrothers relaxing
at Frazer's Hill, Malaysia, in 1975, after he gave a
keynote address at the symposium inaugurating the
Rimba Ilma, still the only scientific botanical garden
in Malaysia. David was fifty years old at the time.

Figure 3. David Fairbrothers leading a tour of the Webb's Mill Bog in the Pinelands Preserve, the day following the symposium in his honor, in June of 2005. David was 80 years old at that time.

Percy Cyril Claude Garnham Tribute

Authored By Francis Edmund Gabriel Cox

Percy Cyril Claude Garnham, known by his intimates as Claude, was born in London on January 15th 1901. He was educated at the Paradise School in London and St Bartholomew's Hospital where he graduated in medicine in 1923. In 1928 he was awarded the degree of MD for which he received the Universality of London Gold Medal. Between 1923 and 1925 he studied at the London School of Hygiene and Tropical Medicine and travelled to Paris, Amsterdam and Rome where he worked with some of the most eminent parasitologists of the time. In 1925 he joined the Colonial Medical Service at the Medical Research Laboratories in Nairobi, Kenya, and in 1928 became Director of the Division of Insect-Borne Diseases where he worked on malaria, plague, yellow fever, leishmaniasis and river blindness and their vectors. In 1947 he joined the staff of the London School of Hygiene and Tropical Medicine first as Reader in Medical Parasitology and later as Professor of Medical Protozoology and Head of the Department of Parasitology. It was during his time at the London School that he, together with Professor Henry Shortt, began his search for the enigmatic stages in the life cycle of the malaria parasite between the injection of sporozoites by a mosquito and the appearance of parasites in the blood. Garnham's interests in malaria focussed on the parasites themselves rather than the disease and he had an encyclopaedic knowledge of the malaria parasites of primates. His understanding of the life cycle of *Hepatocystis kochi* led him to the discovery of the liver stages of this parasite and subsequently those of the primate malaria parasite, *Plasmodium cynomolgi*, and the human parasites, *P. ovale*, *P. vivax* and *P. falciparum*. These discoveries revolutionised the treatment of malaria and opened up the possibility of a vaccine against malaria, an aim still unfulfilled. Garnham retired in 1968 but continued to work as a Senior Research Fellow at Imperial College, London, until 1980.

Cyril Garnham's achievements have been recognised by his election to a Fellowship of the Royal Society in 1964 and his appointment as a Companion of the Order of St Michael and St George (CMG) in the same year. Garnham never sought honours but many were showered on him including fellowship or membership of more than 12 overseas learned societies. He also served parasitology and tropical medicine as President of the Royal Society of Tropical Medicine and Hygiene from 1967-1969, President of the British Society for Parasitology from 1970-1972 and President of the European Federation of Parasitologists in 1971.

Garnham's publication list is impressive, over 400 publications, including his scholarly and classic book, *Malaria Parasites and other Haemosporidia* (1966). He was always in great demand as a speaker at international congresses and was an inspired and generous teacher and many of his students have themselves reached the higher echelons of tropical medicine and parasitology and carried on his tradition.

I first met Cyril Garnham while working as a temporary lab boy at the London School of Hygiene and Tropical Medicine during my university vacations and I must have impressed him because he invited me to work with him after I had graduated. When I was studying for the Diploma in Parasitology and Applied Entomology at the London School he supervised my dissertation and later acted as an informal advisor for my PhD thesis on Host-Parasite Relationships in the Haemosporidia. Afterwards he guided my career with a real paternal interest and I was fortunate to be able to serve with him on the Council of the Royal Society of Tropical Medicine and Hygiene an experience from which I learned a great deal about committee work. He was one of the most formal people I have ever met and it wasn't until I was a professor that he deigned to call me 'Cox' instead of what he regarded to be the more formal Dr Cox. He only once called me Frank!

Cyril Garnham died on December 25th 1994 at and his memorial service at the church of St Bartholomew, appropriately close to St Bartholomew's Hospital Medical School, was crowded with many of the most eminent scholars of tropical medicine and parasitology from the UK and overseas.

Leonard George Goodwin Tribute

Authored By Francis Edmund Gabriel Cox

Leonard George Goodwin, always known as Len or LG, was born in Wood Green in North London on July 11 1915 and went to school at the William Ellis School in London. He read Botany and Zoology at University College London and, after graduation, switched to Pharmacy in which subject he qualified in 1935 and later qualified in medicine. In 1939, shortly before the beginning of the Second World War, he went to work at the Wellcome Bureau of Scientific Research where he investigated the chemotherapy of leishmaniasis, then a serious problem among British troops particularly in Sicily. It was while he was working on the use of pentostam for the treatment of leishmaniasis that he realised that the criteria used for estimating drug dosages were inadequate and began to devise a more rational approach to drug usage. He also worked on the chemotherapy of malaria, sleeping sickness and bilharzia (schistosomiasis) and developed an index to be used for testing drugs. After the war, he remained at the Wellcome Bureau and in 1958 became Director of the Wellcome Laboratories' of Tropical Medicine. In 1964 he was appointed Director of the Nuffield Laboratories of Comparative Medicine attached to the Zoological Society of London and then Director of Science at the Society. He retired in 1980.

Len Goodwin's contribution to science was acknowledged by his election to the Royal Society in 1976 and the award of the Companion of the Order of St Michael and St George (CMG) in 1977. He was President of the Royal Society of Tropical Medicine and Hygiene from 1979-1981 and President of the British Society for Parasitology from 1964-1966.

One little known fact about Len Goodwin is that, while he was working at the Wellcome Laboratories, he pioneered the use of the Syrian, or golden, hamster, *Mesocricetus auratus*, for medical research in the UK. The importance of these animals is that they are all derived from a single litter and are therefore genetically identical. All the pet hamsters in the UK are derived from this first Wellcome colony. He also kept wallabies and was a very good sketch artist.

Len Goodwin was a very private and gentle person who never actively sought any honours or public acknowledgement of his work which was all carried out with the minimum of fuss. He was very much liked by his staff and everyone else who took the trouble to get to know him.

I first met him while serving on the Council of the British Society for Parasitology and later when I was editing the *Wellcome Trust History of Tropical Diseases* where he became a frequent and welcome visitor to my office at the Wellcome Trust Building in Regent's Park and it is with his help and guidance that this book, that he had instigated, came to fruition in 1996.

Len Goodwin died on November 25th 2008.

James E. Gunckel 1914-2011 & Roberta Jean (Longworth) Gunckel 1918-2012

Authored By Alan Knight

Dr. James E. Gunckel, a retired professor of botany at Rutgers University, died September 19, 2011, at Monroe Village, Jamesburg, N.J. He was 97. He was survived by his 93-year companion and wife, R. Jean (Longworth) Gunckel, a son, Fred James Gunckel of Albuquerque, N.M., and daughter, Nancy Gunckel Knight, of Duanesburg, N.Y., as well as two grandsons, Jeffrey A. Knight, Brooklyn, N.Y. and Matthew James Knight, Ithaca, N.Y., and four great-grandchildren. Born in Dayton, Ohio, he graduated from Miami University, Oxford, Ohio, in 1938 and received his doctorate at Harvard in 1946, where he studied and began his research career under Dr. Ralph Wetmore and Dr. Kenneth Thimann. At Rutgers, he chaired what was then called the Botany Department for many years and did pioneering work in two important areas of study: tissue culture (plant cloning) and radiation biology, where he produced benchmark studies on the effect of radiation on a variety of plant species. A prolific publisher of scientific articles, he presided at many national and international botanical meetings, served as the translating editor of the seminal German botanical text General Botany by Wilhelm Nultsch, and edited the textbook Current Topics in Plant Science. A former president of the Torrey Botanical Society, the oldest botanical society in America, he also served many years as editor of The Bulletin of the Torrey Botanical Society, a refereed botanical journal. In 1959/60, having been awarded a Waksman Foundation Fellowship, he did meristem (plant tissue cloning) research at Station Centrale de Physiologic Vegetate, at Versailles, France, under the tutelage of Dr. Georges Morel. His unequaled knowledge of radiation biology, much of it gained through his many summers of research at the Brookhaven National Laboratory, led to his being called upon to provide expert testimony in a legal case pertaining to the Three Mile Island nuclear power plant accident. Dr.Gunckel was particularly proud of the career achievements of his many graduate students at Rutgers. They always wanted to thank him for his commitment to preparing them to assume leading research positions in academia. Professor Gunckel would say, "Don't thank me. Just pass it on to the next generation." His quiet devotion to his home gardens in Somerville and to the Second Reformed Church of New Brunswick, where he was an ordained elder and served in volunteer leadership roles, was well known to his close friends.

Roberta Jean (Longworth) Gunckel

Roberta Jean (Longworth) Gunckel died Friday, April 13, 2012 at Monroe Village, Jamesburg, N.J. She was 94. The daughter of a mining engineer, she was born in British Columbia and grew up in Copper Hill, Tennessee. She attended Duke University and graduated from Miami University (Ohio) with a degree in elementary education. Her career, like her retirement years, was devoted to children. She taught first grade for 23 years in the Highland Park and Bridgewater-Raritan school districts, a career she said she enjoyed every day.

A long-time resident of Bridgewater, N.J, she is survived by daughter, Nancy Gunckel Knight, of Duanesburg, N.Y.; son, Fred James Gunckel, of Albuquerque, N.M.; two grandchildren, Jeffrey Knight, of Brooklyn, N.Y. and Matthew Knight of Ithaca, N.Y. and four great grandchildren. She was predeceased by her husband, James Eugene Gunckel, who passed away in September.

The Team

Jim and Jean Gunckel were the backbone of the Rutgers' Department of Botany during Jim Gunckel's leadership years as professor and chair. The two of them and their children Nancy and Fred hosted multiple events at their magnificent Somerville home for faculty, new faculty and graduate students which often included lodging. These social events included gourmet dinners, dinner parties, bridge tournaments, departmental picnics and receptions. These events promoted strong bonds among the faculty and graduate students which led to the department's premiere national and global reputation. Jean Gunckel enthusiastically shared these leadership responsibilities at Rutgers in addition to her many responsibilities as an educator.

Information Resources:

Information retrieved October 17, 2013 from http://www.legacy.com/obituaries/app/obituary.aspx?pid=157540580Information retrieved October 16, 2013 from http://www.legacy.com/obituaries/mycentraljersey/obituary.aspx?pid=153779072

Roger F. Keller Tribute

Submitted By William R. Sharp

DR. ROGER F. KELLER JR., born in Manchester, NH, passed away at age 89 on December 28, 2011 in Akron, OH. He was preceded in death by his wife, Arline; and his son, John Roger. He is survived by his children, Nancy (Todd) Kislak of Agoura Hills, CA., and Brian (Connie) Keller of Hudson, OH.; grandchildren Heather (Sonny) McClinsey, Michelle and Sarah Kislak, Kendra and Kyle Keller; and great-grandsons, Nicholas Baker and David McClinsey. Roger enlisted in the U.S. Army Oct. 1942, was in the ROTC 3 years at UNH, and was enrolled in Office Candidate School in Ft. Knox, KY., April 1944. He served active duty as a Lieutenant from April 1944 until 1946. He was in the 11th Armored Division in General Patton's 3rd Army during WWII. He received a Purple Heart due to injuries sustained in the Battle of the Bulge. During his time at Walter Reed Army Medical Center in Washington, D.C., where he met the love of his life, his wife Arline. Originally from Milwaukee, WI. Arline served as an occupational therapist at the Walter Reed. Roger received his PhD in Zoology from Michigan State College in 1953. He served as professor and chair at the University of Akron in the Biology Department and provided research guidance to undergraduate and graduate students in the classroom and laboratory until his retirement in 1985. He was a Professor Emeritus of Biology and Professor Emeritus at the Community and Technical College. Roger served on the Board of Trustees at the Akron Zoo for decades. He was a member of the Sons of the American Revolution in his retirement. He was a charismatic and well-respected professor and dearly loved by his family and friends. He unselfishly mentored significant numbers of students seeking medical and graduate degrees.

Dr. Keller provided important leadership in building the University of Akron into a world class research institution. He possessed the uncanny ability to develop cross-campus collaborations in the recruitment of a top ranked research faculty, provide the essential resources for their advancement and develop significant undergraduate and graduate research initiatives.

Information sourced from the *Cleveland Plain Dealer*, Retrieved October 16, 2013 from \http://obits.cleveland.com/obituaries/cleveland/obituary.aspx?pid=15525260

Miodrag Ristic Tribute

Authored By Julius Kreier

Miodrag Ristic was a professor of veterinary pathology and hygiene in the veterinary college of the University of Illinois at Urbana Champaign Illinois. I became his first graduate student shortly after he joined the veterinary college of the University of Illinois. Our relationship persisted for many years after I finished my graduate studies with him and became a professor of microbiology at the Ohio State University.

Dr. Ristic was born in Serbia. He became a prisoner of war when the Germans occupied Serbia before they invaded Russia. He was liberated from the Camp by Canadian troops participating in the Allied invasion of German occupied Europe.

He was quite a linguist. He spoke English, German, French and Russian in addition to his native Serbo-Croatian. An officer of the Canadian troops who he contacted after the camp collapsed recognized his language skills and he remained with the Canadian officer has translator until the war ended. After the war ended he remained in Germany, married a German woman, and with Canadian support he enrolled in a veterinary college. Sometime after he completed the veterinary program he immigrated to the United States where he obtained a position in the veterinary college of the University of Florida at Gainesville. There he developed a successful well-funded research program. At about the same time the veterinary college at the University of Illinois hired a new Dean who wished to bring new faculty active in research and bringing in research grants to enhance research at the school. One of the people he wished to hire was Dr. Ristic.

There arose a problem in his plan when he found that the University of Illinois, unlike the University of Florida, required all faculty to have a PhD, a veterinary degree alone was not sufficient. To solve this problem the new Dean enrolled Dr. Ristic in a PhD program at Illinois. As Dr. Ristic was a well-established research professor he already had enough material for a thesis it just needed to be written up. It still was necessary for him to pass the various examinations required for a PhD degree. He passed the language requirement simply by taking an examination. The various other examinations he also passed when they came up with little trouble. One requirement however required him to spend time on the Illinois campus. He had to take certain courses required of all PhD students. It was because of this requirement that I met him. At the time I was also taking those courses and we ended up being laboratory partners in several of then, I of course at this time knew him only as a fellow graduate student.

After he completed the required courses and passed the various examinations required for his PhD degree he returned to the University of Florida. I did not expect to see him again.

At this time the man who was my advisor left to go to the University of Colorado. When you are a graduate student your advisor is a major factor in your life. I did not know what to do. We had started some research on the development of a vaccine for a viral disease of cattle called shipping fever. All we graduate students did however was to inject a vaccine which our advisor had made but we were not told what the vaccine was nor given much information about the whole project. As a result of the position I was in I became upset and a bit depressed. Fortunately however the chairman of the Department, J.O. Alberts a fine man who had invited me to join his department when I had completed my Master's degree and planned to move to another University assured me that things would work out and that I should continue to work on my degree requirements.

Some months later I was quite surprised when Dr. Ristic walked into my office, told me he was a professor authorized to train graduate students at Illinois and asked me to join him as his student. It will tell you something about him when I tell you that he came into my office in high good humor, said that as we had gotten along well as fellow graduate students he was sure we would get along l well in our new relationship. He then said he had funds for research on anaplasmosis and *Vibrio fetus* and then asked me on which I would prefer to work if I join him. I chose to join him without hesitation and chose to work on the anaplasmosis as it was a disease of the blood and I had been working on hematology in my Master's program. It will tell you something more about him to know that our first joint project was to clean up the mess in the laboratory left by my previous advisor.

In the years that followed. His constant optimism and joy in his work was a pleasure to behold. He informed us all about every aspect of the research going on in the laboratory. He gave help to every student without stint. If a student having problems joined his group they would soon be on the track again to obtaining their degree. It is a tribute to him that in my career I attempted to treat my students as he had treated me.

Maro Ran-Ir Sondahl Tribute

Authored By Antonio Figueira

Submitted By William R. Sharp

I regret to inform the INGENIC Newsletter readership that Dr. Maro Söndahl died early this year in a tragic car accident in Brazil. Maro was not a frequent attendant of cocoa meetings, and probably most of our cocoa research community did not know him well, except for the biotechnologists. Maro was more popular with the coffee community since he dedicated 35 years of his life to this crop, working in many aspects of physiology, breeding and biotechnology.

Maro is widely recognized for his great contributions and pioneering work on tissue culture of various tropical crops. He developed the first protocol for somatic embryogenesis of coffee in the late 1970's, followed by other great achievements in maize, oil palm and roses during the 1980's. He was also a pioneer in cocoa tissue culture. In the 1980's, his team at the DNA Plant Technology Corp. (Cinnaminson, NJ, USA), with support from a chocolate manufacturer, developed the first protocols to obtain cocoa somatic embryos from sporophytic tissues (nucellus and floral parts). Before that, somatic embryos had only been obtained from immature zygotic embryos, with obvious limitations for propagation and genetic transformation. He was granted a US patent for somatic embryogenesis and plant regeneration of cocoa in 1994. His protocol opened the possibility for further developments. In fact, improved somatic embryogenesis protocols were published in 1993 by Nestlé and CIRAD, culminating with the advances developed in the Penn State group in 1998, all derived from Maro's pioneering work.

I first met Dr. Maro Söndahl in 1982 in Rio de Janeiro, Brazil during my last year in college, when he gave a talk about the use of tissue culture in plant breeding. His seminar definitively helped to direct my career to biotechnology. Maro had attended the same school (Brazilian Federal Rural University of Rio de Janeiro), graduating 15 years earlier (1968). He got his Masters degree in 1972 at the Center for Nuclear Energy in Agriculture of the University of São Paulo, where I currently work. His PhD. degree in Cell Biology was from the Developmental Biology Program of Ohio State University (1978).

He started his successful scientific career in 1970 as a researcher at the Agronomic Institute of Campinas, a state owned research center of São Paulo, where he worked mainly with coffee physiology. After concluding his PhD. in the US, he returned to the same Institute, where he became the chairman of the Department of Plant Genetics, and later he was indicated to be the Director of the Biology Division. In 1983, he moved to the US, joining DNA Plant Technology Corporation in Cinnaminson, NJ, as research manager, supervising work with somaclonal variation (coffee, popcorn), protocol development for somatic embryogenesis (cocoa), anther culture (rice, sweet corn) and breeding (sweet corn, popcorn). He became Senior Research Director in 1987 with technical and business responsibilities in cell genetics and breeding on the following crops: coffee, cocoa, oil palm, pineapple, banana, corn, sweet corn, popcorn, rice, oats, watermelon and rose. He later became Director of New Business and Product Development of DNA Plant Technology. In 1993, he started his own company, Fitolink Corp. More recently (1997), he started a new company in Brazil, Bionova (www.bionova- mudas.com.br), working with commercial micropropagation of sugarcane, banana, and pineapple. The tragic accident occurred during a business trip to establish new contracts to provide micropropagated plants to growers in Mossoró. Maro will be remembered for his great contribution to biotechnology of tropical crops and to plant sciences in general.

During his career, Maro was very successful in combining science, publishing important breakthrough articles, with a business oriented entrepreneur perspective. Maro had a great sense of humor and was an entertaining person to have around meetings. He was born in Brazil, but his family was originally from Iceland. He served as an Honorary Consul for Iceland in Curitiba, Brazil since 2000. He was survived by his wife Dr. Clemencia Noriega, who continues to run their business in Brazil, and three children.

Source:
Information retrieved on October 19, 2013 fhttp://ingenic. cas.psu.edu/documents/publications/News/10.pdf

Dr. T. S. Subramanian and Meenakshi Subramanian Tribute

Authored By Geetha Ghai

My parents Dr. T.S. Subramanian (Toppur Seethapathy Subramaian) and Meenakshi Subramanian were my earliest mentors and influencers.

Dr. T.S. Subramanian obtained his high school diploma and Undergraduate degree in Madras (today known as Chennai) India from PS High School and Presidency College, respectively. He obtained his PhD in organic chemistry from Liverpool, England in the 1930s and stayed in Liverpool during the 2nd world war. He was involved in the discovery of DDT scientific and intellectual achievement that went wrong in application. He came back to India in 1945 and was the first Indian director after India obtained freedom from British colonization of the ordinance laboratory in Kanpur India. He brought science to the rural area teaching farmers good agricultural practices along with proper use of herbicides. He then led the Textile research and Jute research. His motto for life was hard work, ethics, helping humanity. He played a major role representing India in FAO, UNESCO, and in establishing science education policies for the country. He played an international role by participating in various Common wealth Conferences, Natick, Canada, Australia and Russia representing Indian science and productivity interests. He led through example and died on September 19, 1985.

Meenakshi Subramanian obtained her undergraduate degree in chemistry from Queen Mary's College Madras, India. Her life was mingled with spirituality and science. She was a magnanimous person filled with compassion for the under privileged. Always providing a helping hand by cooking nutritious meals for the needy children and encouraging them to continue school. Even today when walking on the streets in Chennai (formerly Madras) people walk up to me and state what a wonderful lady she was. She died on July 5, 2005.

My parents Dr. T.S. Subramanian and Mrs. Meenakshi Subramanian 1945 in front of their first house in Kanpur, India

Clara Gertrude Weishaupt

Submitted by William R. Sharp

Clara Gertrude Weishaupt, age 93, died at Greene Memorial Hospital in Xenia, OH 12 August 1991. She was for 22 years an outstanding professor of general botany and local flora in the Department of Botany, The Ohio State University. Simultaneously for 18 years, Dr. Weishaupt provide dedicated leadership as curator of the University Herbarium and conducted research on Ohio flora, culminating in her authoritative book, Vascular Plants of Ohio, 1960, 1968, 1971, and two publications on the grasses of Ohio (1967, 1985).

Born 20 July 1898 to Peter and Elizabeth Barbara (Weisflock) Weishaupt, who lived on a farm west of Lynchburg in Dodson Township, Highland County, OH, Miss Weishaupt was educated there in a one-room elementary school and graduated from the Lynchburg High School (1916). She received a diploma in bookkeeping, shorthand, and typing from Bliss Business College, Columbus, Ohio. At The Ohio State University she completed three degrees, B.S. in Home Economics (1924), M.S. in Botany (1932), and the Ph.D. in Botany (1935).

Her professional career began as a stenographer with the Department of Agricultural Education at The Ohio State University and with the Goodyear Tire and Rubber Company in Akron followed by eight years of teaching biology, mathematics and related subjects in the Lynchburg High School. While at Ohio State University, she was a graduate assistant in the Department of Botany and Plant Pathology (1932-35). Her college teaching career initially was at the State Teachers College, Jacksonville, AL (1935-46), where, while holding the rank of assistant professor and later associate professor of biology, she taught courses in biology, nutrition, field botany, human physiology, industrial arts, and physical science for elementary teachers. At the time she was the only woman on the faculty with a Ph.D. degree. In the Department of Botany and Plant Pathology at The Ohio State University, Dr. Weishaupt served as instructor (1946-51), assistant professor (1951-1960), associate professor (1960-1968), curator of the herbarium (1949-1967), and emeriti associate professor.

Dr. Weishaupt's early interest in the plant sciences was initially fostered in high school while taking an excellent course in botany, but as an undergraduate she developed her education in the areas of home economics and biological chemistry. As a graduate student in the OSU Department of Botany and Plant Pathology, she specialized in plant physiology and completed her master degree thesis on the effects of ultra-violet light on plants, and her Ph.D. dissertation on diffusion of water vapor through multiperforate septa, both completed under the direction of Professor Bernard S. Meyer. While teaching local flora at The Ohio State University, Professor Weishaupt early saw a need for a new field and laboratory manual of Ohio plants that would be useful to the students. Her first effort was a *Guide to Ohio Plants,* co-authored with three other members of the Department. Later she developed her own book, *Vascular Plants of Ohio* 0960), with a revised edition (1968), and a third edition 0970), followed by several subsequent reprinting's. The book is still quite popular and is used by students in local flora classes at various colleges and universities in Ohio and adjacent states.

Not trained as a plant taxonomist and with no experience in herbarium curatorial procedures, Dr. Weishaupt, upon being appointed curator of the OSU Herbarium in 1049, learned quickly the methods necessary to rejuvenate the herbarium. The facility had suffered neglect in the early 1940s during World War II. She brought order to the collection, including the identification of numerous specimens, updating the county distribution maps for the Ohio flora, and conducting extensive field work throughout

Ohio to obtain specimens of species from those counties not well represented in the herbarium

She focused on the State Herbarium, adding to its collection through her own field work and through the contributions of others. Renewed interest in the flora of the state was stimulated by the initiation in 1951 of the Ohio Flora Project, sponsored by the Ohio Academy of Science. The OSU Herbarium was to be the primary resource for this project. During Weishaupt's tenure, two volumes of the Ohio Flora were published by E. Lucy Braun -- *The Woody plants of Ohio* (1961) and *The Monocotyledoneae of Ohio* (1967). Weishaupt's own research on the state flora resulted in the publication of her *Vascular Plants of Ohio* (1960), written for beginning students. Long popular in local flora courses in Ohio and neighboring areas, the book is still in use. Subsequent focus on Poaceae (also called Gramineae or true grasses) a family of obvious agricultural importance, led to her contribution of a treatment for this family to Braun's Monocotyledoneae volume, and to the publication of her *Descriptive Key to the Grasses of Ohio Based upon Vegetative Characters* (1985).

Professor Weishaupt held memberships and offices in many scientific and honorary societies. As a devoted and conscientious professor of research and teaching, Professor Weishaupt will be remembered by many of whom she touched in this capacity. He exciting lecture and demonstration research experiments in the classroom encouraged countless numbers of students to explore careers in the life sciences. She was the recipient of many honors: one of five awarded the Annual Ohio State University Distinguished Alumni Distinguished Teacher Award and the Highland County American Association of Women Distinguished Service Award from the Centennial Honoree Award of the Ohio Academy of Science.

She once said, "I've really has a very ordinary life" Her contributions to the botany of Ohio and the service she gave to so many individuals in the state and the nation are achievements from more than an "ordinary" life.

Information retrieved October 16, 2013 from https://kb.osu. edu/dspace/bitstream/handle/1811/23480/V091N5_221. pdf?sequence=1

Clara G. Weishaupt

AUTHOR CONTRIBUTORS

Volumes I & II

(Author Abbreviated Curriculum vitae listed in
alphabetical order)

Jeff Alder

Jeff Alder, alder.11@osu.edu, Home town: Mount Olive, New
Jersey, Business address:

Bayer HealthCare, 100 Bayer Blvd., PO Box 915, Whippany,
NJ 07981-0915

Academic Institutions: The Ohio State University; B.S., M.S.,
Ph.D., University of Wisconsin (post-doctoral), Journal papers:
approximately 80, Book chapters: 2 Patents: 2, Organizations and
Awards: Contributed to successful development of four antimicro-
bial agents used to treat people with serious bacterial infections;
Chairperson, American Society for Microbiology, Antimicrobial
Chemotherapy, 2012-2014 term; NIH/NIAID and grant Reviewer
for Biodefense, session Chair for Biodefense contract review, NIH/
NIAID 2002 – present; Clinical Laboratory Standards Institute;
Antimicrobial Susceptibility Testing Subcommittee Executive
member; Reviewer, Antimicrobial Agents and Chemotherapy,
Infectious Disease Society of America.

Societies: American Society for Microbiology (ASM); Infectious
disease Society of America (IDSA); European Society for Clinical
Microbiology and Infectious Disease (ESCMID); Various awards

from Abbott Labs, Cubist Pharmaceuticals, and Bayer HealthCare, including the President's Award (Abbott), Scientific mentor of the Year (Abbott), Chairman's Award as employee of the Year (Cubist), and Special Recognition Award (Bayer); Named one of the top 20 notable people in Research and Development; *Research and Development Directions*; February, 2006; Jeff Alder lives with his wife Lisa in Mount Olive, New Jersey. They hope to settle in their "Mountain House" in the Catskills one day. Jeff is the Senior Director, Global Clinical Development for Bayer HealthCare, based in Whippany, NJ.

Henrique V. Amorim

Henrique Vianna de Amorim. Hometown, Piracicaba, SP. Brazil. Graduated in Agricultural Science 1966, ESALQ, the University of São Paulo, Brazil. Master of Science, Ohio State University, Columbus, Ohio – USA, Ph. D. at Univ. São Paulo. From January 1970 to 2001, he was an associate professor in the Biological Sciences Department at University of São Paulo in Biochemistry, Piracicaba, S.P,. Brazil. He launched Fermentec in 1977. Amorim has published over 85 refereed journal articles, book chapters and abstracts, and the authoritative book entitled Alcohol Fermentation Technology: Science and Technology, 2005. **Awards:** Ambassador Medal (2004), Ohio State University, for his achievement in biological Science an entrepreneurship. Entrepreneur of the year 2010, Piracicaba, SP, Brazil. **Professional Membership**: STAB – Brazilian Society sugar and alcohol technicians. IBD – Institute Brewing & Distilling – London UK., **Number of Patents**: 3, **e-mail:** amorim@fermentec.com.br, **Homepage**: www.fermentec.com.br. **Facebook**: Henrique Amorim

Carolyn Brooks

Dr. Carolyn Branch Brooks is a native of Richmond, VA. She received her B.S. degree and the M. S. Degree in Biology from

Tuskegee University and a Ph.D. in Microbiology from The Ohio State University. Dr. Brooks joined the University of Maryland Eastern Shore (UMES) in 1981 and rose through academic and administrative ranks through the years to become a full professor and to serve in the positions of Director, Coordinator, Department Chair, Executive Assistant to the President and Chief of Staff, Research Director of 1890 Land Grant Programs and Dean of the School of Agricultural and Natural Sciences. Since July, 2007 she has served as the Executive Director of the Association of 1890 Research Directors (ARD) which is composed of the research administrators in the food and agricultural sciences at eighteen historically black land grant universities. Among the professional awards she has received are the, **George Washington Carver Public Service Hall of Fame Award**, the **William A. Hinton Award from the American Society for Microbiology,** recognized as **one of Maryland's Top 100 Women**, featured as one of the **100 Distinguished African American Scientists** in *"Distinguished African American Scientists of the 20th Century"*, UMES **National Alumni Association's Faculty Award for Excellence and Achievement, Outstanding Educator Award** from the **Maryland Association for Higher Education, White House Initiative for Historically Black Colleges and Universities - Faculty Award for "Excellence in Science and Technology," the "Woman of the Year Award"** from the Maryland Eastern Shore Branch of the National Association of University Women, **"Chancellor's Research Scholar Award"**, the School of Agricultural Sciences' **Outstanding Faculty Award for Research and the 2005 Spirit of Excellence Award for Community Leadership.** She has published more than 50 journal papers and continues to serve on numerous panels, councils, boards, task forces etc. which has allowed her to serve as a consultant or research and academic program evaluator for universities in California, Michigan, Oregon, Idaho, South Dakota, Florida, New Jersey, New York, Puerto Rico,

Washington State, South Africa, Costa Rica, Honduras, the U.S. Virgin Islands, the Dominican Republic, Tanzania, and Malawi. Her funded research projects allowed her to conduct research in Egypt, Cameroon, Togo, Nigeria and Senegal. Carolyn is extremely active in the Links, Inc., a national service organization of professional African American women. Cbbrooks78@comcast. net

Helanie Carrer

Dr. Helaine Carrer is Associate Professor of Plant Biochemistry and Molecular Biology at the University of Sao Paulo, Agriculture College at Piracicaba (ESALQ). Has a degree in Agronomy Engineering graduated by ESALQ, University of São Paulo in 1983. Obtained her MSc in Agriculture Sciences at The Institute of Nuclear Energy in Agriculture (CENA), University of São Paulo advised by Prof. Otto Jesu Crocomo in 1988. Received her PhD at Rutgers, The State University of New Jersey, USA working with Prof. Pal Maliga at the Waksman Institute on Plastid Transformation Technology in 1994. Actually, she teaches biochemistry and plant molecular biology and conducts research in plant genetic transformation and functional genomics of photosynthetic genes with the goal to develop new sugarcane varieties with drought resistance and higher sugar content in leading a project in the biomass division of BIOEN, Brazil's public consortium for sugarcane to bioenergy R&D. During her career she has published 52 scientific articles, 6 book chapters, participates in 4 patents, advised 17 MSc and 9 PhD students. She received a Medal of Scientific Merit as a leading researcher for the contribution to the DNA Sequencing of *Xylella fastidiosa* plant pathogen bacteria by the governor of the State of Sao Paulo. Current address: Department of Biological Sciences, ESALQ-University of São Paulo. Av. Padua Dias, 11. Piracicaba-SP. 13418-900.

Email: hecarrer@usp.br

Roy Chaleff

Education: Amherst College, B.A., 1968, Yale University, Ph.D., 1972, Brookhaven National Laboratory, post-doctoral fellow, 1972-1974, *Employment:* John Innes Institute, England, Senior Scientific Officer, 1974-1976, Cornell University, Assistant Professor, Depts. of Genetics and Plant Breeding, 1976-1980, E. I. DuPont & Co., Central Research & Development Dept., 1980-1987, American Cyanamid, Director of Plant Biotechnology, 1987-1995, Rutgers University, Professor of Plant Biology, 1995-1998, University of Medicine and Dentistry of New Jersey, Office of the Vice President for Research, Central Administration, Director of Patents and Licensing, and Research Dean for NJ Dental School, 1998-2004, Ben Franklin Technology Partners, 2004-2005, *Currently:* Retired and residing in Pennington, New Jersey.

Frank Cox

Professor Cox is primarily a parasitologist. He graduated in Zoology with Parasitology as his Special Subject at the University of Exeter, UK, and, after postgraduate training at the London School of Hygiene and Tropical Medicine, joined King's College London as Lecturer and Reader in Parasitology in the Department of Zoology and subsequently Professor of Parasite Immunology in the Department of Cell and Molecular Biology. From 1986-1990 he served as Dean of Science in the University of London. He then joined the staff of the London School of Hygiene and Tropical Medicine as a Senior Research Fellow in the Department of Infectious Diseases until his retirement in 2013. He has published over 120 original research papers, reviews and congress proceedings and has authored or co-authored seven books including The *Wellcome Trust Illustrated History of Tropical Diseases* and also an interactive CD, *Six Thousand Years of Tropical Medicine.* Professor Cox has worked on various WHO and other international and national expert committees and has held visiting professorships at UK and overseas universities. He has been Editor of *Parasitology, Trends in Parasitology* and the *Transactions of the Royal Society of Tropical Medicine* and has also served on a number of editorial boards. He holds degrees of PhD and DSc of the University of London. His current interests are in the history of tropical medicine and parasitology. E-mail address: francis.cox1@btinternet.com

Otto J. Crocomo

Full Professor of Biochemistry at University of São Paulo, Rockefeller Fellow with C.C. Delwiche at UC Davis campus, and British Research with L. Fowden, University College, London, visitor professor with D. Boulter, University of Durham, England. Founder of the Center for Agricultural Biotechnology (CEBTEC) at the University of São Paulo, Piracicaba. Lives in Piracicaba, SP, Brasil with his wife Diva and has 5 children and one grandson. **Adolfo Egidio Lovadino Crocomo, Carla Maisa Lovadino Crocomo, Daniel Lovadino Crocomo, Maria Paula Lovadino Crocomo, & Pedro Augusto de Toledo Almeida Crocomo**

Adolfo Egídio graduated in Medicine at the Federal University of the State of Santa Catarina in 1992. He is an ear, nose, and throat specialist. He lives and works in Piracicaba with his wife, Kátia. **Maria Paula** received her B.A. in English Translation at University Ibero-Americana in 1992, in São Paulo. In 1994, she founded in Piracicaba "Interaction-School of Languages" which she directed until 2000. In 2000, she moved to London, UK, where she received her M.A. in Tourism Management from Westminster University. She is currently the Pedagogical Coordinator at "Self School of Languages," in Piracicaba. **Carla Maísa** currently teaches ESL at City College of San Francisco in San Francisco, California. She received her B.A. from Unimep in Piracicaba, where she taught for more than 10 years. In 2000, she decided to move to the San Francisco Bay Area. There she received her M.A. degree in TESOL from San Francisco State University. In addition to teaching, Carla has love for yoga and the arts. **Daniel** lives in Piracicaba. In 1998, he graduated in Tourism at the University Anhembi Morumbi, in São Paulo. In 2004, he received his M.B.A. in Marketing from Unicef in Piracicaba. He is currently a trade representative of high added value agricultural products. He also works with photography, advertisement, and he is a musician.

Pedro Augusto lives in Piracicaba and is Daniel's only son. He is a high-school student, and in his free time, he enjoys playing the guitar and practicing sports.

Joaquim José de Camargo Engler

Joaquim José de Camargo Engler was born in Campinas, State of São Paulo, Brazil. In 1960 he passed the entrance examination for admission in Agricultural Engineering at Escola Superior de Agricultura "Luiz de Queiroz", Universidade de São Paulo (ESALQ/USP) in Piracicaba, São Paulo. Completed college in 1964 and obtained the title of Agricultural Engineer, with specialization in food technology. In 1965 he began his academic career at ESALQ/ USP as a teacher and researcher in the Chair of Economics, current Department of Economics, Administration and Sociology. During his academic career he obtained the titles of "Doctor of Agronomy", "Associate Professor", and "Professor at the University of São Paulo (USP) and completed the degrees "Master of Science" and "Doctor of Philosophy" at The Ohio State University. He was approved in all Teaching Career competitions at USP, hitting the post of full professor. His international experience involves, the Coordination of the Agreement between USP and The Ohio State University for the development of the Research Program on Capital Formation and Technological Innovations in Agriculture; the Coordination of the Agreement between USP, the Ministry of Education and Michigan State University to develop the "Program of Higher Education in Agriculture" (PEAS); the Coordination of the Agreement between USP and the Ford Foundation to develop the Program for Research and Graduate Program in Rural Social Sciences, at ESALQ/USP; the Coordination of Planning and Finance of the Agreement between USP and the Inter-American Development Bank; the Coordination of the International Agreement between USP and European and Latin American Universities (Project UNIBEUR-INFO) member of the International Affairs Committee of the Department of Agricultural Economics and Rural Sociology, The Ohio State University, USA. During his professional activity he received numerous awards and honors, including the Medal "Fernando Costa", the OSU-International Alumni Award-1994, the Medal of Merit in Science and Technology in São Paulo State - 2001; Biography transcribed in the International Directory of Business and Management Scholars and Research of Harvard Business School; the title of "Piracicaba Citizen" in 2007, the Trophy "O Semeador", awarded by ESALQ/USP in 2010 in recognition of significant contribution to education development in the area of Agriculture. He is currently Administrative and Finance Director of the São Paulo State Research Foundation (FAPESP) having been elected and reelected by the Board of Trustees and appointed by the Governor of São Paulo State since 1993, for seven consecutive three-year periods. In addition to the position: And, he serves as Visiting Professor of Economics at Gulbenkian Institute of Science.

Geetha Ghai

Geetha Ghai lives in New Providence, New Jersey, USA. She retired after 13 years as the Associate Director at the Center for Advanced Food Technology Rutgers University, New Brunswick, New Jersey. Prior to this she worked in industry Ciba-Geigy Pharmaceuticals, Summit New Jersey and before that was a faculty member at the Pharmacology Department at the University of Southern Alabama Medical School, Mobile Alabama. She obtained her PhD from the Maharaja Sayajiroa University Vadodra, India in biochemistry and a MBA through the executive program at Rutgers University Business School Newark, New Jersey. She did a postdoctoral fellowship at the State University of New York, Buffalo, New York. She has authored over 57 papers some as primary and others as co-author, a few book chapters and 7 patents to her credit. She was on the grant review board of American Heart Association New Jersey Chapter. She was a member of the American Society of Pharmacology and Experimental Therapeutics and the Federation of American Society of Experimental Biology. For three years she served on the Advisory Board of the State Minority and Multicultural Office an appointment by the Governor of New Jersey. She is a founding member of the South Asian Total Health Initiative based at Rutgers University Medical School, and SKN Foundation both located in New Jersey. Recently, she has been elected to the board of SAGE Eldercare situated at Summit New Jersey. She has been invited to speak at many scientific and non-scientific conferences. Some salient speaking assignments include the motivational keynote speech for Johns Hopkins University Center for Talented Youth, Institute of Food Technology Scientific Lecture series, National Agri-Marketing Association and others. Email: geethaghai@gmail.com

Joseph Hamburger

Joseph Hamburger, Born: Haifa Palestine/Land of Israel, Residence: Jerusalem, Israel, Prof. (research), Member of the Kuvin Center, Hebrew University Hadassah Medical School, P.O.B 12272, Jerusalem 91120, Academic Institutions attended: Hebrew University of Jerusalem (HU), B. Sc., 1960-63, Microbiology and Parasitology, Hebrew University of Jerusalem (HU), M. Sc., 1964-67, Parasitology, Hebrew University of Jerusalem (HU), Ph.D., 1968-73, Parasitology, PROFESSIONAL EXPERIENCE: 1973-1975 Post-Doctoral Fellow at Ohio State University Department of Microbiology - Research subjects: Immunity in rodent malaria, 1975-1976 Research Fellow at Case Western Research University, Department of Geographic Medicine. Research subjects: Identification and characterization of schistosome egg antigens inducing granulomatous hypersensitivity, 1976-1978 Research associate at HU, Division of Helminthology. Research subjects: Characterization of schistosome egg antigens inducing immunopathology. Acquired resistance in schistosomiasis (bilharzia), 1978-1983 Lecturer at the HU, Department of Parasitology. Research subjects: Characterization of schistosome egg antigens inducing immunopathology. Acquired resistance in schistosomiasis (bilharzia), 1984 visiting scientist at Case Western Research University, Department of Geographic Medicine. Research subjects: The use of monoclonal antibodies and recombinant DNA technologies in schistosomiasis research, 1984-1991 Senior Investigator at the HU Department of Parasitology. Research subjects: Characterization of schistosome egg antigens. Identification of infected snails by detecting specific schistosomal antigens. Repetitive sequences in the genome of schistosomes, 1991-1991 Visiting Scientist at Smith College Department of Biological Sciences. Research subjects: Repetitive sequences in the genome of filariae, 1992-to date Prof. (Research) at the HU Department of Parasitology. Research subjects: Molecular markers of infection in schistosomes and filariae- Structure and significance for molecular monitoring of infection in human and intermediate hosts, MEMBERSHIPS: Israel Society of Parasitology (Elected President 1989-90), American Society of Tropical Medicine and Hygiene, The Royal Society for Tropical Medicine and Hygiene, PUBLICATIONS number: 51 and PATENTS number: 1.

Julius P. Kreier

Julius P. Kreier, Professor Emeritus, Department of Microbiology, The Ohio State University, Columbus Ohio; 2047

Iuka Avenue Columbus Ohio, 43201 .Telephone Contact: 614-294-6832, Birthplace: Philadelphia, PA 1926; Education. Philadelphia Public Schools, 1932- 1945. Temple University 1945-1948, University of Pennsylvania, 1949-1953, V.M.D. University of Illinois, 1956-1962, MSc. Ph.D.; Employment. Veterinarian, Cooperative Mexican American Commission for the Eradication of Foot and Mouth Disease, Mexico 1954-1955. Veterinarian, Tuberculosis and Brucellosis Eradication Campaign, 1956 Maryland, USA. University of Illinois, Urbana Illinois, Research Associate 1956-1962. Ohio State University, Assistant Professor, Associate Professor and Full Professor of Microbiology 1962-1989; Professor Emeritus of Microbiology, 1989 to present. Publications 150 in reviewed journals. Books 22; Teaching. Parasitic Protozoology, Rodent Surgery, Introductory Immunology, Infectious Diseases; Mentoring of graduate students: PhD students 24, MSc students 38, Postdoctoral fellows and visiting professors 4; Fulbright Award University of the Republic of Uruguay, invited lecturer, Campinas University Brazil, Veterinary College, Ankara Turkey, Institute of Parasitology Shanghai China, Haryana Agricultural University, Madras, India; My research was primarily on the pathogenesis of infectious diseases of the blood. These were primarily caused by protozoa although some were caused by Rickettsial type organisms. From my work the greatest pleasures I had were from teaching. In fact I also believe that my greatest contribution to science was the students trained. My students are now scattered around the world and most are continuing to carry out scientific investigations.

Jesse Kreier

Jesse Kreier is a lawyer with the Word Trade Organization in Geneva, Switzerland. Since joining the then-GATT Secretariat in 1992, Mr. Kreier has supported multilateral negotiations on international trade issues, has served as legal advisor to numerous dispute settlement panels and has provided technical assistance to assist WTO Members to implement their obligations under the WTO Agreement. From 1987 to 1992, Mr. Kreier practiced law in Washington, D.C., where he specialized in international trade regulation. Previously, he clerked for the Idaho Court of Appeals. Mr. Kreier holds a J .D. degree, *magna cum laude*, from Georgetown University Law Center, a Master of Science in Foreign Service from Georgetown University, and a Bachelor of Arts from Johns Hopkins University. He is admitted to the Bar in California and the District of Columbia USA. Mr. Kreier and his wife Susan Schorr live in the lakeside town of Nyon, Switzerland, a short distance outside Geneva. Their two children, Jacob and Freda, were born in Geneva, and are now in attending college in the United States.

Rachel Kreier

Health economist Rachel Kreier (Rachel.Kreier@gmail.com) lives in Port Jefferson, New York. She received her doctoral degree from Stony Brook University in 2004 at the ripe old age of 48. She has worked as a health rights activist, union staff member, editor, journalist specializing in the health care industry, and professor at Hofstra University in Hempstead, NY. She speaks and teaches frequently about health care reform, and was co-director of the 2010 conference, "New Directions in American Healthcare: Innovations from Home and Abroad." Her published work includes hundreds of newspaper articles, and three peer-reviewed journal articles. Her article, "A dynamic model of health plan choice from a real options perspective," (with B. Sengupta) received the 2011 Best Article Award from the *Atlantic Economic Journal*.

Jerry Ladman

Prior to his retirement in 2007 Jerry served seven years as associate provost for international affairs at The Ohio State University, where he had the broad responsibilities for the University's international programs, area studies, study abroad and international students. As a professor in the Department of Agricultural, Environmental and Development Economics he previously served as director of the Latin American Studies Program and as the director of the Ohio Leadership, Educational and Development Program. In addition, he coordinated the OSU College of Food, Agricultural and Environmental Sciences' program in Mexico and was the founding resident director of the OSU study abroad program held at the Mexican Postgraduate College of Agricultural Sciences. Prior to coming to Ohio State in 1990 he was professor of economics and director of the Center for Latin American Studies at Arizona State University. He has lived twice in Mexico where he was assistant program officer with the Ford Foundation and was a visiting professor at the Postgraduate College. He also spent five years in the Dominican Republic as chief of party of the OSU Agribusiness Partnership Project at the Instituto Superior de Agricultura. He led major projects in Bolivia and El Salvador; was a Fulbright Scholar in Ecuador; and a visiting scholar at the Food Research Institute of Stanford University. During his academic career his research has focused on economic and agricultural development with a special emphasis on rural finance in Latin America, especially in Mexico, Bolivia and Central America. He is author of more than 60 articles and monographs on rural finance, the political economy of Bolivia, the Mexican economy, Mexican migration, the U.S.-Mexican border region and is the editor of two books on Mexican economic and political topics, and one book on Bolivia's political economy. He has received numerous grants totaling more than seven million dollars. The most recent was, during the first five years after his retirement, to work with the Postgraduate College in bringing microfinance to rural areas in Mexico. In the past he served as a member of the Board of Directors of the Research Program on Mexico (PROFMEX),

President of the Pacific Coast Council for Latin American Studies, and President of the Borderlands Scholars Association. He has testified before congress, served as a consultant to a number of organizations, including the World Bank and the United States Agency for International Development. He served as a member of the Executive Committee of the NASULGC Commission on International Programs and was named an Honorary Professor the Catholic University of Bolivia. Since his retirement he has been professor emeritus in the Department of Agricultural, Environmental, and Development Economics and is currently serving as a docent at the Columbus Museum of Art and a member of the Board of Directors of the Columbus Council on World Affairs. He spent most of his youth in Clarion, Iowa. Upon graduation from Clarion high school he studied at Iowa State University, where he received the B.S. degree in Farm Operations and the Ph.D. in economics.

David Lee

David Lee is Emeritus Professor of Biology at Florida International University and resides in Miami, Florida. He was born in Wenatchee in 1942 and raised in Ephrata, on the Columbia Plateau of eastern Washington State. He attended Pacific Lutheran University (B.S. in Biology, 1966) and Rutgers University in New Brunswick for graduate work in Botany (M.S. in 1968, Ph.D. in 1970, in biochemical plant systematic under the direction of David E. Fairbrothers). Following postdoctoral research in plant cell biology and tissue culture with Rod Sharp and Donald Dougall at The Ohio State University (1970-1972), he and his wife Carol moved to Malaysia, where he worked as a Lecturer at the University of Malaya, in Kuala Lumpur, 1973-76. There he developed life-long research interests in the functional ecology of tropical plants. He then worked with Francis Hallé at the University of Montpellier, 1977-78. Following a couple of years of work as a carpenter and landscaper in upstate New York, he moved to Miami to work as Assistant (and then Associate and Full) Professor at the young Florida International University. There he developed a research program in tropical botany, and he helped the institution develop strength in tropical biology, partly through collaboration with local institutions, particularly Fairchild Tropical Botanic Garden. He conducted field work in tropical Asia, Latin America and West Africa. He is best known for discoveries concerning the basis and function of color in vegetative organs (as autumn coloration in temperate trees and structural colors in tropical plants), and the plastic developmental responses of plants to understory shade. This research has resulted in the publication of 87 peer-reviewed articles and book chapters, and 10 books (three edited). His 2007 book, *Nature's Palette* (University of Chicago Press) won the AAP award for scholarly publication in the life and biomedical sciences for that year. He is presently finishing a companion book, *Nature's Fabric,* about leaves, for the same publisher. He received the Bessey Award for Excellence in Teaching by the Botanical Society of America in 2005, and the Alumni Association Outstanding Faculty Award at FIU in 2007. Just prior to his retirement in 2009, he served as Director of The Kampong of the National Tropical Botanical Garden, in Coconut Grove, and recently published a book about its founder, *The World as Garden. The Life and Writings of David Fairchild* (CreateSpace, 2013). Email: leed@fiu.edu.

Raul Machado Neto

Raul Machado Neto, lives and works in Piracicaba, his hometown, and the current institutional address is – Universidade de São Paulo, Escola Superior de Agricultura Luiz de Queiroz (USP/ESALQ), Av. Pádua Dias 11, Piracicaba, São Paulo, CEP 13418260, Brazil. Higher educational training includes BS in Agricultural Science (1973), MS in Animal Science (1977), both at Escola Superior de Agricultura Luiz de Queiroz, the College of Agrculture of Universidade de São Paulo, USP/ESALQ, Piracicaba, São Paulo, Brasil, PhD in Animal Physiology, University of Illinois at Urbana Campaign, USA, in 1980, and Postdoctoral Fellow at Agricultural and Food Research Council-AFRC, Institute for Animal Health, England, (1989-1990). About the professional career, always at USP/ESALQ, started in 1974 in the position Assistant Professor of Universidade de São Paulo/ Escola Superior de Agricultura Luiz de Queiroz, USP/ESALQ USP/ESALQ (1974-1980), Doctor Assistant Professor (1980-1985), Associate Professor (1985), and Full Professor (1997). He received in 2001 the Scientific Merit Medal of State of São Paulo Governor, is currently Research Fellow of CNPq (National Council for Scientific and Technological Development), has delivered numerous lectures and published sixty two papers in scientific journals. His current email is raul.machado@usp.br

Sally Miller

Sally A. Miller, Professor, Department of Plant Pathology, The Ohio State University, Ohio Agricultural Research and Development Center, Wooster, Ohio USA, Email: miller.769@ osu.edu, Twitter: @OhioVeggieDoc, Website: www.oardc.osu. edu/sallymiller, Dr. Sally Miller is a Professor of Plant Pathology in The Ohio State University College of Food, Agriculture and Environmental Sciences. She was born in Canton, Ohio and currently lives in Wooster, Ohio. She received a B. Sc. in Biology from The Ohio State University and M.S. and Ph.D. degrees in Plant Pathology from the University of Wisconsin-Madison. Her career efforts have been centered on plant disease diagnostics and sustainable disease management. Dr. Miller teaches graduate level, laboratory-intensive courses in diagnostic field plant pathology and vegetable disease management, and a short course "Pest and Disease Diagnostics for International Trade and Food Security".

She also serves as State Extension Specialist for vegetable disease management, focusing on integrated disease management in conventional and organic systems. Dr. Miller has published 95 peer-reviewed journal articles, 250 technical reports, 25 book chapters and one co-edited book, and is a co-inventor on seven patents. She is a Fellow of the American Phytopathological Society (APS), served as Director of its Office of International Programs Board from 2007-2013, and was elected APS Vice President in 2013. She was awarded the APS International Service Award in 2002, the OSU Gamma Sigma Delta International Award of Merit in 2007, and the OSU-OARDC Distinguished Multidisciplinary Team Award in 2013.

Mark Mueller

Dr. Mark Muller joined the faculty at The Ohio State University in 1980, initially in the Department of Microbiology. In 1988, along with several colleagues, he helped create the first Department of Molecular Genetics in the country. Dr. Muller was Professor of Molecular Genetics until 2004 when he was recruited to join the Biomolecular Science Program at the University of Central Florida in Orlando. Dr. Muller was involved in forming a new College of Medicine at UCF and joined the Medical School faculty in 2007. He is currently Professor of Medicine at UCF and runs a cancer research group. The laboratory is an established group of researchers working on the molecular biology of cancer and gene regulation. Specific research focus areas include studies on epigenetics, gene silencing, DNA repair and telomerase regulation. Dr. Muller has nearly 100 publications in the fields of cancer and virology and multiple patents (H-index >30). In addition, he has over 400 national and international presentations and abstracts. Dr. Muller currently has national collaborations with researchers at NIH, Mayo Clinic, MD Anderson, as well as international collaborations with the University of Napoli (Naples, Italy) and the

University of Kwazulu-Natal (Pietermaritzburg, South Africa). Dr. Muller has started several for-profit biotechnology companies, including TopoGEN, Inc., Visual Genomics, Inc., Methylation, Ltd., DNA Protein, Ltd., and is active on Scientific Boards in the US and EU. He has worked with Nobel Laureates including Howard Temin (University of Wisconsin) and Michael Smith (University of British Columbia). Dr. Muller is an active member of the American Association for Cancer Research, American Society of Molecular Biology and Biochemistry, and American Association for the Advancement of Science. He reviews manuscripts for multiple journals and grant applications and is a member of the editorial board for The Journal of Plant Pathology and Microbiology. Email contact: Mark.Muller@ucf.edu, Facebook:https://www. facebook.com/mark.t.muller, LaboratoryWebsites:www.biomed. ucf.edu/mtmuller (general public), http://med.ucf.edu/biomed/ directory/profile/dr-mark-t-muller/ (scientific), Current Address: Dr. Mark T. Mueller, Ph.D., UCF, College of Medicine, 6900 Lake Nona Blvd,, Orlando, FL 32127.

John E. Peters

John E. Peters grew up in Dover, Ohio and received his Bachelor's degree from Otterbein College (Westerville, Ohio) and his Masters and Ph.D. degrees from The Ohio State University (Columbus, Ohio). His Master's degree research was in plant tissue culture and his Ph.D. dissertation included the discovery and characterization of a novel proteolytic enzyme from Pseudomonas aeruginosa. His research in these two areas resulted in 15 papers. He served 20 years as an officer in the United States Air Force as a clinical laboratory supervisor and as a faculty member at the United States Air Force Academy Biology Department. Following retirement from the Air Force he began a second career as a faculty member and Biology Department Chair at McHenry County College (Crystal Lake, Illinois). His primary area of expertise has been clinical microbiology education. He is a member of the American Society for Microbiology and the American Society of Clinical Pathologists (ASCP-MT),

Email-johnpampeters@hotmail.com,Facebook–john.e.peters.9

Robert Pfister

I joined the faculty at the Ohio State University at the end of summer in 1966. I was hired as an assistant professor of microbiology and after seven years rose to the rank of Professor. I became chairman of the department of microbiology in 1973 and resigned from the chairmanship in 1985. I retired from the Ohio State University in 1993. The courses that I taught during my tenure as a professor were introductory microbiology, microbial cytology, electron microscopy, general biology, and microbial seminar. During the time that I was in the department, I was also chairman of the graduate committee.

I published 70 scientific articles in various professional journals, and presented numerous papers at society meetings. I developed a "numerical taxonomy" program using "Fortran" for the IBM 1620 computer at Lamont Geological Observatory while working in the laboratory of Dr. Paul Burkholder. We studied and identified various microorganisms In the Antarctic using this numerical taxonomy program. With Dr. Burkholder's help I started a sponge collecting program in Puerto Rico to look for drugs from the sea. I was one of the early scientists to recognize microbial forms in ancient sediments, and to study the movement of toxic heavy metals in native soils using the interaction of microbes, visualizing them in both transmission and scanning electron microscopy. I also had numerous cooperative programs with fellow scientists.

I mentored 65 doctoral students and about 80 Masters degree students during my tenure at the Ohio State University. Most importantly, I had a wonderful wife and raised four wonderful children during my scientific career. My life has been blessed, yours can also be if you try.

James A. Quinn

Jim Quinn graduated from high school (Valedictorian) in Guymon, Oklahoma, was active in 4-H (State President, Oklahoma Hall of Fame), and received a B.S. in Crops and Soils at Oklahoma Panhandle State University (Valedictorian, Student Body President). He received a M.S. (Range Science) and a Ph.D. (Botanical Science) at Colorado State University. Immediately after the receipt of his Ph.D., he joined the faculty at Rutgers University (Assistant Professor, 1966-71; Associate Professor, 1971-77; Professor, 1977-2000; Professor Emeritus, 2001-Present). He taught undergraduate and graduate courses in biology, botany, and ecology, serving on 176 graduate student committees and as Chairperson of the graduate committees for 12 M.S. and 13 Ph.D. degree recipients. He was an author of 167 publications and published abstracts of papers at meetings (62 articles in referred journals), and was an invited symposium speaker at national (4) and international (4) meetings. Other recognitions and awards include External Examiner for Ph.D. dissertations in Canada (2) and Australia (4) and for three university graduate programs, Alumni Hall of Fame at Oklahoma Panhandle State University, and a Torrey Botanical Society Life Member Award for "long and dedicated service." A. QuinnProfessional memberships are the Botanical Society of America (Vice-Chairman, 1977 and Chairman, 1978, of the Ecological Section; Editorial Board, Amer. J. Bot., 1980-82), Ecological Society of America (Certified "Senior Ecologist", 1990-2011), Nature Conservancy, NJ Conservation Foundation, NJ Academy of Science (Council, 1972-76; Treasurer and Executive Committee, 1976-80), Pinelands Preservation Alliance, Sigma Xi (Chapter Vice-President, 1985-86; President-Elect, 1991-92; President, 1992-93), Society for

Range Management, and Torrey Botanical Society (Council, 1970-79; Director, 1980-81; President, 1982-83; Associate Editor, Bull. Torrey Bot. Club, 1983-85; Director, 1998-2000). Email: (quinn@aesop.rutgers.edu)

Rosa Shine Raskin

Rosa Shine Raskin has a background in technical and business information analysis, and clinical medicine. She has worked in academic medical centers throughout Ohio, the federal government, and industries in the Fortune 100 for over 30 years. She has held positions as a Research Assistant, Environmental Biologist and Technical Information Specialist in Materials Science, Tires, Engineered Products, Pneumatic Tools, Airplane Safety, Food Science, and Instructor in Clinical Medicine. She co-founded a start-up company and is the principal consultant at Rosa S. Raskin & Associates LLC, empowering companies with the information they need to succeed. She participates in the publication of articles and books in medicine and psychology. She serves as Contributing Author for the leading international trade journals on coatings in the Asia Pacific Region, Middle East, Africa, and Europe. Her articles often appear on the cover of the journals for which she writes. She published a book on Amazon.com about her experience as a Displaced Person and the continued search for her lost sister entitled, *Walk Forward,* a different Schindler's List, book's trailer is at ***www.youtube.com/watch?v=Zp7uQap6p2M***

Her company website is at *www.raskinfo.com* and she writes three blogs published both on Amazon.com and on the web, including "Information Specialist Secrets" at *preciousinformationspecialist. blogspot.com,* "Precious Cooking" at *preciouscooking.blogspot.com,* and "Most Precious Memories" at *mostpreciousmemories.blogspot.com.* She has an M.L.S. in Library Science from Kent State University, an M.S. in Microbiology and a B.S. in Biology from The Ohio State University.

Contact information, *rosaraskin@gmail, rosa@raskinfo.com,*
Skype: rosa.raskin1,
Twitter handle: *RosaSRaskin,*
Linkedin Profile: *www.linkedin.com/profile/view?id=15895388,*
Pinterest: *www.pinterest.com/rosaraskin/*
Personal Facebook page: *www.facebook.com/rosa.s.raskin,*
Facebook page for her book, *Walk Forward: www.facebook.com/ walkforwardbook*

Thomas Seed

Paterson, NJ. <u>Current residence</u>: Bethesda, MD, USA. <u>Communication address</u>: tmseed@verizon.net. <u>Education</u>: *1964-1968* – University of Connecticut, Storrs, CT, BSc (Bacteriology); *1968-1972* – Ohio State University, Columbus, OH, MSc (1969 - Microbiology), PhD (1972 – Microbiology; *1972-1973* – Case Western Reserve University, Institute of Pathology, Cleveland, OH, (Postdoctoral Fellowship in Cellular Ultrastructure). <u>Positions & Institutions</u>: *1998-present* – Consultant, Tech Micro Services Co., Bethesda, MD; *1995-1997* - Associate Chief of Research, Radiation Effects Research Institute, Hiroshima, Japan; *1993-1995* – Research Professor, Vitreous State Laboratory, Department of Physics, Catholic University of America, Washington, DC; *1996- 2003* – Senior Scientist/Group Leader, Radiation Casualty Management, Armed Forces Radiobiology Research Institute, Bethesda, MD; *1975-1995* – Research Biologist & Group Leaders for Radiation Hematology (1982-1995), Radiation Leukemogenesis (1979-1981), Cellular Indicators (1977-1978), Electron Microscopy (1975-1995), Divisional of Mechanistic Biology and Biotechnology, Argonne National Laboratory, Argonne, IL; *1973-1975* – Assistant Biologist & Head, Ultrastructure Group, Blood Research Laboratory, American National Red Cross, Bethesda, MD. <u>Professional works</u>: Journal articles (107 as per 'Pub Med' listing); Total cited publications, including books, book chapters, guidance documents, etc. (~186 publications cited ~2800 as per Google Scholar listing); Patents: 3 patents- 'Radiation Countermeasures' (US Patent US 7,919,525 B2; US Patent 7,665,694 B2; EP Patent EP 1,767,215); Awards/honors (select listing): Appointed, The Ohio State University Research Fellowship, 1971-1972; Appointed Head, US Delegation, NATO Research Study Group-23, 1996-1998; Elected, Chairman, NATO Research Task Group TG-006 (1999-2001); Awarded,

Distinguished Seminar Speaker (2006); Elected, Council member (2005-2010) & Consociate member (2010-present), National Council on Radiation Protection and Measurements; Professional affiliations: American Association for the Advancement of Science (AAAS- emeritus); Microscopy Society of America (MSA- emeritus); American Society for Microbiology (ASM-emeritus); International Society of Experimental Hematology (ISEH- emeritus); Radiation Research Society (RRS- retired member). General research interests: Nature and mechanisms of radiation injuries; Medical countermeasures; Hematopoiesis (structure/function/pathologic mechanisms); Leukemogenesis (nature/mechanisms); Low level radiation/chemical toxicity.

William R. Sharp

Dr. Sharp, birthplace, Akron, Ohio, has a background in biotechnology, translation of science into business ideas, spawning start-up companies and extensive technology transfer experience in the Americas and Asia. He has authored over seventy refereed research papers, abstracts and eight books in the field of plant cell biology including the five volume series entitled the Handbook of Plant Cell Culture. He previously held the positions: Dean of Research, Cook College (Now SEBS – The School of Environmental & Biological Sciences) and Director of Research, New Jersey Agricultural Experiment Station, Rutgers University; Executive Vice-President, DNA Pharmaceuticals, Inc.;

Executive Vice-President for Research, DNA Plant Technology, Corp; Research Director, Pioneer Research, Campbell Institute for Research & Technology, the Campbell Soup Company; Full Professor, Ohio State University; and Fellow, Argonne National Laboratory. He was a Fulbright Grantee during 1971 and 1973. Dr. Sharp holds a Ph.D. in Plant Cell Biology from Rutgers University. He is the recipient of the title Eminent Professor and the Luiz Queiroz Distinguished Service Medal award from the University of Sao Paulo and more recently, The Ohio State University Board of Trustees Honorary Services Award.

Jeffrey William Sharp

Jeffrey Sharp, an award-winning producer and publishing entrepreneur, is President/CEO of Story Mining & Supply Co., a Los Angeles-based production company committed to acquiring, developing, and financing and producing multi-platform premium content through unique access to quality material. SMS combines deep financial resources, strong creative relationships and an executive team that has produced award winning and commercially successful films. Sharp has produced a series of Academy Award winning and Golden Globe nominated films over the past ten years like Boys Don't Cry, You Can Count on Me, Nicholas Nickleby and Proof. With those movies as well as other renowned adaptations, including A Home at the End of the World, The Night Listener, and Evening, Sharp has worked to develop new models for integration of the publishing and film industries. Prior to Story Mining, Sharp co-founded the digital publisher Open Road Integrated Media with former HarperCollins CEO Jane Friedman. Story Mining and Open Road have a strategic partnership to develop and co-produce feature films and television shows based upon Open Road titles including: Lie Down in Darkness, based on the novel by William Styron, written and to be directed by Scott Cooper (Crazy Heart) and Cocoa Beach, a television series adapted by Andre and Maria Jacquemetton (Mad Men) based on the life of NBC news reporter Jay Barbree's recollections of the early days of the U.S. space program in Cape Canaveral, Florida. Sharp holds an MFA from Columbia University and a BA from Colgate University. In 2005, the Columbia University School of the Arts honored him with the Andrew Sarris award for his contribution to independent film. Sharp currently serves as Chairman of the Hamptons International Film Festival Advisory Board, member of the Executive Board of Literacy Partners, Special Advisor for the Book Meets Film Forum at the Taipei International Book Expo as well as a member of the Advisory Board of BookExpo of America. He is a member of BAFTA (British Academy of Film and Television Arts).

Harry E. Sommer

Harry E. Sommer, Associate Professor, D.B. School of Forest Resources, University of Georgia; Education: B.S. in Agriculture with honors, University of Vermont, Honors Topic: Quinone Occurrence in Hemlock, 1963; M.S. in Biochemistry, University of Maine, 1966, Thesis Title: "Purification and Characterization of Fructose 1, 6-diphosphatase"; Ph.D. Ohio State University 1972, Dissertation: "Influence of 2, 4 Dichlorophenoxyacetic Acid on Nitrate Reductase and Protein in Wild Carrot (Dacucus carota L.) Tissue Culture; Selected Publications: Sommer, H.E., C.L. Brown and P.P. Kormanik, 1975, Differentiation of Planets in Longleaf Pine (Pinus palustris Mill) Tissue culture in vitro, Bot. Gaz, 136: 196-200; Brown, C.L and H.E. Sommer, 1975, An Atlas of Gymnosperms Cultured in vitro, 1925-1974, Georgia Research Council, Macon, Georgia, p 271.; Sommer, H.E. (1975) Differentiation of Adventitious Buds on Douglas Fir Embryos in vitro, Proceedings of Imperial Plant Prop. Soc. 25; 125-127.; Sommer, H.E. and C.E. Brown, 1979, Application of Tissue Culture to Forest Tree Improvement in W.R. Sharp, P.O.

Larsen, E.F. Paddock and V. Raghavan, eds. Plant Cell and Tissue Culture: Principles and Applications, pp 461-491.; Sommer, H.E. Organogenesis in Angiosperm Trees, Bull. Soc. Bot. Fr. Actual Bot. 130: 79-81; Brown, C.L. and H.E, Sommer, 1983, Shoot Growth and Histogenesis of Trees Possessing Diverse Patterns of Shoot Development, Amer. Journal Bot. 79: 335-346.

Judy Lyman Snow

Judith Lyman Snow, Basking Ridge, NJ, Judysnow99@gmail.com, *Employment*

Rutgers University, Cook College--1985-2011, Biotechnology Center for Agriculture and the Environment (AgBiotech); Rockefeller Foundation, Agricultural Sciences--1980-1984, *Education,* Cornell University, MS in Horticulture--1976, PhD in Plant Breeding—1980 Duke University, BA in Botany, Magna cum Laude—1974, *Publications:* Author, annual reports and work plans for the AgBiotech Center at Rutgers for over 20 years, Author/co-author/editor of 2 books, 4 book chapters and 7 journal articles, Editor, Cook College *Grants Alert* weekly email newsletter 1993-96, *Awards:* Junior Science & Humanities Symposium Service Award, Rutgers University—2002 Cook College/NJAES Individual Impact Award, Rutgers University—1996 Rutgers University Merit Award—1994, *Associations/Affiliations*: Missouri Botanical Garden, William L. Brown Center for Plant Genetic Resources: Secretary of the Advisory Board—2002-2008, *Diversity* journal for plant genetic resources: Secretary/Treasurer—1996-2002.

Dan Tomas Spira

Born: 1932. Nitra Czechoslovakia; Emigrated to Israel 1949; Address: 35 Ben Zvi Blv. Jerusalem 96260; Tel: +972 (0)544 599 576. E-mail dant@mail.huji.ac.il; Institution: Hebrew University Hadassah Medical School P.O.B. 12272 Jerusalem 91120 ISRAEL; Study: The Hebrew University in Jerusalem; 1954 Begin. 1959 M.Sc.: Thesis: Blood Loss and Replacement in Malarious Rats.

1966. Ph.D.: Thesis Antigenic Analysis of Plasmodia; Awards: 1975 The Royal Society, London, Bruno Mendel Fellow at National Institutes for Medical Research Mill Hill, England; 1983 Minerva Fellowship, DFG. Germany at: Dept. of Physiological Chemistry, Eberhard Karls University, Tubingen Germany; Publications: about 100, the first one: Spira D. and Zuckerman A. 1962. Science 137: 356-357; two books edited; Professional activities: 1997 elected President of International Congress of Protozoology and appointed Chairman of The XI congress of ICOP. Retired: October 1, 2001; Prof. Emeritus Dan T. Spira; The Kuvin Centre for the Study of Infectious and Tropical Disease; Hebrew University School of Medicine, Jerusalem, ISRAEL; Tel. +972-54-4599576.

Donald Styer

Donald J. (Chip) Styer II, Systems Developer, The Ohio State University, Ohio Agricultural Research and Development Center, Wooster, Ohio USA, Email styer.21@osu.edu, Dr. Chip Styer is a Systems Developer at the Ohio Agricultural Research and Development Center, The Ohio State University College of Food, Agriculture and Environmental Sciences. He was born in El Paso, Texas, and currently lives in Wooster, Ohio. He received a B. Sc. in Microbiology from The Ohio State University and Ph.D. in Plant Pathology from the University of Wisconsin-Madison. Dr. Styer's doctoral research was in phytobacteriology, and as a research scientist at DNA Plant Technology Corporation, he developed plant tissue culture bioreactor systems. His principle interest at DNAP was in the development and implementation of data management systems for plant biotechnology applications. After moving to Ohio, he worked in computer hardware and software marketing and later joined OSU as a Systems Specialist. He was responsible for implementation of a College-wide online faculty reporting system. His current efforts include database management, web development, and data compilation and analysis.

ABOUT THE EDITORS

(Editors Abbreviated *Curriculum vitae* Listed in
Alphabetical Order)

Otto J. Crocomo

Full Professor of Biochemistry at University of São Paulo,
Rockefeller Fellow with C. C. Delwiche at UC Davis campus, and
British Research with L. Fowden, University College, London,
visitor professor with D. Boulter, University of Durham, England.
Founder of the Center for Agricultural Biotechnology (CEBTEC)
at the University of São Paulo, Piracicaba. Lives in Piracicaba, SP,
Brasil with his wife Diva and has 5 children and one grandson.

Julius P. Kreier

A Short Not Completely Academic Biography of Julius Kreier
was born in Philadelphia in 1926. I attended the public school system there. In high school, I followed the academic program but

with a supplemental program in agriculture. At 18, in 1944, I was called up for the draft but rejected for service because of severe arthritis of my left hip joint caused by a bacterial infection as a young child. After high school, I enrolled in Temple University, with a major in biology. Actually the program, I followed was the premedical program.

During the third year in Temple University, I applied for the College of Veterinary Medicine at the University of Pennsylvania. I was accepted and entered the veterinary college in the fall of next year. I therefore did not get a Master's degree from Temple University.

In 1953, I graduated from veterinary school and got a job with the US department of Agriculture to work on foot and mouth disease eradication project in Mexico. When the program ended successfully in 1955, I transferred to the eastern shore of Maryland and worked there on tuberculosis and brucellosis control program. Just before reporting to the job in Maryland, I married Ruth Casten, a woman, I met during my second year in veterinary college. We had maintained a relationship during the five years in which I completed my veterinary degree and worked in Mexico. Despite our frequent geographic separation during the time we continued contact. After our marriage, we remained together until her death in 2010. In 1957, I enrolled in a graduate program at the University of Illinois. There I received a Master of Science degree and a PHD. Perhaps of more significance during these years, my wife presented me with a little girl and then a little boy. I was hired by the department of microbiology at the Ohio State University in 1963. I retired from the position on December 31, 1988. I am now a professor emeritus of microbiology. I occupy myself now with making wooden bowls on a lath, sculpturing animal and human figures in clay and other materials and tending my garden greenhouse. I visit my children as often as possible and go to movies and ballet with friends. I am now also involved in producing this book with my good friend Rod Sharp. I do all of these things to help fill the void created by the absence of my wife for 55 years. It has been said that the paths of glory lead but to the grave. This is true but it is also true that all the other paths lead to the same place. What is important is the journey not the end. While we are making the journey, we should do our best to make it pleasant not only for ourselves but also for those with whom we travel it. This last consideration is of importance particularly to those of us who are to be teachers because our behavior may be imitated by our students.

William R. Sharp

Dr. Sharp has a background in biotechnology, translation of science into business ideas, spawning start-up companies and extensive technology transfer experience in the Americas and Asia. He has authored over seventy original research papers, abstracts and books in the field of plant cell biology including the five volume series entitled the Handbook of Plant Cell Culture. He previously held the positions: Dean of Research, Cook College and Director of Research, New Jersey Agricultural Experiment Station, Rutgers University; Executive Vice-President, DNA Pharmaceuticals, Inc.; Executive Vice-President for Research, DNA Plant Technology, Corp; Research Director, Pioneer Research, Campbell Institute for Research & Technology, the Campbell Soup Company; Full Professor, Ohio State University; and Fellow, Argonne National Laboratory. He was a Fulbright Grantee during 1971 and 1973. Dr. Sharp holds a Ph.D. in Plant Cell Biology from Rutgers University. He is the recipient of the title Eminent Professor and the Luiz Queiroz Distinguished Service Medal award from the University of Sao Paulo.

APPENDIX

Academic and Scientific Cooperation between the ESALQ/USP and OSU

Joaquim José de Camargo Engler

The academic and scientific collaboration between the College of Agriculture "Luiz de Queiroz", Universidade de São Paulo (ESALQ/USP) and the College of Food, Agricultural and Environmental Sciences at The Ohio State University (FAES/OSU) began formally in 1964 with the signing of an Institutional Agreement between the two entities, with financial support from the United State Agency for International Development.

The relationship between these two academic institutions already existed, but individually and informally. The said agreement was aimed at institutional improvement of higher education and research in agriculture in Brazil. This general objective should be achieved through integration of teaching, research and extension and development of a postgraduate program of high level, as well as research that could support government policies aimed at economic and social development of Brazil.

A number of long-term and mid-term projects were developed to achieve the goals of the Program of Cooperation between the research and teaching faculty who participated as consultant collaborators with the faculty at ESALQ/USP. The projects included development and offering of new graduate courses, improvement of existing facilities with emphasis on expanding and strengthening relationships in education and research exchanges with institutions in Brazil and abroad.

OSU Research and Teaching Faculty assigned to Piracicaba, as participants in the Cooperation Agreement were the following, with their respective areas:

- Allen Steinhauer, Entomology
- Alvin Moxon, Animal Science
- Claire Young, Agricultural Education
- Clyde Allison, Plant Pathology
- David O.Hansen, Rural Sociology
- Donald Larson, Agricultural Economics
- Eva Wilson, Home Economics
- Fred Deatherage, Food Technology
- John Parsons, Agronomy
- John Sitterley, Agricultural Economics
- Kelso Wessel, Agricultural Economics
- Olen Leonard, Rural Sociology
- Paul Clayton, Poultry Science
- Richard L.Meyer, Agricultural Economics
- Robert Welsh, Agricultural Economics
- Roger Williams, Entomology
- Trevor Arscott, Agronomy
- Walter Slatter, Food Technology

In order to expand and strengthen the competence of its faculty, ESALQ/USP, with the active participation of OSU, emphasized their training activities in human resources with emphasis on Graduate Research and Teaching. In this sense 29 ESALQ faculty members were sent to American universities including OSU and other institutions for the implementation of programs of Master of Sciences (MS) and 24 programs for Doctor of Philosophy Degree (Ph.D.). Other professors participated in collaborative research as post-doctoral fellows and visiting professors.

For obtaining evidence concerning the MS or Ph.D. program at the American universities during the Agreement the following ESALQ faculty served as counterparts in their respective areas of expertise:

- Adilson Paschoal, Zoology
- Antonio Galvao, Forestry
- Arare Pedroso, Entomology
- Avany Santos Correa, Home Economics
- Gaius Octavius Nogueira Cardoso, Plant Pathology
- Roberto Cássio Melo Godoy, Statistics
- Cesario Pires, Forestry
- Cyro Paulino da Costa, Horticulture
- Delmar Antonio Marchetti, Agronomy
- Diva Resende, Home Economics
- Elke Jurandy B. Nogueira Cardoso, Plant Pathology
- Fernando Perez Curi, Agricultural Economics
- Geraldo Tosello, Agronomy
- Gilberto Casadei Baptist Entomology
- Helena Teixeira Martins, Home Economics
- Henrique Vianna de Amorim, Plant Physiology
- Humberto de Campos, Statistics
- Ignacio Dal Fabbro, Agronomy
- Iracema de Sa, Home Economics
- Irenaeus Umberto Packer, Animal Science
- John Nunes Nogueira, Food Technology
- Joaquim José de Camargo Engler, Agricultural Economics
- Joaquim Oliveira, Food Science & Nutrition
- Jose Molina Son, Rural Sociology
- Keigo Minami, Horticulture
- Luiz Antonio Balastreire, Agricultural Engineering
- Dulce Maria Bergamin Bandeira, Psychology
- Max Lázaro Vieira Bose, Animal Science
- Mitsue Hironaka, Home Economics

- Moacyr Corsi, Agronomy
- Murilo Graner, Food Technology
- Oriovaldo Fall, Rural Sociology
- Paul F. C. Araujo, Agricultural Economics
- Paulo Roberto Cantarelli, Agricultural Biochemistry
- Paul Martin Soder, Ecology
- Randolph Custodio, Poultry Science
- Raul Dantas d'Arce, Animal Science
- Ricardo Shirota, Agricultural Economics
- Roberto Cobbe, Agricultural Journalism
- Roberto Dias de Moraes e Silva, Poultry Science
- Rose Higaki, Home Economics
- Rubens Valentini, Agricultural Economics
- Ruy Caldas, Biochemistry
- Sergio Brandt, Agricultural Economics
- Sérgio Paranhos, Agronomy
- Toshiaki Kinjo, Agronomy
- Valdomiro Bittencourt, Agronomy
- Vidal Pedroso de Faria, Animal Science
- Violet Coast, Home Economics
- Walter de Paula Lima, Agronomy
- Wilson Roberto Soares Mattos, Animal Science
- Zilda Matos, Agricultural Economics
- Zilmar Mark Ziller, Agronomy

Although the coordination of activities for improvement of the ESALQ faculty in the USA were made by OSU, they were held in various Universities and Institutions aiming to offer the best working conditions in their respective areas of expertise.

In addition to creation of formal programs of graduate training, several short-term activities were developed, including seminars and exchanges of members of ESALQ and OSU with each other, in order to increase collaboration and competence in teaching, research, the extension service and university administration. Among these activities, the "Agricultural Marketing and Food Technology Project," in which 14 students and five teachers of the newly created area of specialization in Food Technology from ESALQ attended conferences and visits to Ohio State University, California, Louisiana and Florida, and USDA research laboratories and industrial production equipment and food in the United States of America as a way to complement the offerings at ESALQ.

Another collaborative research activity between ESALQ and OSU was entitled the "Capital Formation and Technological Innovations in Agriculture in Developing Countries". This was held in the period 1971/73. Upon completion of the Agreement ESALQ OSU-USAID-funded academic-scientific collaboration program between these institutions, This program continued, including the realization of joint programs, such as the PICA (Program of Assistance to the Inter-University for Agricultural Sciences), which aimed at Inter-University cooperation between U.S. and Brazilian institutions in partnerships, in which ESALQ and OSU worked with the College of Agriculture of Para for improving their teaching and research programs.

Another ESALQ-OSU project occurred as part of "Higher Agricultural Education Program (PEAS), developed by the Ministry of Education, in which important joint teaching and research efforts were conducted. The partnership between OSU and ESALQ including the realization of research projects together with funding agencies to American and Brazilian, as the National Council for Scientific and Technological (CNPq) and Foundation for Research Support of São Paulo (FAPESP), especially in the areas of Agricultural Engineering, Soil Science, Seed Biology, Agricultural Economics.

ESALQ today annually receives students under the auspicines of the Tripartite Research and Teaching Program from OSU as well as at Rutgers University for cooperative programs.

The collaboration between the academic and scientific faculty of ESALQ / USP and FAES / OSU had a great impact on both institutions. The ESALQ expanded and improved its program of Graduate Studies and Research at OSU and the internationalization of its teaching and research. Thus both institutions continued to maintain this fruitful partnership.

www.ingramcontent.com/pod-product-compliance
Lightning Source LLC
Chambersburg PA
CBHW051436170526
45166CB00001B/4